Lecture Notes in Physics

For information about Vols. 1–131, please contact your bookseller or Springer-Verlag.

Vol. 132: Systems Far from Equilibrium. Proceedings, 1980. Edited by L. Garrido. XV, 403 pages. 1980.

Vol. 133: Narrow Gap Semiconductors Physics and Applications. Proceedings, 1979. Edited by W. Zawadzki. X, 572 pages. 1980.

Vol. 134: $\gamma\gamma$ Collisions. Proceedings, 1980. Edited by G. Cochard and P. Kessler. XIII, 400 pages. 1980.

Vol. 135: Group Theoretical Methods in Physics. Proceedings, 1980. Edited by K. B. Wolf. XXVI, 629 pages. 1980.

Vol. 136: The Role of Coherent Structures in Modelling Turbulence and Mixing. Proceedings 1980. Edited by J. Jimenez. XIII, 393 pages. 1981.

Vol. 137: From Collective States to Quarks in Nuclei. Edited by H. Arenhövel and A. M. Saruis. VII, 414 pages. 1981.

Vol. 138: The Many-Body Problem. Proceedings 1980. Edited by R. Guardiola and J. Ros. V, 374 pages. 1981.

Vol. 139: H. D. Doebner, Differential Geometric Methods in Mathematical Physics. Proceedings 1981. VII, 329 pages. 1981.

Vol. 140: P. Kramer, M. Saraceno, Geometry of the Time-Dependent Variational Principle in Quantum Mechanics. IV, 98 pages. 1981.

Vol. 141: Seventh International Conference on Numerical Methods in Fluid Dynamics. Proceedings. Edited by W. C. Reynolds and R. W. MacCormack. VIII, 485 pages. 1981.

Vol. 142: Recent Progress in Many-Body Theories. Proceedings. Edited by J. G. Zabolitzky, M. de Llano, M. Fortes and J. W. Clark. VIII, 479 pages. 1981.

Vol. 143: Present Status and Aims of Quantum Electrodynamics. Proceedings, 1980. Edited by G. Gräff, E. Klempt and G. Werth. VI, 302 pages. 1981.

Vol. 144: Topics in Nuclear Physics I. A Comprehensive Review of Recent Developments. Edited by T.T.S. Kuo and S.S.M. Wong. XX, 567 pages. 1981.

Vol. 145: Topics in Nuclear Physics II. A Comprehensive Review of Recent Developments. Proceedings 1980/81. Edited by T. T. S. Kuo and S. S. M. Wong. VIII, 571-1.082 pages. 1981.

Vol. 146: B. J. West, On the Simpler Aspects of Nonlinear Fluctuating. Deep Gravity Waves. VI, 341 pages. 1981.

Vol. 147: J. Messer, Temperature Dependent Thomas-Fermi Theory. IX, 131 pages. 1981.

Vol. 148: Advances in Fluid Mechanics. Proceedings, 1980. Edited by E. Krause. VII, 361 pages. 1981.

Vol. 149: Disordered Systems and Localization. Proceedings, 1981. Edited by C. Castellani, C. Castro, and L. Peliti. XII, 308 pages. 1981.

Vol. 150: N. Straumann, Allgemeine Relativitätstheorie und relativistische Astrophysik. VII, 418 Seiten. 1981.

Vol. 151: Integrable Quantum Field Theory. Proceedings, 1981. Edited by J. Hietarinta and C. Montonen. V, 251 pages. 1982.

Vol. 152: Physics of Narrow Gap Semiconductors. Proceedings, 1981. Edited by E. Gornik, H. Heinrich and L. Palmetshofer. XIII, 485 pages. 1982.

Vol. 153: Mathematical Problems in Theoretical Physics. Proceedings, 1981. Edited by R. Schrader, R. Seiler, and D.A. Uhlenbrock. XII, 429 pages. 1982.

Vol. 154: Macroscopic Properties of Disordered Media. Proceedings, 1981. Edited by R. Burridge, S. Childress, and G. Papanicolaou. VII, 307 pages. 1982.

Vol. 155: Quantum Optics. Proceedings, 1981. Edited by C.A. Engelbrecht. VIII, 329 pages. 1982.

Vol. 156: Resonances in Heavy Ion Reactions. Proceedings, 1981. Edited by K.A. Eberhard. XII, 448 pages. 1982.

Vol. 157: P. Niyogi, Integral Equation Method in Transonic Flow. XI, 189 pages. 1982.

Vol. 158: Dynamics of Nuclear Fission and Related Collective Phenomena. Proceedings, 1981. Edited by P. David, T. Mayer-Kuckuk, and A. van der Woude. X, 462 pages. 1982.

Vol. 159: E. Seiler, Gauge Theories as a Problem of Constructive Quantum Field Theory and Statistical Mechanics. V, 192 pages. 1982.

Vol. 160: Unified Theories of Elementary Particles. Critical Assessment and Prospects. Proceedings, 1981. Edited by P. Breitenlohner and H.P. Dürr. VI, 217 pages. 1982.

Vol. 161: Interacting Bosons in Nuclei. Proceedings, 1981. Edited by J.S. Dehesa, J.M.G. Gomez, and J. Ros. V, 209 pages. 1982.

Vol. 162: Relativistic Action at a Distance: Classical and Quantum Aspects. Proceedings, 1981. Edited by J. Llosa. X, 263 pages. 1982.

Vol. 163: J. S. Darrozes, C. Francois, Mécanique des Fluides Incompressibles. XIX, 459 pages. 1982.

Vol. 164: Stability of Thermodynamic Systems. Proceedings, 1981. Edited by J. Casas-Vázquez and G. Lebon. VII, 321 pages. 1982.

Vol. 165: N. Mukunda, H. van Dam, L.C. Biedenharn, Relativistic Models of Extended Hadrons Obeying a Mass-Spin Trajectory Constraint. Edited by A. Böhm and J.D. Dollard. VI, 163 pages. 1982.

Vol. 166: Computer Simulation of Solids. Edited by C.R.A. Catlow and W.C. Mackrodt. XII, 320 pages. 1982.

Vol. 167: G. Fieck, Symmetry of Polycentric Systems. VI, 137 pages, 1982.

Vol. 168: Heavy-Ion Collisions. Proceedings, 1982. Edited by G. Madurga and M. Lozano. VI, 429 pages. 1982.

Vol. 169: K. Sundermeyer, Constrained Dynamics. IV, 318 pages. 1982.

Vol. 170: Eighth International Conference on Numerical Methods in Fluid Dynamics. Proceedings, 1982. Edited by E. Krause. X, 569 pages. 1982.

Vol. 171: Time-Dependent Hartree-Fock and Beyond. Proceedings, 1982. Edited by K. Goeke and P.-G. Reinhard. VIII, 426 pages. 1982.

Vol. 172: Ionic Liquids, Molten Salts and Polyelectrolytes. Proceedings, 1982. Edited by K.-H. Bennemann, F. Brouers, and D. Quitmann. VII, 253 pages. 1982.

Lecture Notes in Physics

197

Quarks and Nuclear Structure

Proceedings of the 3rd Klaus Erkelenz Symposium Held at Bad Honnef, June 13–16, 1983

Edited by K. Bleuler

Springer-Verlag
Berlin Heidelberg GmbH 1984

Editor

K. Bleuler
Institute for Theoretical Nuclear Physics
University of Bonn, Nußallee 14–16, D-5300 Bonn 1

Organized by
K. Holinde, H. R. Petry, D. Schütte, K. Bleuler
Institute for Theoretical Nuclear Physics
University of Bonn, Nußallee 14–16, D-5300 Bonn 1

This conference was sponsored by the "Stiftung Volkswagenwerk" in Hannover. This generous financial help is greatly acknowledged by the organizers.

ISBN 978-3-540-12922-6

Library of Congress Cataloging in Publication Data. Klaus Erkelenz Symposium (3rd: 1983: Bad Honnef, Germany) Quarks and nuclear structure. (Lecture notes in physics; 197) 1. Quarks–Congresses. 2. Nuclear structure–Congresses. I. Bleuler, Konrad, 1912-. II. Title. III. Series. QC793.5.Q2522K57 1983 539.7'21 84-5386

ISBN 978-3-540-12922-6 ISBN 978-3-540-38807-4 (eBook)
DOI 10.1007/978-3-540-38807-4

Originally published by Springer-Verlag Berlin Heidelberg New York Tokyo in 1984

Printing and binding: Beltz Offsetdruck, Hemsbach/Bergstr.

TABLE OF CONTENTS

INTRODUCTION AND SURVEY

K. Bleuler (Bonn)

During the past years we took part in a most important and successful development of particle physics: It was realized that all fundamental interactions have to be derived from a local gauge principle. In particular, this viewpoint represents the basis of the Salam-Weinberg theory of electro-weak interactions which led recently to these wonderful verifications. In very much the same way quantum chromodynamics (QCD) was successfully checked in the realm of high-energy particle physics. It is, however, of decisive importance that QCD serves as a basis for all strong interactions, i.e., for nuclear physics in general. In fact, the underlying SU(3)-color gauge principle yields a well-defined interaction scheme between the gauge (or gluon) field and the Dirac wave function which describes the quarks, i.e., the fundamental constituents of all hadrons. Within this framework, this interaction appears to be the new and fundamental starting point for research in nuclear theory. This means that - in principle - the various, partly phenomenological parameters of the conventional theory (i.e., the masses and interaction properties of the relevant hadrons, so far expressed by coupling constants and form factors) should be derived from one single constant (i.e., the Λ-parameter within the expression for the "running" coupling constant) appearing in the gauge-invariant coupling scheme. This amounts, of course, to an enormous and far-reaching program which, as will be immediately realized, can be reached only through various intermediate steps and, for the time being, a large number of simplifying assumptions. The main scope of this symposium is to give an account of present day research within this general frame.

The various contributions may be ordered - according to rather different, and so far feasible, ways of approach - into three groups:

(i) Calculations based directly on the fundamental laws of QCD (Chaps.1 and 2); for the time being these attempts lead to the determination of the masses and - in a very restricted sense - to the inner structure of the relevant light hadrons, i.e., nucleons and bosons.

(ii) Special models representing these bound quark-gluon systems in simplified (parametrized) form (Chaps.3 and 4). The corresponding, partly phenomenological approaches (mainly so-called bags) are - in view of the enormous extension of the QCD-calculations - of greatest practical importance.

(iii) Special applications leading to corrections (or improvements) with respect to the conventional theory (Chaps.5 and 6). In view of the relatively large diameters of the various hadrons, these investigations are mostly concerned with the short-range part of nuclear interactions.

The first chapter is thus entirely devoted to the fundamental question of a determination of the masses of the relevant hadrons from first principles, with the help of the so-called lattice gauge theory (lectures by M. Bander, G. Schierholz, J. Engels, and F. Palumbo). Although this mathematical approximation scheme is far from being ideal, the results appear to be most promising and represent, in fact, a decisive step in our understanding of heavy particles in general. Apart from reducing the empirical masses of the relevant light hadrons to one single strong coupling constant, one obtains, at the same time, an idea of the inner structure of the building stones of nuclear matter. Different approximation schemes with very much the same scope were discussed in the second chapter. It was of decisive importance to see that dispersion theoretical methods yield similar results and can also be used for dealing with the problem of the interaction between hadrons (lectures by J. Namyslowski and, in a somewhat different sense, by G. Miller). A most important part of the general theory represents the fundamental relation between the principle of symmetry breaking of the QCD-vacuum state and the characteristic properties of the pion (lecture by W. Weise and R. Brockmann). In this respect it was of the greatest interest to realize that this "ground state" may be treated with the help of methods which had been developed within the framework of a general theory of nuclear matter (D. Schütte).

It is clear that the most complicated lattice-theoretical calculations cannot, at present, be extended to the determination of nuclear structure as a whole and not even to a detailed interpretation of the nucleonic properties (magnetic moments, charge distribution, etc.). For this reason, it is of decisive importance to represent the structure of the light hadrons with the help of simplifying models, i.e., the well-known bags, which, so to speak, represent an overall parametrization of the detailed quark-gluon structure of hadrons (Chap.3 with lectures by G. Miller, J. Wambach, and A. Szymacha; L. Wilets proposes an interesting "dynamical" enlargement by replacing the walls of the bags by the action of a "soliton field"). In view of the fact that these viewpoints by no means lead to a unique prescription, it appeared interesting to compare the different models proposed so far. In this context it was appealing to discuss a special model which yields nuclear structure as a whole (Chap.4). As it turns out that the bag diameters are, in fact, not small with respect to characteristic nuclear dimensions, and that the internal excitation energy is comparable to nuclear binding energies, it is clear that nuclear structure as a whole might differ in an essential way from the conventional N-nucleon picture. It was thus proposed (lecture by H.R. Petry) that the bag concept be extended (i.e., by introducing a so-called "overall" bag) in order to represent spherical nuclei as a whole. It was very encouraging to realize that, in this way, characteristic nuclear properties were reproduced in a rather natural and relatively simple way. (In this context a field-theoretical viewpoint was taken by E. Hilf.) Under these circumstances it was, of course, of greatest importance that the quark distribution in finite nuclei had been measured experimentally, and that it was shown to differ in an essential way from the distribution expected from unchanged inner structures of the individual nucleons (lecture on the EMC-effect by K. Rith). An interesting generalization of nuclear structure stems from inserting strange quarks into the system in order to represent empirical results obtained through embedding hyperons into nuclear matter (J. Pirner).

The last part of the conference dealt with the problem of nuclear forces in the light of QCD and with a comparison with the conventional theory (Chaps.5 and 6). The lectures of A. Fässler, M. Harvey, and P. Sauer thus contained quark-theoretical interpretations of the short-range part of nuclear forces as well as natural extensions to larger distances. In this connection it is of special interest to study and to analyze the forces between heavy quarks within the framework of the empirical results on the Y/J-particle (lecture by A. Martin). In order to have a basis for a general comparison, R. Machleidt represented a detailed and, to some extent, complete analysis of boson exchange in the conventional sense. Taking internal excitation of the nucleons as well as the finite extension of the sources into account, it was shown that all empirical NN-scattering phases could be reproduced in a satisfactory way. Within this frame, it turned out that the numerical values of the source extensions (replacing approximately the effect of an inner quark structure) played a decisive role. It was, therefore, of interest to see that gluon exchange yields to some extent similar results for the inner part (K. Holinde). Eventually, boson exchange was discussed from a more general standpoint by M. Bolsterli, whereas A. de Swart discussed the fundamental nucleon-antinucleon problem, both from the bosonic and from the quark-theoretical viewpoint.

A short, tentative survey of these contributions yields, in fact, various promising attempts toward a deeper understanding of nuclear structure with the help of quark-theoretical viewpoints. It became clear, however, that only a long and difficult way might lead to the final goal of full quantum chromodynamical interpretation. In particular, it was realized that there still exists a large gap between the direct lattice and dispersion theoretical determination of hadronic properties and the (phenomenological) description through various bag models. A more detailed analysis should lead to an interpretation of the important empirical results on the inner quark-gluon structure and - hopefully - to an understanding of the (fundamental but unsolved) confinement problem. A similar situation occurs in the determination of the structure of finite nuclei. Again, we have two extreme viewpoints, i.e., a conventional system of nucleons bound by boson exchange (enlarged by quark-theoretical interpretations of the short-range part) and an overall quark structure based on a natural extension of the bag concept. The gap should, again, be filled by a much more detailed description which should, among other things, describe the (unknown) behavior of the inner structure of individual nucleons within matter and - if possible - clarify the confinement problem (maybe in connection with the EMC-effect). The lectures thus suggested new directions for further (urgently needed) research which might also lead to a deeper insight into the underlying field theoretical concepts.

This meeting was held for the third time, in commemoration of the untimely death of Klaus Erkelenz 10 years ago; it was, again, supported by the "Volkswagenstiftung" in a most generous way.

Konrad Bleuler
Bonn, January 1984

RECENT DEVELOPMENTS IN LATTICE GAUGE THEORIES

Myron Bander
Service de Physique Théorique
CEN-SACLAY, 91191 Gif sur Yvette, Cedex, France
and
Department of Physics
University of California
Irvine, California 92717, U.S.A.

I. Introduction

Since 1980 most of the work on QCD has involved Monte Carlo simulations of lattice gauge theories. In this report I will review these works, emphasizing the most recent and/or the most accurate ones. The basic concepts of OCD, as well as a description of prior developments can be found (with all modesty) in my review [1]. This talk will be divided into three parts.

(i) Results from theories with no dynamical quarks. These include a determination of the string tension and the spectra of quarkless mesons (glueballs).

(ii) Inclusion of quarks in the quenched approximation. The topics discussed include chiral symmetry breaking, hadronic spectra and static properties of hadrons.

(iii) Caveats. Some of the studies on the validity of the quenched approximation and the problems brought on by the practical restrictions on lattice sizes are presented.

Of course, time limiations prevent a complete, self contained discussion ; this report should be viewed as a synopsis of results and as a guide to the literature. This discussion will be restricted to color SU(3) and zero temperature.

Before proceeding we shall established some notation. The continuum QCD action is

$$S = \int d^4x \left\{ \frac{1}{2g^2} \mathrm{Tr}\, F_{\mu\nu}F_{\mu\nu} + \sum_f \bar{q}^{(f)} \, (i\not{D} - m^{(f)}) q^{(f)} \right\} \tag{I.1}$$

with

$$F_{\mu\nu} = \partial_\mu A_\nu - \partial_\nu A_\mu + [A_\mu, A_\nu] \tag{I.2}$$

$$A_\mu = A_\mu^\alpha \frac{\lambda^\alpha}{2} \quad (\lambda^\alpha \text{ are the SU(3) generators}).$$

$q^{(f)}$ is the quark field of flavour f and the covariant derivative D_μ is

$$D_\mu = \partial_\mu - iA_\mu \, .$$

The coupling constant g depends implicitely on a cut-off Λ (or renormalization point μ). This dependence is obtained from the renormalization group equation

$$-\Lambda \frac{\partial g}{\partial \Lambda} = \beta_o g^3 + \beta_1 g^5 + \dots \tag{I.3}$$

For SU(3)

$$\beta_o = \frac{11}{16\pi^2} \quad ; \quad \beta_1 = \frac{102}{(16\pi^2)^2} \tag{I.4}$$

On the lattice, instead of the gauge potentials A_μ, we introduce link variables $U_{x,\varepsilon}$ defined on the link starting at point x and extending to $x+\varepsilon a$, with a the lattice parameter and ε a unit vector. For a pure gauge theory the lattice version of the action is

$$S = \beta \sum_P \mathrm{Tr} \prod_P U \quad ; \; \beta = 6/g^2 \tag{I.5}$$

The summation refers to all fundamental plaquettes of the lattice, and the product is over the U's defined on the four links bounding a plaquette.

Should this theory have a continuum limit, then renormalization group arguments indicate that for small g and a the following mass scale

$$\Lambda_o = \frac{1}{a} [\beta_o \, g^2(a)]^{-\beta_1/2\beta_o^2} \exp[-1/2\beta_o \, g^2(a)] \tag{I.6}$$

should be a independent. In fact, in the absence of other scales, such a quark masses, all dimensionful *physical* quantities should be expressible as pure numbers multiplied by appropriate powers of Λ_o. For example, we believe that the large distance quark-antiquark potential is linear

$$V(R) \sim K\, R. \tag{I.7}$$

The string tension K should, as we vary g, behave as

$$\sqrt{K} = A\, \Lambda_o \tag{I.8}$$

with A a number to be determined.

II. Theories without dynamical quarks

As QCD without matter fields is a non trivial interacting theory, we may study some of its consequences. These include a determination of the string tension K, (in fact, dynamical quarks would destroy such a linear term) and the spectrum of hadrons made exclusively out of gauge potential degrees of freedom ; these are commonly refered to as glueballs.

A. String tension

The string tension is most commonly determined by evaluating the expectation vlaue of

$$W[C] = \mathrm{Tr} \prod_{\ell \in C} U_\ell \tag{II.1}$$

where C is a large contour in the lattice. It A(C) is the minimal area of some surface bounded by C, we expect

$$W[C] \sim \exp[-KA(C)] \tag{II.2}$$

K is the same coefficient that appears in Eq.(I.7). The purpose of measuring K is threefold :

(i) We wish to check (I.6) and (I.8) thus determining whether the lattice theory approaches for small a what we believe the perturbative continuum theory to be.

(ii) If the above point is satisfied, we can use (I.6) and (I.8) together with the phenomenologically determined value of K

$$\sqrt{K} \sim 440 \text{ MeV} \tag{II.3}$$

to set the value of the lattice constant a for each value of the coupling constant g. In turn we will be able to use a to determine the values of other physical quantities, such as hadron masses.

(iii) The cut-off Λ_o can be related to a continuum cut-off which can in principle be determined from the phenomenology of high energy, high momentum transfer collisions.

A recent Monte Carlo measurement of the string tension [2] is presented in Fig.1

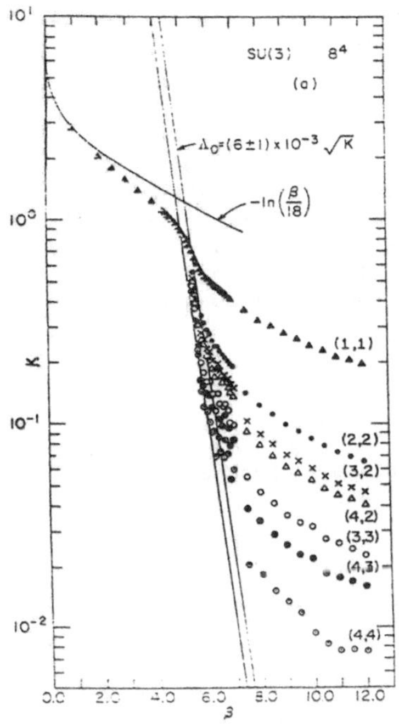

Fig.1 : String tension as a function of the coupling constant (From Ref. 2).

In this figure we are supposed to look at the envelope of curves for different size contours entering Eq.(II.1). To the extent this envelope is a straight line (for small g^2 or large β). With a slope determined by (I.6), to that extent is point (i) satisfied. The intercept of this line determines the constant A in Eq.(I.8). From Ref.2 we obtain A = 170 $\pm$ 30. A recent [3] evaluation of K on a large ($10^3\times 20$) lattice, however restricted to one value of g^2 ($g^2=1$), gives a value of A about twice as large. Using (I.8) and the phenomenologically determined string tension we obtain :

$$\Lambda_o \sim \begin{cases} 2.6 \text{ MeV (Ref.2)} \\ 1.3 \text{ MeV (Ref.3)} \end{cases} \qquad \text{(II.4)}$$

In order to compare these results to other ways of determining Λ we must relate this lattice cut-off Λ_o to a continuum one. This has been done [4] and we find $\Lambda_{MOM} = 83.5\ \Lambda_o$ yielding Λ_{MON} to lie between 100 MeV and 200 MeV in agreement with recent phenomenological determinations. The reason for the difference between the results of Ref.2 and Ref.3 may be due to the fact that the loops considered are not large enough. More comments will be made in the last section.

B. Gluonic Mesons

Hadronic masses may be obtained by evaluating the average of

$$\sum_{\vec{x}} \langle O_H(\vec{x},t)\ O_H(\vec{x},0)\rangle \sim e^{-m_H t} \qquad \text{(II.5)}$$

where O_H is an operator that creates the hadron of interest out of the vacuum. For example, for a gluonic meson with quantum numbers 0^{++} we could use $O_H = \mathrm{Tr}\ F^2$ while for 0^{-+}, $O_H = \mathrm{TR}\ F\ \tilde{F}$ would do, etc. In Monte Carlo evaluations we use lattice analogous of these operators. The masses are given in terms of the inverse lattice constant and we must use (I.6) and (I.8) to obtain then in physical units. In Table 1 we present two sets of results for gluonic meson masses, based on $\Lambda_o = 2.4$ MeV (cf. Eq.(II.4))

TABLE 1

Glueball Masses (MeV)

	Ref.5	Ref.6
0^{++}	740 ± 40	730 ± 100
0^{-+}	1400 ± 200	
2^{++}	1600 ± 100	(1100 - 2200)
1^{-+}	1750 ± 200	

The wide variation for 2^{++} meson in the calculation of Ref.6 is a reflection of the fact that this mass depends on the value of the gauge coupling constant. *It has not yet reached the scaling limit.* It is interesting to note that the 0^{-+} and 2^{++} masses are in the region of recently discovered resonances [7]. Most interesting is the prediction for the 1^{-+} meson which is a state that cannot be made out of a quark and an antiquark.

III. Calculations with Dynamical Quarks

A quark field may be accomodated on the lattice by adding to the action the term

$$S_j = \sum_{\vec{r}} \bar{\psi}_{\vec{r}} \psi_{\vec{r}} - K \sum_{\hat{\varepsilon}} [\psi_{\vec{r}+\hat{\varepsilon}a} \, U_{\vec{r},\varepsilon} \, (r+\gamma_\mu) \, \psi_{\vec{r}} + h.c.] \tag{III.1}$$

$$= \sum_{\vec{r},\vec{s}} \psi_{\vec{r}} \, Q_{rs} \, \psi_{\vec{s}}$$

The constant r plays no role in the continuum limit and is introduced, for technical reasons in order to eliminate a multiplication of fermionic degrees of freedom. The parameter K is related to the bare quark mass by

$$(r\,K)^{-1} = 8 + 2am_q^{(o)} \tag{III.2}$$

It is refered to as the "hopping parameter".

In the case where r=1 the fermions are refered to as Wilson fermions ; where r=0 and the various spin components are placed on different lattice sites the fermions are refered to as Susskind fermions. In the latter case we still have some doubling of degrees of freedom but the chiral limit is easier to control.

As the quark fields are anticommuting variables we cannot perform a usual Monte Carlo calculation on them. As the fermion fields appear quadratically, a formal integration may be performed. The term that appears is a determinant of the operator Q. Most calculations performed up till now have set this determinant equal to one (quenched approximation). This corresponds to the neglect of closed fermion loops in a Feynman diagram expansion. Justification for this approach comes from the large N expansion of QCD where such loops do not contribute, and from the smallness of the ρ-ω mass difference, which is due to such loops. In the next section we shall discuss some further tests of this approximation.

A. Chiral Symmetry Breaking

We expect that QCD will spontaneously break chiral symmetry. This may be studied by examining the expectation value, $\langle\bar{\psi}\psi\rangle$ in the limit of vanishing quark mass. Figure 2 shows a recent evaluation (with Susskind fermions) of this quantity showing that for $\beta > 5.5$ we are in a scaling region

B. Hadron Spectrum

Analogous to the evaluation of the gluonic meson spectra we may evaluate the masses of hadrons containing quarks, both mesons and baryons. In addition to fixing the scale of the lattice spacing we have one more parameter to fix, namely K (cf. Eq.(III.1) and Eq.(II.2)) or equivalently the quark mass. Ideally the procedure one should follow (for Wilson fermions) is :

1. Fix β (presumably in the scaling region)

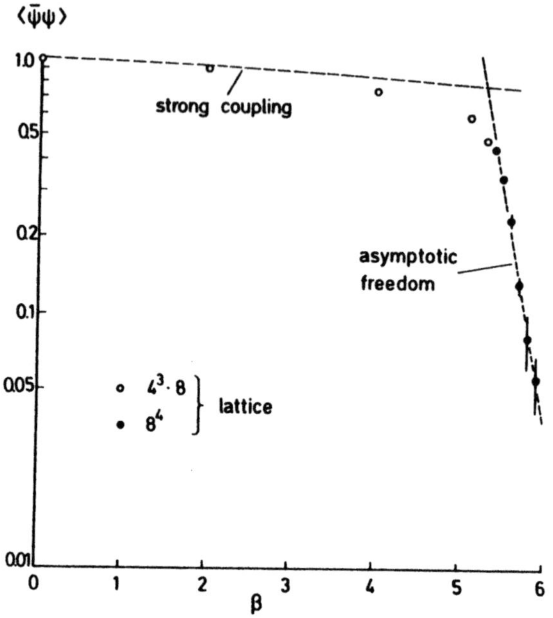

Fig.2 : Chiral symmetry parameter $\langle\bar{\psi}\psi\rangle$.

2. Adjust K so that the pion is masless (chiral limit) and then tune it in order to obtain the correct pion mass
3. Measure other physical quantities
4. Repeat the above for a different value β in order to check scaling.

The second step is usually done by extrapolation from small values of K, while step four has not been performed extensively. Recent results are summarized in Table 2.

TABLE 2

Hadron Masses (MeV)

	Ref.9 (Susskind)	Ref.10 (Wilson)
ρ(770)	730 ± 90	820 ± 30
A_1(1270)	1190 ± 90	
N(940)	920 ± 100	1250 ± 150
Δ(1230)		1350 ± 200
f_π(130)	134	(100-200)
K(490)		500 ± 30
K^*(840)		930 ± 40

C. Static Hadrons Properties

By placing a baryon in an external magnetic field one may measure its magnetic moment. A recent work gives [11]

$$g_p = 2.7 \pm 1.0 \qquad (2.79 \text{ exp})$$

$$g_n = -1.6 \pm 0.5 \quad (-1.91 \text{ exp})$$

A quantity related to the decay rate of $\rho \to \pi\pi$ has also been determined [12]

$$g_{\rho\pi\pi} = 3.0 \pm 1.0 \qquad (6.1 \pm .1 \text{ exp})$$

IV. Caveats and Checks

The effects of two approximations must be checked. One concerns the finite size of the lattice and the other is the setting of the fermion determinant to one (quenched approximation). In addition to the effects of finite lattice size the scaling behavior of physical quantities, as measured on the lattice, should be verified by approaching the scaling limit as close as possible. Unfortunately these two effects work against each other. Going to smaller couplings, equivalently smaller lattice constants has the effect reducing the overall size of the lattice. Many calculations have been performed on lattices of the order of 10^4 at $\beta = 6/g^2 = 1$. At this coupling the lattice parameter is ~ 0.1 F. The hadrons contained therein are somewhat squeezed [13]. The effects of increasing the lattice to 16^4 and decreasing g^2 to 0.95 is discussed in Ref.14.

Improvements on the quenched approximation have been studied by two methods. The first consists of evaluating the fermion determinant by an expansion in the hopping parameter K (cf. Eq.III.1) [15]. Inclusion of fermion loops for one fermion flavor changes the physical value of the lattice spacing by 10%. In the second method, the dynamical fermions are included by considering an analogous dynamical boson problem (pseudofermion method) [16]. Fermion loops change the value of $\langle\bar{\psi}\psi\rangle$ by about 5% for each fermion species.

I wish to express my thanks to Dr. E. Marinari, Dr. A. Morel and Dr. S. Meshkov for useful discussions.

References

[1] M. Bander, Phys. Rep. C 75 (1981) 205

[2] R. Ardill, M. Creutz and K. Moriarty, Brookhaven Report BNL-32377 (1982) ; M. Creutz and K. Moriarty, Phys. Rev. D 26 (1982) 2166

[3] G. Parisi, R. Petronzio and F. Rapuano, Phys. Lett. 128 B (1983) 418

[4] A. Hasenfratz and P. Hasenfratz, Phys. Lett. 93 B (1980) 165

[5] K. Ishikawa, A. Sato, G. Shierholz and M. Teper, DESY report 83-061 (1983)

[6] B. Berg and A. Billoire, Nucl. Phys. B 221 (1983) 103 and B 226 (1983) 405

[7] D. Hitlin, report to the "International Symposium on Lepton and Photon Interactions at High Energies" (1983) and California Institute of Technology Report-68-1071

[8] I. Barbour, P. Gibbs, J. Gilchrist, H. Schneider, G. Shierholz and M. Teper, DESY Report 83-093 (1983)

[9] J. Gilchrist, H. Schneider, G. Shierholz and M. Teper, DESY Report 83-094 (1983)
[10] H. Lipps, G. Martinelli, R. Petronzio and F. Rapuano, Phys. Lett. 126 B (1983) 250
[11] C. Bernard, T. Draper and K. Olynyk, Phys. Rev. Lett. 49 (1982) 1076
[12] S. Gottlieb, P. Mackenzie, H. Thacker and D. Weingarten, Fermilab Report 83/60-THY (1983
[13] P. Hasenfratz and I. Montvay, Phys. Rev. Lett. 50 (1983) 309
[14] P. Hasenfratz and I. Montvay, DESY Report 83-072 (1983) and Ref.10
[15] I. Montvay, DESY preprint 83-074 (1983)
[16] H. Hamber, E. Marinari, G. Parisi and C. Rebbi, Phys. Rev. 124 B (1983) 99

THE HADRON SPECTRUM IN LATTICE QCD

G. Schierholz

Deutsches Elektronen-Synchrotron DESY, Hamburg

Abstract: I give a brief introduction to lattice QCD and discuss some of the recent calculations of the hadron mass spectrum. I also address the question of spontaneous chiral symmetry breaking which most obviously influences the character of the hadron spectrum.

I. Introduction

Recent months have seen considerable progress in understanding the hadron mass spectrum. This has required two developments: Firstly the construction of a theoretically appealing and phenomenologically reasonable fundamental field theory, QCD, and secondly the development of nonperturbative methods, the lattice Monte Carlo techniques, to solve and hence to give a real meaning to the QCD Lagrangian.

In this talk I shall introduce you to "lattice QCD" and discuss some of the recent calculations of the hadron mass spectrum.

The plan of the talk is as follows: After the introduction in this section I shall, in section 2, first discuss that part of the hadron spectrum that is composed primarily of the gluon degrees of freedom. These hadrons, the glueball states, are very much a prediction of QCD, and the absence of specific experimental information on these states provides us with the opportunity of making quantitative predictions that have no experimental bias. The dynamics that most obviously influences the character of the quark states is that of chiral symmetry and its spontaneous breaking. If QCD is to be a successful candidate theory it must reproduce this phenomenon. If it does not then the meson and baryon spectrum predicted by QCD will possess parity doubling and no very light pions, in damaging contrast to the experimental situation. In section 3 I shall convince you that QCD indeed breaks its (zero quark mass) chiral symmetry dynamically. This is largely a question of the lattice size. In section 4 finally I will present very recent results of a Monte Carlo calculation of the meson and baryon mass spectrum on a $10^3 \times 16$ lattice which we have checked to be large enough to accommodate the dynamics of spontaneous chiral symmetry breaking and the hadrons themselves. I shall conclude with some remarks in section 5.

Lattice gauge theory[1]

For the sake of simplification let us for the moment regard only the pure gauge theory. The starting point is the Feynman path integral representation[(2)] in Euclidean space-time

$$Z = \int [dA_\mu(x)]\, e^{-\frac{1}{2g^2}\int d^4x \mathrm{Tr}(F_{\mu\nu}F_{\mu\nu})} \quad , \qquad (1)$$

where the field strength tensor is

$$F_\mu = \partial_\mu A_\nu - \partial_\nu A_\mu + [A_\mu, A_\nu]. \qquad (2)$$

Note that the coupling has been absorbed into the matrix-valued vector field $A_\mu(x)$:

$$A_\mu(x) = g\, A^a_\mu(x) T^a, \qquad (3)$$

where the $A^a_\mu(x)$ are the gauge potentials and the T^a are a representation of the SU(3) Lie algebra.

As it stands the representation (1) is incomplete. Any attempt at a calculation will encounter ultraviolet infinities. The conventional way to deal with this problem is to add to (1) a regularization prescription. That is, the theory is first mutilated by the arbitrary removal of the offending high momentum / short distance components of the fields, and then these components are systematically reintroduced while varying the coupling in such a way as to keep the theory finite. That means that beyond a certain momentum the reintroduction of the even higher momentum components will not affect the low energy physics. To implement the regularization procedure one must specify a momentum p_0 (distance a) such that momentum components higher than p_0 (distances smaller than a) are to be thrown away, and then the bare coupling g^2 in (1) is effectively the coupling on the size scale p_0^{-1} (a). We then take $p_0 \to \infty$ $(a \to 0)$ varying $g^2(p_0)$ $(g^2(a))$ so as to keep the low-energy (large distance) physics unchanged. This necessarily breaks the scale invariance of the theory.

The standard regularization procedure in perturbative calculations is dimensional regularization[(3)]. This is however of no use for the calculation of nonperturbative quantities such as the hadron masses. An alternative procedure to remove high momenta/small distances is by discretising space-time. If we put the theory onto a lattice of spacing a we automatically restrict all the components of the momentum to be less than or equal to

$$p_0 = \frac{\pi}{a} \quad . \tag{4}$$

To recover the full theory we let $a \to 0$. Once a is much less than the dynamical lengths in the theory, characterized by some correlation length ξ, the physics of the theory should remain unchanged as $a \to 0$ as long as we vary g^2 with a according to the usual (two-loop) renormalization group formula

$$g^2(a) = \frac{16\pi^2}{11\ln(a\Lambda)^{-2} + \frac{102}{11}\ln\ln(a\Lambda)^{-2}} \quad , \tag{5}$$

which (to two loops) is the same for all regularization procedures apart from the Λ parameter which can be related to the more familiar Λ_{mom} via a perturbative calculation. For calculational convenience we also wish to confine space-time to a finite volume, say L lattice spacings in each direction. Then if $\xi << La$ the physics should be the same as for the infinite volume limit. The purpose of this discussion is to make the point that lattice QCD is not some kind of approximation to "true" QCD but, when treated in the appropriate region of parameters

$$a << \xi << La \quad , \tag{6}$$

precisely is "true" QCD.

The basic geometric facets of the lattice are the sites, the links, the plaquettes, etc. A vector gauge field belongs most naturally to the links joining neighbouring sites

$$A_\mu(x) \to U_\mu(n) = e^{iaA_\mu} \approx 1 + iaA_\mu \quad . \tag{7}$$

For small a the U_μ are just the elementary line integrals of the continuum gauge potential. A tensor field could naturally be associated with a plaquette. With fermions the situation is less straightforward, and this contributes to the difficulties in calculating with them.

The partition function (1) then becomes

$$Z = \int \prod_{n,\mu} [\, dU_\mu(n)]\ e^{-S(U)} \quad , \tag{8}$$

where the measure is the Haar invariant measure over the gauge group SU(3). The action S(U) must reproduce the usual continuum Euclidean action (1-3) in the limit $a \to 0$. An example is the Wilson action

$$S(U) = \beta \sum_{n} (1 - \tfrac{1}{3} \mathrm{Tr}\,\square), \quad \beta = \frac{6}{g^2}, \tag{9}$$

where Tr $\square$ means taking the trace of the matrix obtained by multiplying together the four matrices U_μ on the links forming the plaquette.

Monte Carlo solution

Formulated in Euclidean space-time the lattice regulated gauge theory looks like a statistical mechanical system with a Boltzmann factor e^{-S} which suggests(4) using methods developed for solving such systems. The method that interests us here is the Monte Carlo technique(5).

Suppose we want to calculate the vacuum expectation value

$$\langle O(U) \rangle = \frac{1}{Z} \int \prod_{n,\mu} [dU_\mu(n)]\ O(U)\ e^{-S(U)} \quad . \tag{10}$$

The essence of the Monte Carlo method then is to generate a sequence of (typical) gauge field configurations according to the distribution $\prod_{n,\mu} [dU_\mu(n)]\ e^{-S(U)}$. Suppose we have N such configurations

$$\{U_\mu\ (n)\}_{i=1,\ldots,N} \quad . \tag{11}$$

Then the vacuum expectation value (10) is given by

$$\langle O(U) \rangle \simeq \frac{1}{N} \sum_{i=1}^{N} O(\{U_\mu(n)\}_i) \tag{12}$$

which becomes exact as $N \to \infty$.

2. Glueball Mass Spectrum (6)

We turn now to the calculation of the masses of the glueball states. All (SU(3)) calculations so far have been done in the quenched approximation, that is without fermion loops, using the Wilson action(9).

We first form wave-functionals with the desired J^{PC} quantum numbers. Since the theory is confining the glueballs are colour singlets, and an (over)complete set of colour singlet operators is given by the set of all closed loops. To make the operators have the required J^{PC} properties we take suitable linear combinations of rotations and reflections and real or imaginary parts (the real part having C = +1, the imaginary part having C = -1). In Fig. 1 I show for example the 2x2 loop and the linear combinations of this loop that project onto $J^{PC} = 0^{++}$ and $J^{PC} = 2^{++}$.

Calculational procedure

The by now standard technique[7] for calculating glueball masses proceeds roughly as follows[8]. Take the trial glueball wave-functional, $\Phi(\vec{n},t)$, constructed as above with the desired J^{PC} properties and centred on the site $(\vec{n},t)$. Make the $\vec{p}=0$ translation invariant sum

$$\Phi(\vec{p}=0,t) = \sum_{\vec{n}} \Phi(\vec{n},t) \tag{13}$$

and measure the correlation function

$$\begin{aligned}\Gamma_t &= \langle\Phi(\vec{p}=0,t)\ \Phi(\vec{p}=0,0)\rangle \\ &= \langle\Phi(\vec{p}=0,0)\ e^{-Ht}\Phi(\vec{p}=0,0)\rangle \\ &= \sum_{n=0}^{\infty} e^{-m_n t}\,|\langle\Omega|\Phi|n\rangle|^2 \quad ,\end{aligned} \tag{14}$$

where $|\Omega\rangle$ is the vacuum. Vary Φ over a class of wave-functionals to obtain a large enough projection onto the desired glueball state so that Γ_t is dominated by the lowest mass glueball propagator for $t \geq a$. Then obtain the lowest glueball mass from

$$ma = \ln \left\{\frac{\Gamma_{(n-1)a}}{\Gamma_{na}}\right\} . \tag{15}$$

The choice of n will depend on the purity of the wave-functional (as to the admixture of higher excited states). Generally n = 2 or n = 3

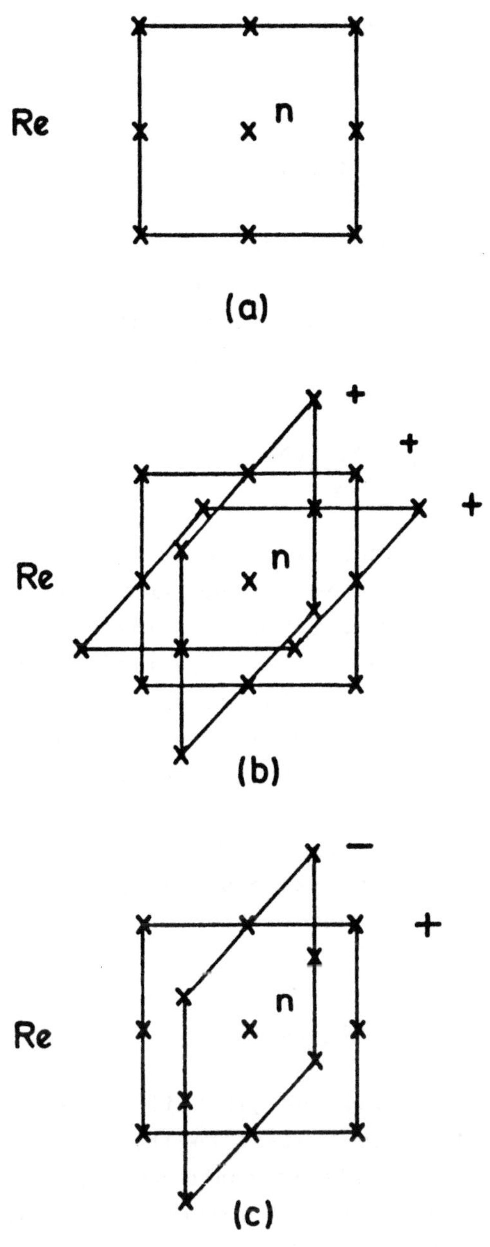

Fig.1: (a) A 2x2 loop of link matrices, (b) a 0^{++} combination of 2x2 loops, (c) a 2^{++} combination of 2x2 loops.

will be large enough at present values of β to make m_1, m_2, ... die away in equ. (14).

When are we in the continuum limit?

As I have indicated in the introduction we will have to do our calculations at sufficiently small lattice spacings such that the masses we obtain are indeed those of the continuum theory. In practice one assumes that the lattice spacing is small enough to be given by the perturbative formula

$$a(\beta) = \frac{83.5}{\Lambda_{mom}} \exp\left(-\frac{4\pi^2}{33}\beta\right) \left(\frac{8\pi^2}{33}\beta\right)^{\frac{51}{121}} , \qquad (16)$$

i.e. the inverse of (5), where the factor 83.5 comes from having replaced the lattice Λ parameter by the more conventional Λ_{mom} (9). A criterion for being in the continuum region then is that if we redo the calculation at larger values of β (smaller lattice spacings a), the dimensionless quantity ma (equ.(15)) we calculate should scale like a(β) . In other words m should be the same in units of Λ_{mom}. The problem with this criterion is that there are possibly higher order corrections to (16) which might become visible as the errors of the Monte Carlo calculations decrease with time. An alternative criterion which avoids (16) is that the ratio of masses or the ratio of mass to (string tension)$^{\frac{1}{2}}$ is independent of β.

The lattice spacing in physical units

The final step in calculating the glueball masses is to express these masses in MeV units to make contact to the real world (i.e. we must know a in MeV^{-1}). In principle we could use (16) with Λ_{mom} from deep inelastic experiments. However such values of Λ_{mom} are notoriously ill-determined. The absence of any firm experimental glueball candidates (though recently two glueball candidates have been announced (10)) rules out to normalize any one of the calculated glueball masses to experiment. Instead one can calculate (say) the ρ mass and use its experimental value to set the scale of a(β), provided that the calculated hadron spectrum comes out reasonably well. Such a calculation exists[11] (see below) and we shall use it. One could also normalize to calculations of the string tension[12], the gluon condensate[13] $<\frac{\alpha_s}{\pi} F_{\mu\nu}F_{\mu\nu}>$ or the chiral condensate[14] $<\bar{\psi}\psi>$. Since our mass calculations will be predominantly from $\beta \sim 5.7$ we shall center our analysis around this value (which should be borne in mind as the two-loop perturbative formula (16) is not likely to be exactly valid). Using (16) we obtain

from	$\Lambda_{mom}(\beta{\sim}5.7)$ MeV
rho mass[11], m_ρ	210 ± 30
string tension[12], $\sqrt{K}$ = 400 MeV	230 ± 30
gluon condensate[13], $<\frac{\alpha_s}{\pi} F_{\mu\nu}F_{\mu\nu}>=(336\pm10\text{MeV})^4$ [15]	180 ± 20
chiral condensate[14], $<\bar{\psi}\psi>_{inv}=(225\pm25\text{MeV})^3$ [16]	195 ± 25

This gives us consistently

$$\Lambda_{mom} \approx 200 \text{ MeV} \tag{17}$$

and

$$a(\beta{=}5.7) = 1.37\text{GeV}^{-1} = 0.28 \text{ fermi}. \tag{18}$$

Results

On a hypercubic $L_s^3 \times L_t$ lattice (of $L_s(L_t)$ lattice spacings in the spacial (time) direction the inequality (6) reads in a more graphic language

$$a << D << \tfrac{1}{2} \min (L_s a, L_t a) \tag{19}$$

where (here) D is the glueball diameter. The factor $\frac{1}{2}$ comes from using the periodic boundary conditions to reduce the obvious finite volume effects. In order for the calculation of hadron properties to be credible we also require the temperature ($= 1/L_t a$) to be much smaller than the deconfinement temperature of ≈ 200 MeV[17]. To realize (19) we need at least a rough estimate of the glueball size.

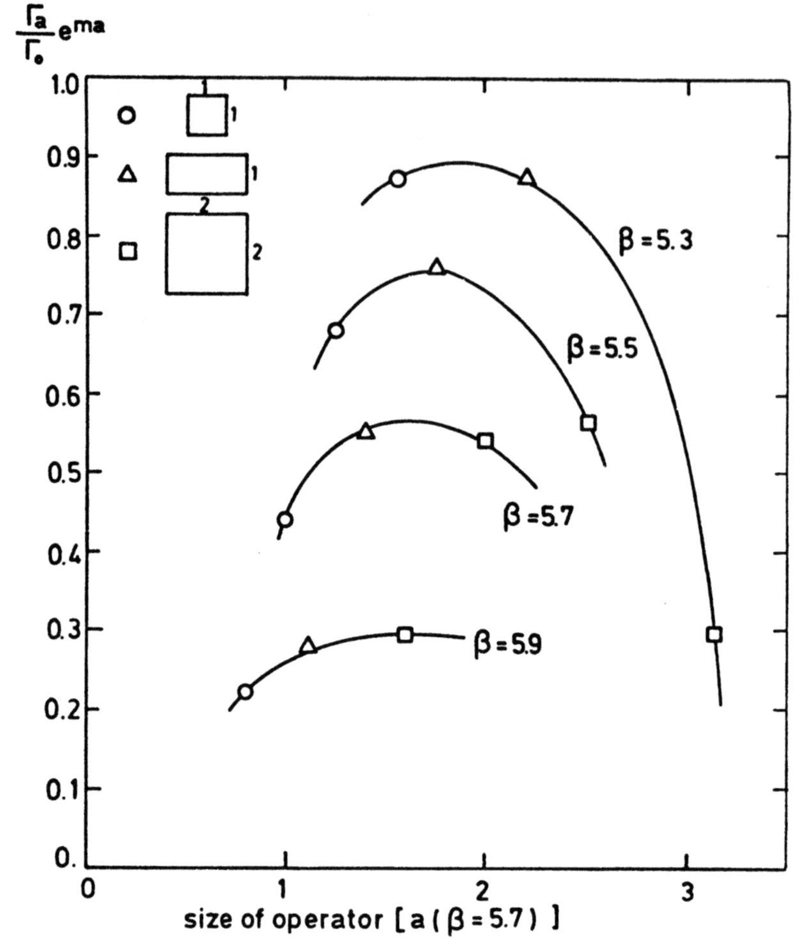

Fig.2: Γ_a/Γ_o for various operators versus the size of the operators in units of the ß = 5.7 lattice spacing.

This may be found in ref.18 where we have calculated Γ_a/Γ_0 for several basic loops of differing sizes. It should have a maximum at approximately the glueball diameter. The result of the calculation done at several values of β I have shown in Fig. 2, where the geometrical loop sizes have been expressed in terms of the lattice spacing a(β=5.7) using (16). We find

$$D \approx (1.5\text{-}2.0)\ a(\beta=5.7) \approx 0.5 \text{ fermi.} \tag{20}$$

From (20) we conclude that (19) is more or less satisfied for $5.3 \lesssim \beta \lesssim 5.9$ (albeit with an optimistic interpretation of the strong inequalities in (19)) for $L_s, L_t = 4$ and accordingly for larger lattices. If we now pick $L_t = 8$ then the temperature varies from ≈ 60 to ≈ 120 MeV in this range of β values which is also well below the deconfinement temperature.

The calculation of glueball masses requires high (Monte Carlo) statistics. For this reason almost all of the work has been on lattices as small as possible, typically $L_s^3 \times L_t = 4^3 \times 8$. To be sure that finite size effects are small it is important to also have a glueball calculation on a much larger lattice. Such a calculation now exists[19], showing that the results on an 8^4 lattice do not differ very much from those on $4^3 \times 8$ lattices as we shall see.

We have systematically calculated[18,20] the masses of the supposedly low-lying glueball states. We find four light glueballs, the 0^{++}, 2^{++}, 0^{-+} and the oddball 1^{-+}. In addition to our work there have also been estimates for the bulk of the glueball spectrum by Berg and Billoire[21,22] and a 0^{++} estimate by Michael and Teasdale[23].

0^{++}:

In Fig.3 I have plotted the statistically best $L_s=4$ results[18] together with some results on a $5^3 \times 8$ lattice[22] and those on an 8^4 lattice[19]. All the estimates come from Γ_{2a}/Γ_a except for β=5.9 where Γ_{3a}/Γ_a has been used. We observe that any systematic variation of $m(0^{++})$ between β=5.3 and β=5.9 is less than 20%. In this same range of **β** the lattice spacing (16) decreases by a factor of ≈ 2, which leads us to conclude that we are seeing evidence for the calculated 0^{++} mass being that of the continuum theory.

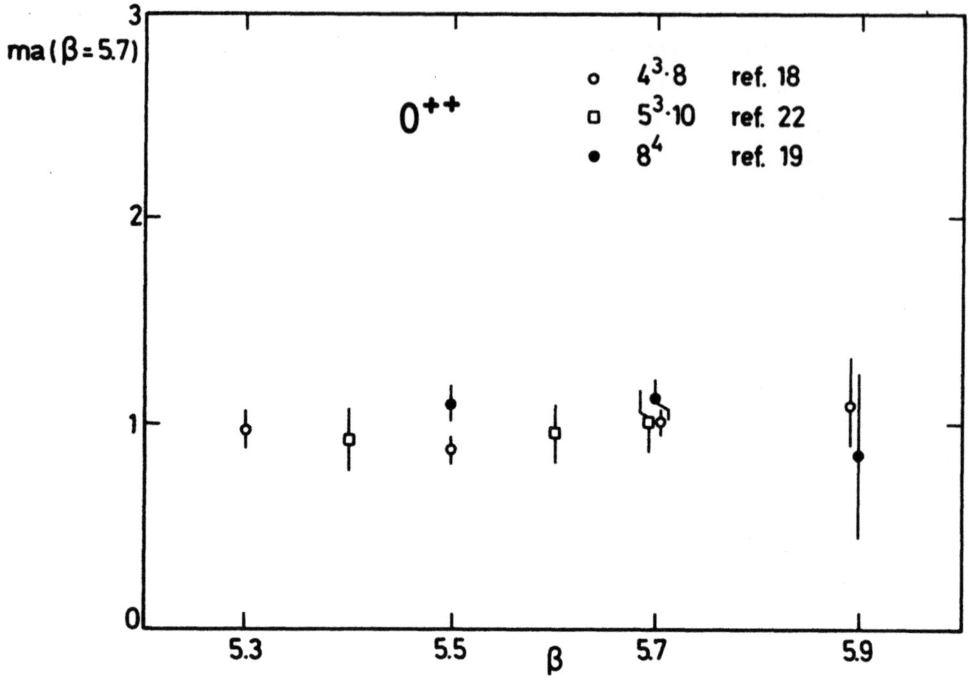

Fig.3: 0^{++} mass on 4^3x8, 5^3x8 and 8^4 lattices in units of a^{-1}(ß=5.7).

If we compare $m(0^{++})$ obtained on $4^3 \times 8$ and 8^4 lattices we find that any changes are small. The largest change is at $\beta = 5.5$ which might be associated with the diminishing of the specific heat peak(24) with increasing lattice size(13).

2^{++}:

In Fig.4 I have plotted our results obtained on $4^3 \times 8$(19) and 8^4 (18) lattices. The systematic variation of $m(2^{++})$ between $\beta = 5.3$ and $\beta = 5.9$ is $\approx$ 30% which again is much smaller than the $\approx$ 100% decrease of the lattice spacing in this range of β. This gives us confidence that also the calculated 2^{++} mass is that of the continuum. The masses in Fig.4 show no finite size effects outside the errors.

The 2^{++} glueball mass has also been calculated by Berg and Billoire(22). They claim to see no continuum scaling behaviour. For a possible resolution of this discrepancy I refer the interested reader to refs. 6, 18.

0^{-+} and 1^{-+}:

There are claims(18) for a light pseudoscalar 0^{-+}glueball and a light 1^{-+} oddball. Oddballs are glueballs which have quantum numbers not accessible to a quark-antiquark pair. So far there is no numerical test of continuum behaviour or of finite size effects for either state. The presumption is that finite size effects will be small and that the masses have reached their continuum limit which needs however to be checked.

Summary of masses:

The calculated masses of the low-lying glueball states are

$$\begin{aligned} m(0^{++}) &= 770 \pm 40 \text{ MeV}, \\ m(0^{-+}) &= 1420 {}^{+240}_{-170} \text{ MeV}, \\ m(2^{++}) &= 1670 {}^{+110}_{-90} \text{ MeV}, \\ m(1^{-+}) &= 1730 \pm 220 \text{ MeV}. \end{aligned} \tag{21}$$

The scale used is $\Lambda_{mom} = 200$ MeV as I have discussed above. For the experimental status of the search for glueballs see ref. 25.

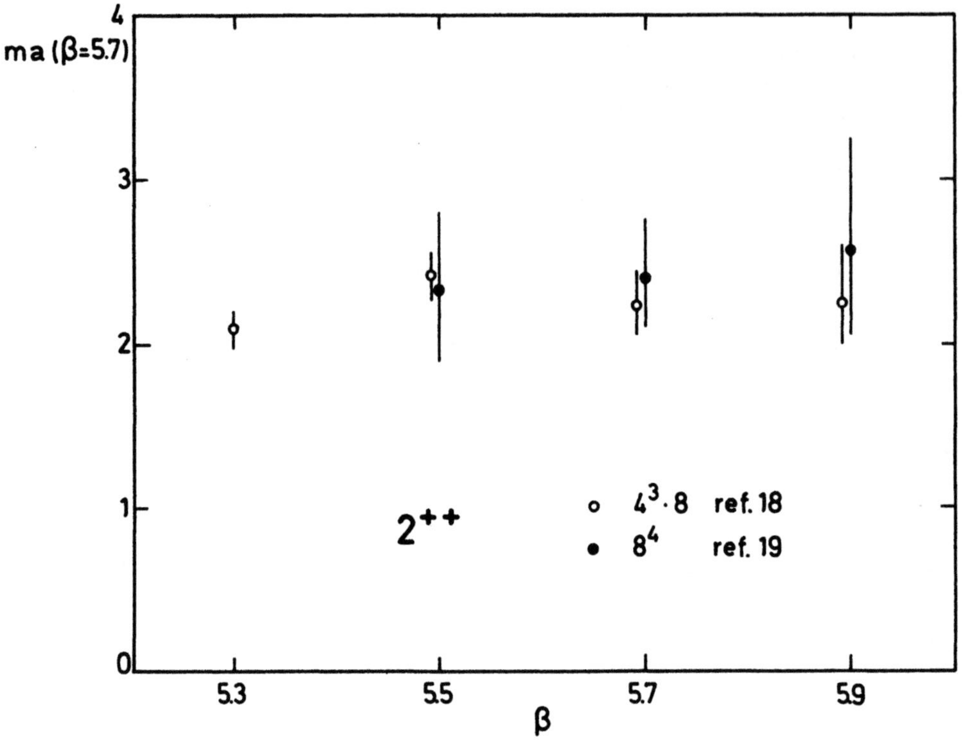

Fig.4: 2^{++} mass on 4^3x8 and 8^4 lattices in units of a^{-1}(ß=5.7).

3. Evidence for Spontaneous Chiral Symmetry Breaking?

Until recently it was not established whether QCD breaks its chiral symmetry dynamically or not. This problem is made difficult by its nonperturbative nature and required to go new roads to resolve it: the lattice Monte Carlo method.

Fermions on the lattice

So far we have dealt with the pure gauge theory only. To put quarks onto the lattice is less straightforward, and I refer the reader for a more detailed discussion to the literature(26).

For the fermions we want an action which explicitly possesses a continuous chiral symmetry on top of reproducing the usual continuum Euclidean action in the limit $a \to 0$. The Wilson fermion action(27) breaks all chiral symmetries by irrelevant operators, and there is no evidence that it has recovered chiral symmetry (enough to break it spontaneously) in the range of couplings numerically accessible at present. The naive action

$$S(\bar{\psi}, \psi, U) \equiv -\bar{\psi}[M(U)+2ma]\psi \tag{22}$$

$$= -\sum_{n,\mu} \bar{\psi}(n)\gamma_\mu \,[U_\mu(n)\,\psi(n+\mu) - U_\mu^+(n-\mu)\,\psi(n-\mu)] - 2ma\sum_n \bar{\psi}(n)\psi(n)$$

has an obvious chiral symmetry when $m = 0$ but describes 16 "flavours" instead of one. Unfortunately a local, explicitly chiral invariant formulation of the fermion action on the lattice without species doubling does not exist(28). The next best choice are Kogut-Susskind staggered fermions(29) which are identical to naive fermions if one only excites one of their 4 decoupled modes and describe 4 flavours. This relationship is most clearly seen if we decompose the (anti-hermitean) matrix M into

$$M = \Gamma \mathcal{M} \Gamma^+ , \tag{23}$$

where Γ is a unitary block diagonal matrix containing all the γ matrices(30), and write

$$\psi = \Gamma\chi \quad . \tag{24}$$

Then the action (22) becomes

$$S(\bar{\chi}, \chi, U) = -\bar{\chi}[M(U)+2ma]\chi \quad , \tag{25}$$

where the 4 Dirac components of χ decouple. This allows us to work with a single component fermion field , thereby reducing the number of degrees of freedom by a factor of 4. The action (25) reduced to a single component χ field is the action for staggered fermions. The interpretation as 4-flavoured lattice fermionic fields[31] follows by dividing the lattice into hypercubes and defining a new field (of 4 Dirac components x 4 flavours) to be $\chi(n)$ times an appropriate phase factor at each of the 16 corners of the hypercube. The matrix M is a NxN sparse complex matrix where $N = 3\ L_s^3 \times L_t$. We shall use antiperiodic boundary conditions for the fermion fields.

The action for staggered fermions has a U(4) x U(4) symmetry for m=0 in the continuum limit. At finite lattice spacing the U(4) chiral symmetry is explicitly broken down to U(1). The important point is that there is an explicit continuous chiral U(1) symmetry for which

$$\langle\bar{\psi}\psi(m)\rangle = \frac{3}{N} \ \langle Tr(M+2ma)^{-1}\rangle \tag{26}$$

is an order parameter, and this U(1) is not the anomaly afflicted piece of U(4) but belongs to the SU(4) [32].

In what follows we shall have to restrict ourselves to the quenched approximation as in case of the glueball calculation. That means the vacuum expectation values (cf. equs. (10) - (12)) of interest are defined by averaging over gauge field configurations which are in equilibrium with the pure gauge field action (9). So far nobody has mastered to include fermion loops in generating "snapshots" of the vacuum in an efficient way.

$\langle\bar{\psi}\psi\rangle$

The usual criterion for chiral symmetry breaking is that the chiral condensate (26) should not vanish as $m \to 0$. Since a finite lattice (being a system with a finite number of degrees of freedom) will, given enough time, rotate through all the degenerate minima of the effective potential so that one obtains $<\bar{\psi}\psi(m\to 0)> = 0$ even if the symmetry is dynamically broken, the correct criterion is that

$$\lim_{m\to 0}\ \lim_{\text{volume}\to\infty} <\bar{\psi}\psi(m)> \neq 0. \qquad (27)$$

To check whether (27) holds or not one must calculate (26) for several different lattice volumes. The condensate should tend to an envelope which has a nonzero intercept at $m = 0$. The largest volumes should be such that the mass at which $<\bar{\psi}\psi(m)>$ breaks away from the envelope to collapse to zero is much smaller than the gauge theory scale, $O(\Lambda_{mom})$, to demonstrate that this is a kinematic rather than a dynamic effect. It is clear that such a calculation requires great accuracy at very small quark masses.

To show that the spontaneous breaking is a continuum property one must do the calculation over a range of β and demonstrate the correct continuum renormalization group behaviour for (27).

The numerical problem we are faced with in (26) is that of calculating the inverse or the eigenvalues of a very large complex matrix. The best method we know at present for obtaining (selected rows of) the inverse is the conjugate gradient method(33), and the best method we know for obtaining the eigenvalues is the Lanczos algorithm(33). In the present context there are good reasons for preferring the latter method.

Results(14)

In Fig.5 I have plotted the calculated $<\bar{\psi}\psi(m)>$ for $4^3 \times 8$, $6^3 \times 8$ and 8^4 lattices at $\beta=5.7$. Also shown is $<\bar{\psi}\psi(m)>$ at 3 mass values for a $10^3 \times 16$ lattice. These values come from a hadron spectrum calculation(11) using the conjugate gradient algorithm. We observe that a $4^3 \times 8$ lattice shows no significant signal of spontaneous chiral symmetry breaking. As we increase the lattice we see a clear envelope developing, and this envelope has a nonzero intercept at zero quark mass. Comparing our 8^4 and $10^3 \times 16$ results we see that for this

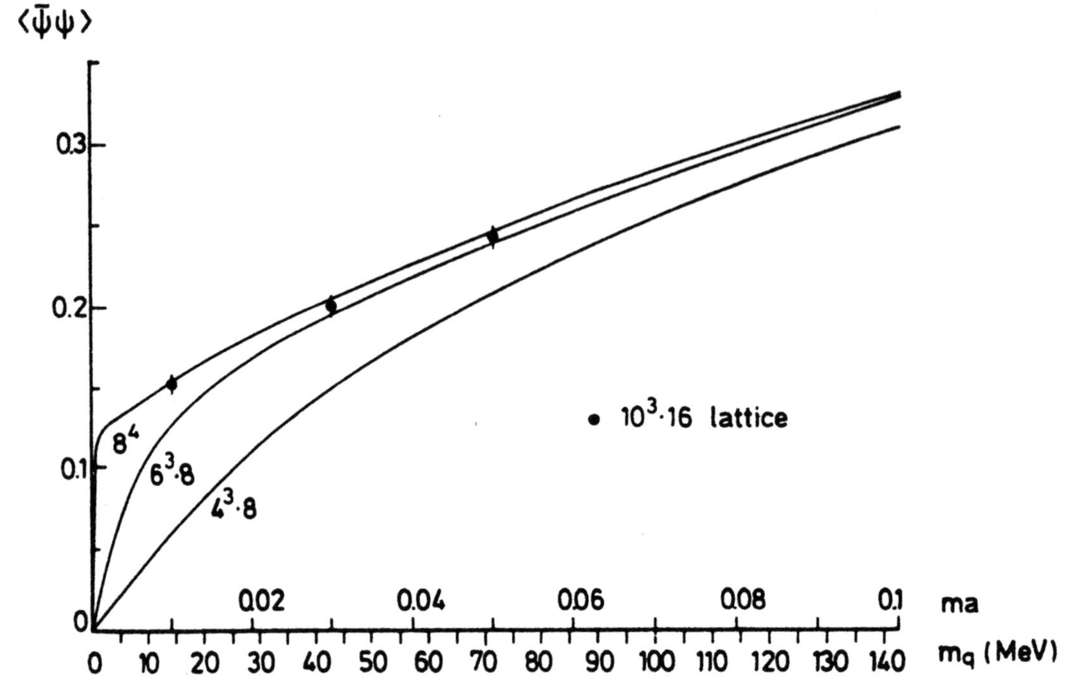

Fig.5: $\langle\bar{\psi}\psi\rangle$ as a function of the bare (renormalization group invariant) quark mass ma (m_q) obtained on the 4^3x8, 6^3x8 and 8^4 lattice at ß = 5.7. The solid circles come from the hadron spectrum calculation on the 10^3x16 lattice to be described later on.

physics an 8^4 lattice at $\beta\sim5.7$ is effectively of infinite volume down to $ma \approx 0.01$. The break-away occurs at a quark mass $ma = O(0.002)$ where it is obviously a kinematic effect. The bare quark mass m is related to the more physical renormalization group invariant quark mass[16] m_q also indicated in Fig.5 by

$$m_q = \alpha_{mom}^{-\frac{4}{\pi}} m, \quad \alpha_{mom} = \frac{g^2{}_{mom}}{4\pi} . \tag{28}$$

To demonstrate that this is a continuum effect we have calculated $<\bar{\psi}\psi(m)>$ for values of β from 0 to 5.9. We find that we have roughly a linear dependence on ma for $ma \lesssim 0.1$ until the break-away to zero occurs. We obtain $<\bar{\psi}\psi(0)>$ by extrapolating this linear dependence to ma=0. Our values for $<\bar{\psi}\psi(0)>$ are plotted in Fig.6. We can see the strong coupling behaviour[34] at low β, a transition region between β = 4 and 5.1 and very clear continuum (asymptotic freedom) behaviour for $\beta > 5.1$.

Our calculated (dimensionless) $<\bar{\psi}\psi>$ can be expressed in terms of the continuum (dimensionful) renormalization group invariant $<\bar{\psi}\psi>_{inv}$ by

$$<\bar{\psi}\psi> = 2\alpha_{mom}^{-\frac{4}{\pi}} a^3 <\bar{\psi}\psi>_{inv} \tag{29}$$

(the normalization uses 4 flavours). Using (16) and Λ_{mom} = 200 MeV (see (17) and above) I have plotted $<\bar{\psi}\psi>^{1/3}_{inv}$ as a function of β in Fig.7. We see that we have good numerical agreement with the experimental value[16] for $<\bar{u}u>$.

4. Meson and Baryon Masses

We may hope now that the Kogut-Susskind fermionic lattice QCD action (25) will also provide a realistic description of the meson and baryon spectrum.

The fact that the promise of the early calculations[35] on very small lattices was not fulfilled by later, more detailed work[36] emphasized for us the dangers of using Wilson fermions. This was widely interpreted at the time as a problem of finite size effects[37], with the cure to be found by increasing the lattice size. That finite size

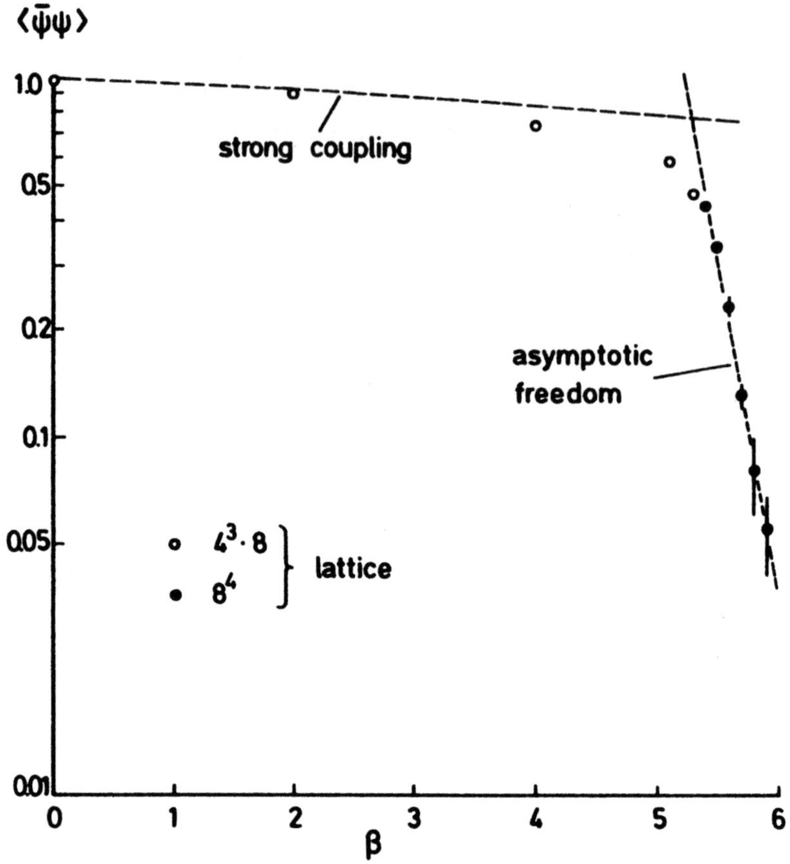

Fig.6: The intercept $\langle\bar{\psi}\psi(0)\rangle$ for various values of ß together with the O(1/d) strong coupling and scaling (asymptotic freedom) curves.

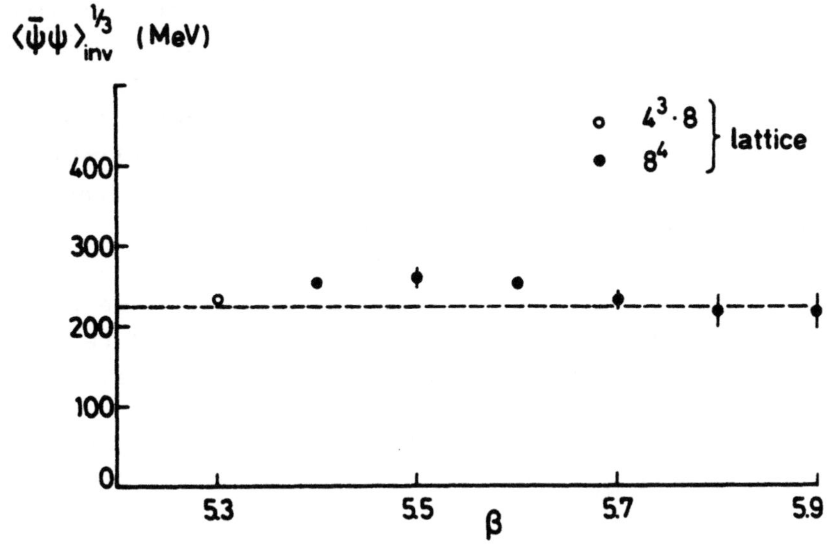

Fig.7: The renormalization group invariant $\langle\bar{\psi}\psi\rangle_{inv}^{1/3}$ for 5.3 = ß = 5.9. The dashed line represents the experimental value $\langle\bar{u}u\rangle$ = 225 ($\pm$25) MeV.

effects were part of the problem was clear. There is however also the possibility that these effects are being magnified by the forced extrapolation of mass calculations performed at large pion masses to zero pion mass in a theory that has not yet developed a chiral symmetry that can be spontaneously broken. Recent calculations with Wilson fermions on large $10^3 \times 20$ [38] and 16^4 [39] lattices have in fact shown that the problems are still there.

In continuation of our work on chiral symmetry breaking we have calculated the meson and baryon masses on a $10^3 \times 16$ lattice [11]. We have performed our calculation at $\beta=5.7$. There the spatial extent of our lattice is (cf. equ. (18)

$$10a(\beta=5.7) = 2.8 \text{ fermi} \gtrsim 2D, \tag{30}$$

where D is a typical meson or baryon diameter. This seems a reasonably large space for the hadron on a lattice (with antiperiodic fermionic boundary conditions).

I shall not go into details of the flavour structure of staggered fermions. All what we need for calculating meson and baryon masses are operators that project onto meson and baryon states made out of the desired combination of flavours. The local operators

$$\bar\psi(n)\gamma_5\psi(n),\bar\psi(n)\gamma_\mu\psi(n),\bar\psi(n)\gamma_5\gamma_\mu\psi(n),\bar\psi(n)\psi(n),\bar\psi(n)\sigma_{\mu\nu}\psi(n) \tag{31}$$

and

$$\varepsilon_{ABC}(\psi^A(n)C\gamma_5\psi^B(n))\psi^C(n),\ \varepsilon_{ABC}(\psi^A(n)C\gamma_\mu\psi^B(n))\psi^C(n) \tag{32}$$

will do this job with $\psi(n) = \Gamma(n)\chi(n)$, $\chi(n)$ being a single component fermion field (cf. equ.(25)). The operators (31) project onto the[32] π, ρ, A_1, ε (i.e. the $(I,J)^{PC} = (0,0)^{++}$singlet) and B, respectively, while (32) project onto the nucleon (N) and Δ and their negative parity partners. This then leads us to evaluate the propagators ($\Gamma(0) = 1$)

$$M_{PS}(t) = \sum_{\substack{\vec{n} \\ t=n_4 a}} \tfrac{1}{4}\, Tr(\gamma^+(n)\gamma_5\Gamma(n)\gamma_5)\ \langle\bar{\chi}(n)\chi(n)\bar{\chi}(0)\chi(0)\rangle$$

$$= \sum_{\substack{\vec{n} \\ A,B}} \langle|G^{AB}(n)|^2\rangle = C_\pi(e^{-m_\pi t} + e^{-m_\pi(16a-t)}) + \ldots,$$

etc. (33)

where

$$G^{AB}(n'-n) = [(M+2ma)^{-1}]^{AB}_{nn'} \quad . \tag{34}$$

We have used the conjugate gradient algorithm[33] to calculate the quark propagators (34). As before we work in the quenched approximation. That is the meson and baryon propagators (33) are sampled over a set of gauge field configurations in equilibrium with the action (9).

The result we obtain for the pion mass is shown in Fig.8. We find that $m_\pi^2 \sim m_q$, which is what one expects for a Goldstone pion. The dashed line in Fig.8 is parameterized by

$$(m_\pi a) = 7.60\ ma. \tag{35}$$

Using Λ_{mom} = 200 MeV again we infer from (35) that the pion takes its experimental mass value for ma ≈ 0.005, which corresponds to a renormalization group invariant quark mass of

$$m_q \approx 7\ \text{MeV} \quad . \tag{36}$$

This is the mass value one also obtains from current algebra[16]. The results for m_ρ, m_{A_1}, m_ε and the nucleon mass m_N are shown in Fig.9. The calculated hadron masses can be interpolated by the linear mass fits

$$m_\rho a = 0.98 + 4.10\ ma, \tag{37}$$
$$m_{A_1} a = 1.61 + 3.30\ ma,$$
$$m_\varepsilon a = 0.88 + 5.70\ ma,$$
$$m_N a = 1.21 + 9.00\ ma,$$

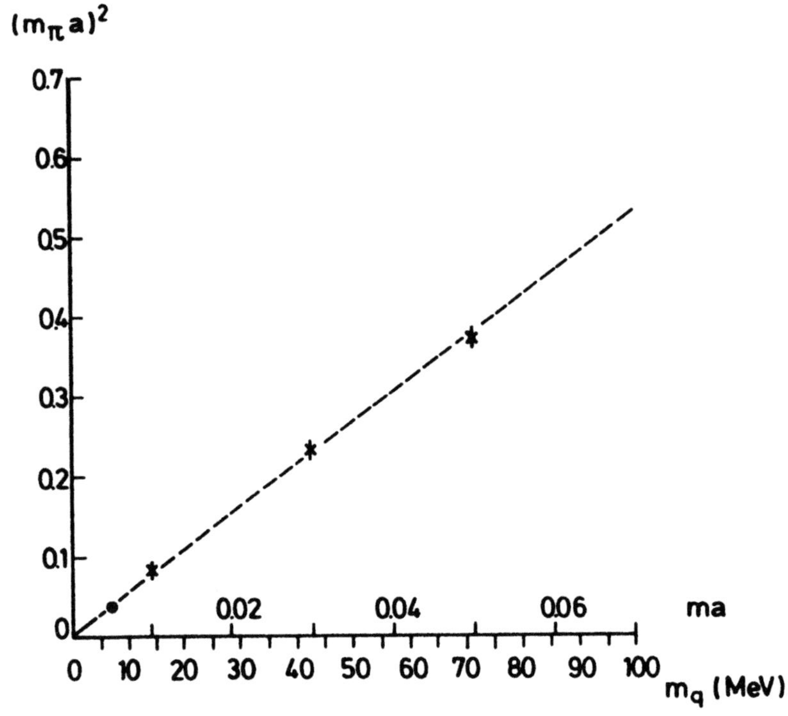

Fig.8: $(m_\pi)^2$ as a function of ma (m_q). The solid circle on the linear extrapolation curve marks the position of the physical pion, from which we can read off the value of the quark mass m_q.

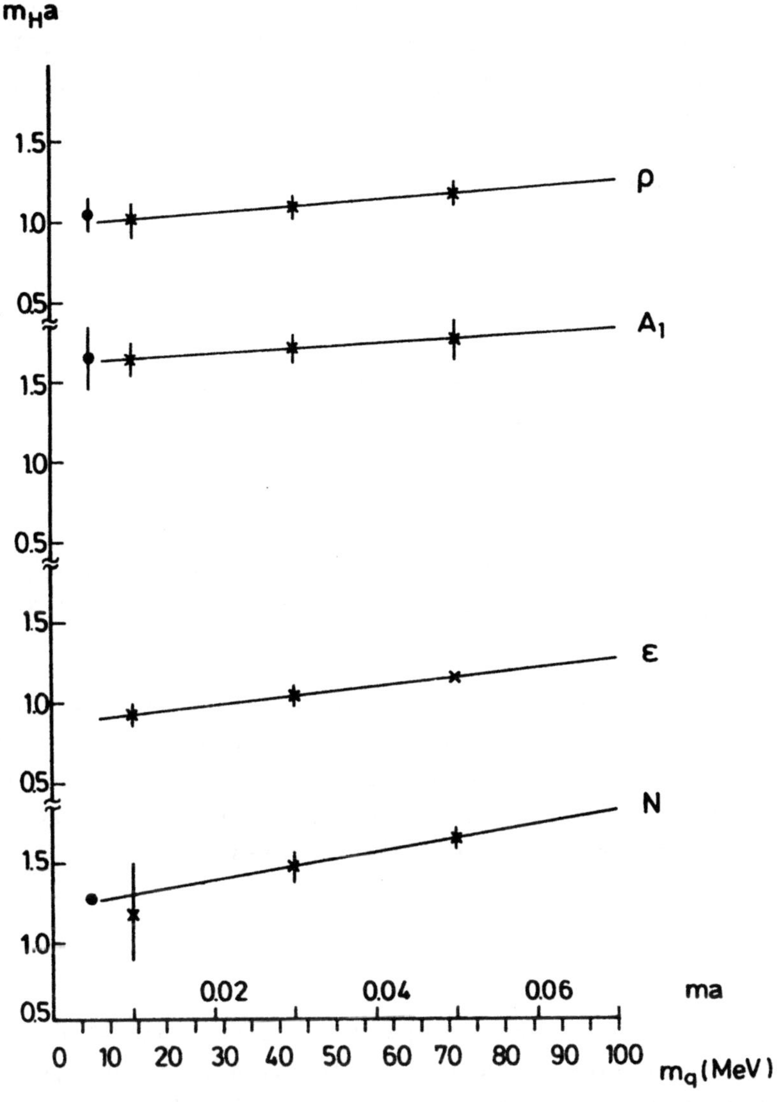

Fig.9: The hadron masses, $m_H a$, as a function of ma (m_q). The solid circles at the physical quark mass indicate their experimental values.

which correspond to the solid lines in Fig.9. We shall use (37) now to evaluate m_ρ, m_{A_1}, m_ε and m_N at the physical quark mass (36). For our usual Λ_{mom} = 200 MeV and taking the statistical errors porperly into account we obtain

$$\begin{aligned} m_\rho &= (730 \pm 90)\ \text{MeV}, \\ m_{A_1} &= (1119 \pm 90)\ \text{MeV}, \\ m_\varepsilon &= (660 \pm 50)\ \text{MeV}, \\ m_N &= (920 \pm 100)\ \text{MeV}. \end{aligned} \qquad (38)$$

This fits the experimental low-lying hadron masses quite well. From the slope of (35) we also extract

$$f_\pi \approx 134\ \text{MeV} \qquad (39)$$

using $m_\pi^2 f_\pi^2 = 4m_q\langle\bar\psi\psi\rangle_{inv}$ and our calculated value of $\langle\bar\psi\psi\rangle_{inv}$. This is to be compared to the experimental value of [16] f_π = (131.9 ± 0.1)MeV.

We have repeated our calculation at β=5.4 and 5.9 to check for continuum scaling behaviour[40]. We find that the hadron masses scale within the statistical errors, except for the slope of the pion (35) at β =5.4 which however has a natural explanation. We are also calculating the masses of various other states.

5. Concluding Remarks

We may say that altogether we begin to obtain a consistent picture of quark and gluon physics in quenched QCD, which is encouraging and more than we could expect on these rather coarse lattices and from working at the "edge" of the continuum region.

In the near future we will be able to double the sizes of our lattices, and we will have accumulated higher statistics. This will enable us to reduce the systematic uncertainties and the statistical errors. We are also beginning to make progress in including fermion loops.

Acknowledgements

I like to thank my colleagues Ian Barbour, Philip Gibbs, Philip Gilchrist, Kenzo Ishikawa, Heinz Schneider and Mike Teper for a most prolific collaboration and for continuous discussions on the subject of this talk.

References

1. K. Wilson: Phys. Rev. D10, 2445 (1974)

2. R.P. Feynman: Rev. Mod. Phys. 20, 267 (1948);
   R.P. Feynman and A.R. Hibbs: Quantum Mechanics and Path Integrals (McGraw-Hill, 1965)

3. G. 't Hooft and M. Veltman: Diagrammar, CERN report 73-9 (1973)

4. K. Wilson and J. Kogut: Phys. Rep. 12C, 75 (1974)

5. M. Creutz: Phys. Rev. Lett. 43, 553 (1979);
   K. Wilson: Cargèse lectures (1979);
   M. Creutz, L. Jacobs and C. Rebbi: Phys. Rev. D20, 1915 (1979)

6. For a recent review see:
   M. Teper: talk at the Brighton conference (1983), and LAPP preprint TH-91 (1983)

7. M. Falcioni, E. Marinari, M.L. Paciello, G. Parisi, F. Rapuano, B. Taglienti and Zhang Yi-cheng: Phys. Lett. 110B, 295 (1982);
   K. Ishikawa, G. Schierholz and M. Teper: Phys. Lett. 110B, 399 (1982);
   B. Berg, A. Billoire and C. Rebbi: Ann. Phys. 142, 185 (1982)

8. K. Ishikawa, G. Schierholz and M. Teper: Zeitschr. f. Phys. C19, 69 (1982)

9. A. Hasenfratz and P. Hasenfratz: Phys. Lett. 93B, 165 (1980)

10. D.L. Scharre et al.: Phys. Lett. 97B, 329 (1980);
    C. Edwards et al.: Phys. Rev. Lett. 48, 458 (1982);
    ibid. 49, 259 (1982)

11. J.P. Gilchrist, G. Schierholz, H. Schneider and M. Teper: DESY preprint 83-094 (1983), to be published in Phys. Lett. B

12. For recent calculations on large lattices see:
    M. Fukugita, T. Kaneko and A. Ukawa: KEK preprint TH 63 (1983);
    F. Gutbrod, P. Hasenfratz, I. Montvay and Z. Kunszt: CERN preprint TH-3591 (1983)

13. G. Schierholz and M. Teper: in preparation;
    see also ref.6

14. I.M. Barbour, P. Gibbs, J.P. Gilchrist, G. Schierholz, H. Schneider and M. Teper: DESY preprint 83-093 (1983), to be published in Phys. Lett. B

15. G. Launer, S. Narison and R. Tarrach: CERN preprint TH-3712 (1983)

16. J. Gasser and H. Leutwyler: Phys. Rep. 87C, 78 (1982)

17. J. Engels, F. Karsch, H. Satz and I. Montvay: Nucl. Phys. B205, 545 (1982);
    J. Kogut, M. Stone, H.W. Wyld, W.R. Gibbs, J. Shigemitsu, S.H. Shenker and D.K. Sinclair: Illinois preprint TH-82-39 (1982)

18. K. Ishikawa, A. Sato, G. Schierholz and M. Teper: DESY preprint 83-061 (1983), to be published in Zeitschr. f. Phys. C

19. G. Schierholz and M. Teper: DESY preprint 83-107 (1983), to be published in Phys. Lett. B

20. K. Ishikawa, G. Schierholz and M. Teper: Phys. Lett. 116B, 429 (1982);
K. Ishikawa, A. Sato, G. Schierholz and M. Teper: Phys. Lett. 120B, 387 (1983)

21. B. Berg and A. Billoire: Phys. Lett. 113B, 65 (1982); ibid. 114B, 324 (1982); Nucl. Phys. B221, 109 (1983)

22. B. Berg and A. Billoire: Nucl. Phys. B226, 405 (1983)

23. C. Michael and I. Teasdale: Nucl. Phys. B215, 433 (1983)

24. B. Lautrup and M. Nauenberg: Phys. Rev. Lett 45, 1755 (1980)

25. E. Bloom: talk at the Paris conference (1982)

26. See e.g. the reprint collection:
C. Rebbi: Lattice Gauge Theories and Monte Carlo Simulations (World Scientific, 1983)

27. K. Wilson: Erice lectures (1975)

28. L.H. Karsten and J. Smit: Nucl. Phys. B183, 103 (1981);
H.B. Nielsen and M. Ninomiya: Nucl. Phys. B185, 20 (1981)

29. J. Kogut and L. Susskind: Phys. Rev. D11, 395 (1975);
L. Susskind: Phys. Rev. D16, 3031 (1977);
T. Banks, J. Kogut and L. Susskind: Phys. Rev. D13, 1043 (1976);
T. Banks, S. Raby, L. Susskind, J. Kogut, D.R.T. Jones, P.N. Scharbach and D. Sinclair: Phys. Rev. D15, 1111 (1976);
H.S. Sharatchandra, H.J. Thun and P. Weisz: Nucl. Phys. B192, 205 (1981)

30. N. Kawamoto and J. Smit: Nucl. Phys. B192, 100 (1981)

31. A. Duncan and R. Roskies: Pittsburgh preprint PITT-82-6 (1982);
P. Becher and H. Joos: Zeitschr. f. Phys. C15, 343 (1982);
T. Banks, Y. Dothan and D. Horn: Phys. Lett. 117B, 413 (1982);
H. Kluberg-Stern, A. Morel, O. Napoly and B. Petersson: Saclay preprint DPh/83/29 (1983);
J. Kogut, M. Stone, H. Wyld, S. Shenker, J. Shigemitsu and D. Sinclair: Illinois preprint DOE/ER/01545-335 (1983)

32. H. Kluberg-Stern, A. Morel, O. Napoly and B. Petersson: ref.31

33. I.M. Barbour, J.P. Gilchrist, G. Schierholz, H. Schneider and M. Teper: Phys. Lett. 127B, 433 (1983), and references therein

34. T. Jolicoeur, H. Kluberg-Stern, M. Lev, A. Morel and B. Petersson: Saclay preprint SPhT/83-70 (1983)

35. H. Hamber and G. Parisi: Phys. Rev. Lett. 47, 1792 (1981); E. Marinari, G. Parisi and C. Rebbi: Phys. Rev. Lett. 47, 1795 (1981); D. Weingarten: Phys. Lett. 109B, 57 (1982); A. Hasenfratz, P. Hasenfratz, Z. Kunszt and C.B. Lang: Phys. Lett. 110B, 289 (1982)

36. D. Weingarten: Nucl. Phys. B215, 1 (1983); F. Fucito, G. Martinelli, C. Omero, G. Parisi, R. Petronzio and F. Rapuano: Nucl. Phys. B210, 407 (1982); R. Gupta and A. Patel: Caltech preprint 68-966 (1982); A. Hasenfratz, P. Hasenfratz, Z. Kunszt and C.B. Lang: Phys. Lett. 117B, 81 (1982)

37. P. Hasenfratz and I. Montvay: Phys. Rev. Lett. 50, 309 (1983); G. Martinelli, G. Parisi, R. Petronzio and F. Rapuano: Phys. Lett. 122B, 283 (1983); R. Gupta and A. Patel: Phys. Lett. 124B, 94 (1983); C. Bernard, T. Draper and K. Olynyk: Phys. Rev. D27, 227 (1983)

38. H. Lipps, G. Martinelli, R. Petronzio and F. Rapuano: CERN preprint TH-3548 (1983)

39. P. Hasenfratz and I. Montvay: DESY preprint 83-072 (1983)

40. J.P. Gilchrist, G. Schierholz, H. Schneider and M. Teper: in preparation

FINITE TEMPERATURE QCD

J. Engels
Fakultät für Physik
Universität Bielefeld
Germany

Abstract:

A review of finite temperature lattice calculations for quantum chromodynamics is given. We show how the thermodynamic quantities can be evaluated by Monte Carlo methods, once finite temperature field theory has been formulated on a lattice. The existing results for chemical potential zero and in quenched approximation are discussed. They exhibit a clear first order transition for SU(3) lattice QCD and probably a second order transition for SU(2) lattice QCD. The chiral and deconfinement transitions are coinciding in the quenched approximation.

I. Introduction

The composite nature of hadrons, as assumed by the quark model or quantum chromodynamics (QCD), leads to the immediate conjecture, that at sufficiently high temperatures or densities strongly interacting matter undergoes a phase transition from a state of interacting hadrons to a quark-gluon plasma[1]. The phase transition may eventually proceed in two distinct steps[2]: one, where the quarks and gluons bound in hadrons deconfine into massive quarks and gluons and a second transition, in which chiral symmetry is restored and the quarks and gluons become massless.

Since, with quantum chormodynamics we hope to have the basic theory of strong interactions, we should be able to show that QCD indeed leads to a phase diagram for strong-interaction matter like the one depicted in fig. 1. It is still unclear, however, whether the chiral and deconfinement transitions occur at different critical temperatures and if so, which one is lower.

The Monte Carlo evaluation of lattice QCD[3] gives us in principal the possibility to cover the whole range of physical temperatures T (and chemical potentials μ). Because of technical limitations at present only parts of the phase diagram are investigated. There exist detailed calculations for zero chemical potential[4-11] - in the "quenched"

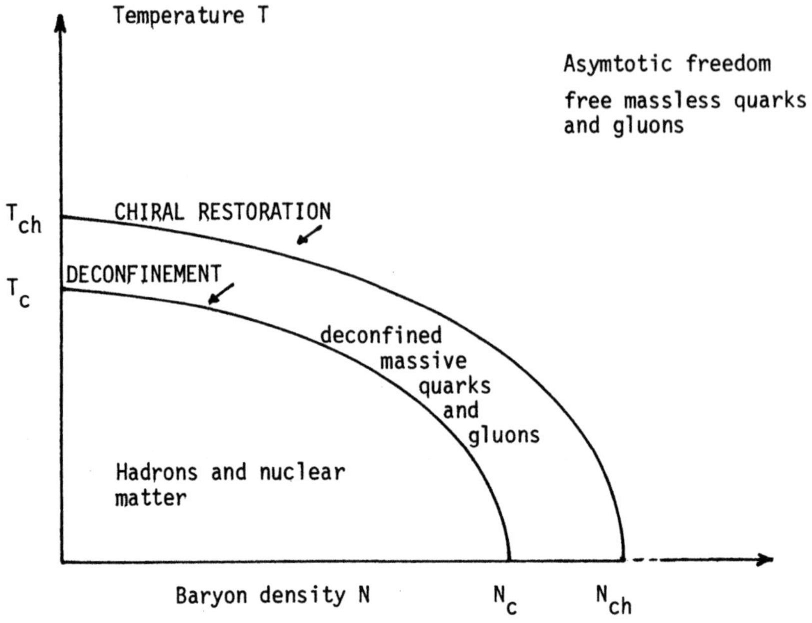

Figure 1 : The phase diagram for strongly interacting matter: confined, deconfined with spontaneously broken chiral symmetry, and deconfined with restored chiral symmetry.

approximation, where virtual quark loops are neglected, the essential finite temperature work is done. We shall report on these results. For $\mu \neq 0$ first, exploratory lattice calculations have been carried out by Kogut et al.[12] confirming the general picture.

The order of a transition and the transition temperature can be determined in two ways:

i) the thermodynamic quantities like energy density, specific heat, pressure etc. are calculated as a function of temperature. Discontinuous or singular behaviour of any of these variables is then a sign for a transition[7];
ii) one calculates order parameters associated with the corresponding transition, e.g. the thermal Wilson loop $\langle L \rangle$ is an order parameter for the deconfinement transition in pure Yang-Mills theory[4,5,6], $\langle \bar{\psi}\psi \rangle$ is an order parameter for the chiral transition in QCD[8,9,13,14].

In the next section, we shall review the relevant equations for finite temperature field theory and their application to QCD on the lattice. The following section presents the results for the pure Yang-Mills system, i.e. QCD without fermions. The contribution of the fermions and the chiral transition are then discussed in section IV. In section V we give a summary and outlook on possible future progress in this field.

II. QCD for Finite Physical Temperatures on the Lattice

The formulation of quantum field theory for finite physical temperatures starts with the definition of the partition function

$$Z = \mathrm{Tr}\; e^{-\beta(H-\mu N)} \quad , \tag{1}$$

where β is the inverse of the temperature, $\beta = T^{-1}$, μ the chemical potential, H the Hamiltonian of the system and N the particle density (the difference of the quark and antiquark densities). For simplicity and because we do not report on non-zero μ results we omit in the following all terms proportional to μ . The partition function can also be written as a Feynman path integral[15)]

$$Z = \int[d\varphi]\, \exp\,\{\int_0^\beta d\tau \int_V d^3x\, \mathcal{L}(\varphi, i\dot{\varphi})\} \quad , \tag{2}$$

where $\mathcal{L}$ is the Lagrangian density, $\tau = it$ and φ denotes all kinds of classical fields. Because of the trace operation in eq. (1) only periodic (antiperiodic) configurations

$$\varphi(\vec{x},0) = \pm\varphi(\vec{x},\beta) \tag{3}$$

for Bose (Fermion) fields are to be considered in the functional integral. Since the functional-integral formalism does not include normal ordering, results obtained from eq. (2) have to be corrected for possible ground state contributions. In terms of the Euclidean action

$$S = -\int_0^\beta d\tau \int_V d^3x\, \mathcal{L} \quad , \tag{4}$$

the partition function becomes

$$Z(\beta,V) = \int[d\varphi]e^{-S} \quad . \tag{5}$$

The Euclidean form thus contains a four-dimensional integral - the action, which is asymmetric in the space and temperature directions.

To obtain a momentum cut-off in the theory and to perform the $x-\tau$ integrations in eq. (4) we introduce a lattice[16)]. It is an asymmetric one, with N_σ sites and spacing a_σ for all spatial directions and N_β sites and spacing a_β for the thermal direction. Moreover, the lattice must be periodic in the thermal direction. Though it is not required, one usually takes periodic boundary conditions in the spatial directions too. Volume and temperature are then given by

$$V = (N_\sigma a_\sigma)^3 \quad , \qquad \beta = N_\beta a_\beta = T^{-1} \quad . \tag{6}$$

Only the choice of asymmetric lattices allows for independent variations of volume and temperature. This is necessary for two reasons:

i) the integration boundaries in the action integral are different for thermal and spatial directions and

ii) at fixed N_σ and N_β, i.e. for a lattice with a fixed number of points, one wants to take partial derivatives with respect to V and β, because the energy density is given by

$$\varepsilon = -\frac{1}{V}\frac{\partial \ln Z}{\partial \beta}\bigg|_V \tag{7}$$

and the pressure by

$$P = \frac{1}{\beta}\frac{\partial \ln Z}{\partial V}\bigg|_\beta \quad . \tag{8}$$

In the actual calculations[7] these derivatives are replaced by derivatives with respect to a_σ and the asymmetry variable $\xi = a_\sigma/a_\beta$. Then one obtains e.g. instead of eq. (7)

$$\varepsilon = \frac{\xi^2}{N_\sigma^3 N_\beta a_\sigma^4}\frac{\partial \ln Z}{\partial \xi}\bigg|_{a_\sigma} \quad . \tag{9}$$

Let us now go from the general case to QCD. The corresponding Lagrangian density is

$$\mathcal{L} = -\frac{1}{4} F^a_{\mu\nu} F^{\mu\nu}_a + \bar{\psi}(i\not\partial - g\not A_a \lambda^a)\psi = \mathcal{L}_{YM}(A) + \mathcal{L}_F(A,\psi,\bar{\psi}) \quad , \tag{10}$$

with

$$F^a_{\mu\nu} = \partial_\mu A^a_\nu - \partial_\nu A^a_\mu - g\, f^a_{bc}\, A^b_\mu\, A^c_\nu \quad . \tag{11}$$

The f^{abc} are the $SU(N_c)$ structure functions, g the coupling constant and the colour indices a,b,c run from 1 to $N_c^2 - 1$. On the lattice the functional integration variables A, the gauge fields, are substituted by link variables

$$U_{xy} = \exp\{-i(x-y)^\mu A_\mu(\tfrac{x+y}{2})\} \quad , \tag{12}$$

where x and y denote adjacent sites on the lattice and $A_\mu = A^a_\mu \lambda_a$, the λ_a are the generators of the $SU(N_c)$ group. The spinor fields are attached to the sites of the lattice, they become site variables ψ_S, $\bar{\psi}_S$. The partition function can then be written as

$$Z = \int \prod_{\text{links}} dU \int \prod_{\text{sites}} d\psi\, d\bar{\psi}\, \exp\{-S^G(U) - S^F(U,\bar{\psi},\psi)\} \quad , \tag{13}$$

where in thermal direction the bosonic integration variables U obey periodic boundary conditions and the spinor fields $\bar{\psi},\psi$ antiperiodic boundary conditions. The action $S^G(U)$ describes the pure Yang-Mills part, $S^F(U,\bar{\psi},\psi)$ the fermionic part of the full QCD action. There are now many possibilities for the dependence of the action on the lattice variables. They only requirement to be satisfied is, that in the continuum limit, where

$$N_\sigma, N_\beta \to \infty \quad ; \quad a_\sigma, a_\beta \to 0$$

with V and β fixed, one has to recover the correct classical continuum action. The most widely used action for the pure gauge field part is the one of Wilson[16]:

$$S^G(U) = 2N_c \, [K_\sigma^G \sum_{\{P_\sigma\}} (1 - \frac{1}{N_c} \mathrm{Re\,Tr}\, U_{ij} U_{jk} U_{kl} U_{li})$$

$$+ K_\beta^G \sum_{\{P_\beta\}} (1 - \frac{1}{N_c} \mathrm{Re\,Tr}\, U_{ij} U_{jk} U_{kl} U_{li})] \quad . \qquad (14)$$

Here the sum $\{P_\sigma\}$ runs over all plaquettes with only spatial links, the sum $\{P_\beta\}$ over all plaquettes with two spatial and two temporal links.

As a consequence of the asymmetry of the lattice spacings two coupling constants g_σ and g_β appear in the action[17], namely via

$$K_\sigma^G = \frac{1}{g_\sigma^2} \xi^{-1} \quad ; \quad K_\beta^G = \frac{1}{g_\beta^2} \xi \quad . \qquad (15)$$

One needs two of them for $a_\sigma \neq a_\beta$ $(\xi \neq 1)$, because then independent variations in a_σ and a_β or ξ can be compensated by changing g_σ and g_β correspondingly, such that physically measurable quantities remain the same. So, g_σ and g_β become functions of a_σ and ξ. For $\xi = 1$, i.e. $a_\sigma = a_\beta = a$, there is, of course, only one coupling constant

$$g(a) = g_\sigma(a_\sigma, 1) = g_\beta(a_\sigma, 1) \quad ; \quad K_{\sigma,\beta}^G(\xi = 1) = g^{-2} \quad . \qquad (16)$$

Choosing Wilson fermions[16], the fermionic part of the action is (for one quark flavour)

$$S^F = \sum_{n,m} \bar{\Psi}_n Q_{nm} \Psi_m \quad , \qquad (17)$$

where

$$Q_{nm} = 1\!\!1 - K_\beta^F M_{0,nm} - K_\sigma^F \sum_{\mu=1}^{3} M_{\mu,nm} \quad , \qquad (18)$$

$$M_{\mu,nm} = (1 - \gamma_\mu) U_{nm} \delta_{n,m-\hat{\mu}} + (1 + \gamma_\mu) U_{nm}^+ \delta_{n,m+\hat{\mu}} \quad . \qquad (19)$$

In the last three equations colour and spinor indices have been suppressed. Because of the antiperiodic boundary conditions for fermion variables in temperature direction at the upper boundary $\delta_{n,m\pm\hat{0}}$ has to be replaced by $-\delta_{n,m\pm\hat{0}}$. Notice that, as S^G, also the fermionic part S^F depends on two different coupling constants (here called hopping parameters) K_σ^F and K_β^F for asymmetric lattice spacings:

$$K^F_\sigma = \frac{4}{3+\xi} k_\sigma(a_\sigma,\xi) \quad ; \quad K^F_\beta = \frac{4\xi}{3+\xi} k_\beta(a_\sigma,\xi) \quad . \tag{20}$$

For $\xi = 1$ we have

$$K^F_{\sigma,\beta}(\xi=1) = k(a) \quad , \tag{21}$$

which is the usual hopping parameter[16].

The integral in eq. (13) over the fermionic degrees of freedom $\bar{\Psi},\psi$ can be performed and yields[18]

$$Z = \int \prod_{\text{links}} dU \, e^{-S^G(U)} \det Q \quad . \tag{22}$$

Evaluating formula (9) leads then to two terms for the energy density[14]

$$\varepsilon = \varepsilon^G + \varepsilon^F \quad , \tag{23}$$

with

$$\begin{aligned} \varepsilon^G a^4_\sigma = & - \frac{2N_c\xi^2}{N^3_\sigma N_\beta} Z^{-1} \int \Pi \, dU \, e^{-S^G} \det Q \, \times \\ & \times \{ \frac{\partial K^G_\sigma}{\partial \xi} \sum_{\{P_\sigma\}} (1 - \frac{1}{N_c} \mathrm{Re\,Tr\,} UUUU) \\ & + \frac{\partial K^G_\beta}{\partial \xi} \sum_{\{P_\beta\}} (1 - \frac{1}{N_c} \mathrm{Re\,Tr\,} UUUU) \} \end{aligned} \tag{24}$$

for the gauge field contribution, and

$$\begin{aligned} \varepsilon^F a^4_\sigma = & - \frac{2N_c\xi^2}{N^3_\sigma N_\beta} Z^{-1} \int \Pi \, dU \, e^{-S^G} \det Q \, \times \\ & \times \{ \frac{\partial K^F_\sigma}{\partial \xi} \sum_{\mu=1}^{3} \mathrm{Tr}(M_\mu Q^{-1}) + \frac{\partial K^F_\beta}{\partial \xi} \mathrm{Tr}(M_o Q^{-1}) \} \end{aligned} \tag{25}$$

for the fermion contribution.

Compared to the pure Yang-Mills theory the computational problem introduced by the fermions is to determine $\det Q$ and Q^{-1} ; e.g. in ε^G the only influence of the fermions is contained in the factor $\det Q$, which essentially renormalizes g . Since the calculation of the determinant of Q presents the most serious difficulties, one usually takes the socalled quenched approximation[19], which consists in setting in all equations

$$\det Q \equiv 1 \quad . \tag{26}$$

This approximation corresponds to the neglect of all virtual quark loops. The results, which we show in the following, were all obtained using eq. (26).

III. Results for the Yang-Mills System

The Monte Carlo evaluation of the above formulae has to be carried out on a spatial lattice as large as possible - one would like to come close to the thermodynamic limit $V \to \infty$ $(N_\sigma \to \infty)$. In the cases reported here, $\xi = 1$, i.e. $a_\sigma = a_\beta = a$ was used. For the Yang-Mills system the only parameter is then the coupling constant g . In the continuum limit, g and the lattice spacing a are related by the renormalization group equation (RGE)

$$a\,\Lambda_L = \exp\left\{-\frac{24\pi^2}{11N_c g^2} - \frac{51}{121}\ln\frac{11N_c g^2}{48\pi^2}\right\} \quad , \tag{27}$$

where Λ_L is the lattice scale parameter. All physical units on the lattice are appropriate powers of a or Λ_L^{-1} . The lattice scale parameter can be expressed in conventional units (MeV), once a known physical quantity (usually the string tension[3]) is measured on the lattice. At fixed g^2 , the temperature is then obtained from

$$T = 1/(N_\beta a(g^2)) \quad . \tag{28}$$

The thermodynamic average of a quantity X is given by

$$\langle X \rangle = \int \Pi\, dU\, e^{-S(U)}\, X(U) \Big/ \int \Pi\, dU\, e^{-S(U)} \quad , \tag{29}$$

$$S(U) = S^G(U) + \ln \det Q(U) \quad . \tag{30}$$

For the pure Yang-Mills case or for full QCD in quenched approximation S is equal to S^G .

The calculation of the energy density ε^G from eq. (24) essentially requires the determination of plaquette averages

$$\bar{P} = 1 - \frac{1}{N_c}\,\mathrm{Re}\langle \mathrm{Tr}\, UUUU\rangle \quad , \tag{31}$$

because for $\xi = 1$ we have

$$\varepsilon^G a^4 = 6N_c\,\Big\{ g^{-2}(\bar{P}_\sigma - \bar{P}_\beta) - \frac{\partial g_\sigma^{-2}}{\partial \xi}\Big|_{\xi=1}(\bar{P}_\sigma - \bar{P}_{sym}) - \frac{\partial g_\beta^{-2}}{\partial \xi}\Big|_{\xi=1}(\bar{P}_\beta - \bar{P}_{sym})\Big\} \quad . \tag{32}$$

The $\bar{P}_{\sigma,\beta}$ denote the averages of spatial and thermal plaquettes, $\bar{P}_{sym}$ is the average plaquette value on a large symmetric $(N_\sigma = N_\beta)$ lattice. Subtraction of $\bar{P}_{sym}$ in eq. (32) just takes away the vacuum contribution, since a large symmetric lattice simulates a zero temperature system. The derivatives of $g_{\sigma,\beta}^{-2}$ with respect to ξ are known constants in the limit $g^2 \to 0$ [20].

What is the expected behaviour as a function of temperature of the gauge field energy density? Below the critical temperature T_c one expects to find the energy density of a gas of hadrons. For a pure Yang-Mills system these hadrons must be glueballs and their exitations. At T_c a phase transition - the deconfining transition - should occur, signalled by a discontinuity for a first order transition or a singularity in the specific heat for a second order transition. For large temperatures ε^G should approach the energy density of a gas of free gluons, which is given by the Stefan-Boltzmann law

$$\varepsilon^G_{SB} = \frac{\pi^2}{30}\, 2(N_c^2 - 1)\, T^4 \quad . \tag{33}$$

As can be seen in figs. 2 and 3 these expectations are in fact borne out by actual Monte Carlo simulations of the Yang-Mills system[6,7].

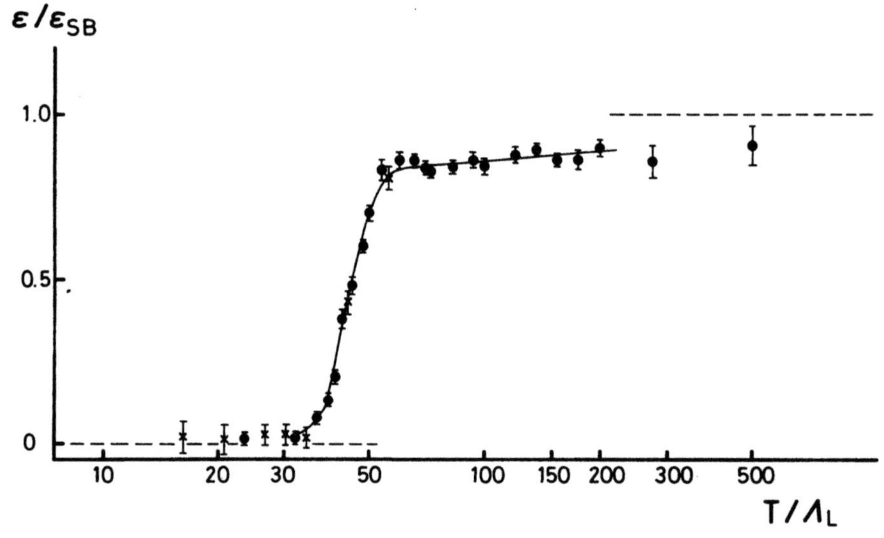

Figure 2 : The ratio $\varepsilon^G/\varepsilon^G_{SB}$ for the pure SU(3) Yang-Mills system calculated on a $10^3 \times 3$ lattice versus temperature from ref.7.

Obviously, both for SU(2) and SU(3), there is a phase transition, probably of second order for SU(2). First results[6] for SU(3) suggest a first order transition in that case. However, to confirm this assumption a more accurate and detailed investigation of the transition region is necessary. To this end one evaluates the thermal Wilson loop L

$$L = \frac{1}{N_c} \mathrm{Tr} \prod_{\tau=1}^{N_\beta} U_{\vec{x};\tau,\tau+1} \quad . \tag{34}$$

It is related to the free energy F_q of an isolated colour charge[4)]

$$\langle L \rangle \sim e^{-\beta F_q} \quad . \tag{35}$$

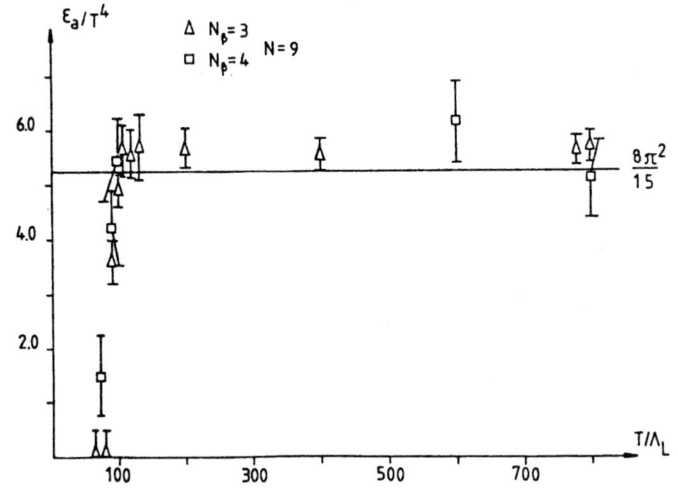

Figure 3 : The quantity ε_a^G/T^4 for the pure SU(3) Yang-Mills system calculated on lattices with $N_\sigma = 9$, $N_\beta = 3,4$ [6]; ε_a^G contains only the main contribution to ε^G , which is proportional to g^{-2} . The horizontal line is the Stefan-Boltzmann result.

On an infinite lattice $<L>$ vanishes in the confined phase, is non-zero above the deconfinement temperature and therefore an order parameter for the deconfinement transition. If we have a first order transition, then there must be a critical coupling g_c^2 , where two different states of the lattice exist, one of them having a non-zero $<L>$-value. We can make use of the Monte Carlo (MC) (Metropolis) method itself to obtain these distinct states. A normal MC calculation would start with either a completely ordered state of the lattice, i.e. all links being equal to the unit matrix (that corresponds to $g^2 \to 0$, high temperature) or a random (hot) configuration of links (corresponding to $g^2 \to \infty$, low temperature). During the MC procedure the lattice will thermalize, i.e. reach its thermal equilibrium. At the critical coupling the ordered start will then lead to the non-zero $<L>$-value, the hot start to a close to zero $<L>$-value. If the spatial volume V of the lattice is too small, there will be flips between the two phases, but when the spatial volume is large enough the two distinct states will persist for a large number of iterations. This enables one to determine the critical coupling g_c (and with eq. (28) T_c) with great accuracy. In fig. 4 such a search for the first order transition in SU(3) Yang-Mills theory on an $8^3 \times 3$ lattice is shown[10]. At $6/g^2 = 5.5531$ clearly two coexistent states are found. The energy densities of the two states are of course also different; the complete energy density as a function of

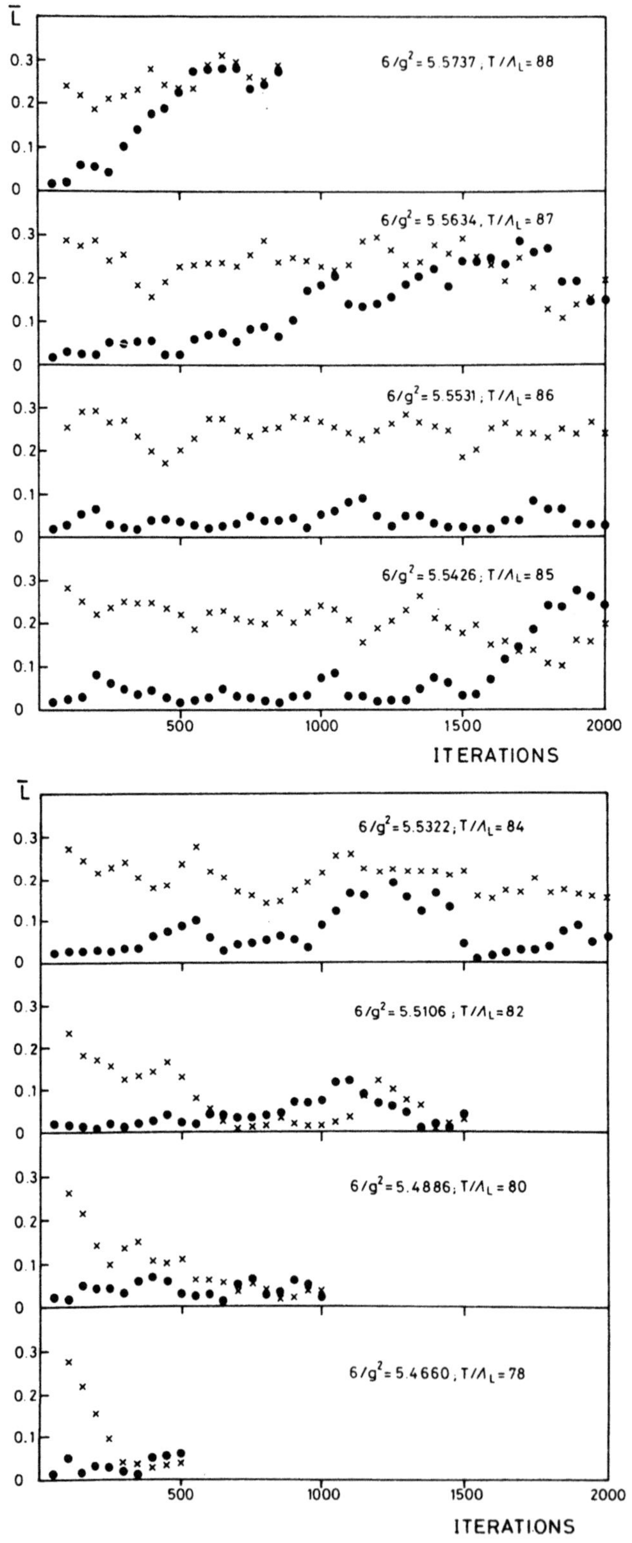

Figure 4 :
The average $\overline{L}$ over 50 successive iterations of the order parameter L for the SU(3) Yang-Mills system, as a function of the total number of iterations after ordered (x) and random (o) starts, calculated on an $8^3 \times 3$ lattice[10)] for various values of the coupling $6/g^2$; also shown is the associated temperature, using the RGE, eq. (27).

temperature is shown in fig. 5.

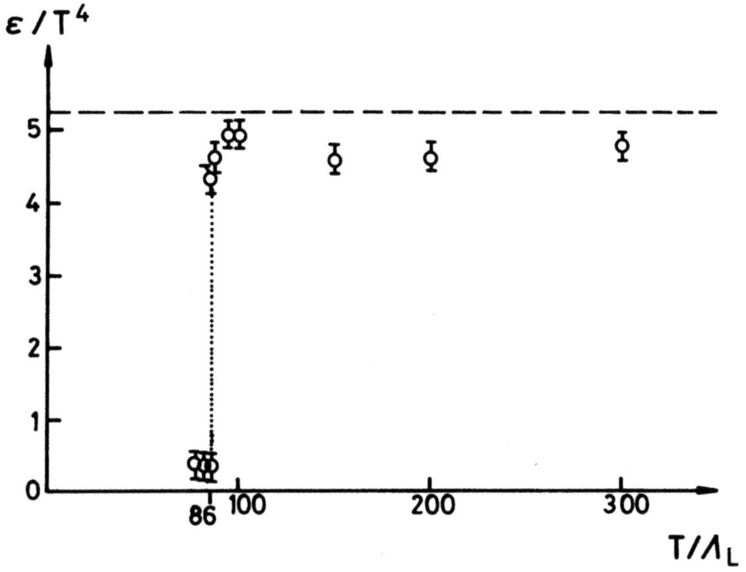

Figure 5 : The energy density ε^G/T^4 for the SU(3) Yang-Mills system as function of temperature, calculated on an $8^3 \times 3$ lattice[10].

In table 1 the results for the deconfinement temperature as obtained from calculations on lattices with different N_β are compared[10].

Table 1: The deconfinement temperature

N_β	$6/g_c^2$	$T_c[\Lambda_L]$	$T_c[\sqrt{\sigma}]$
2	5.11 ± 0.01	78 ± 1	
3	5.55 ± 0.01	86 ± 1	0.519 ± 0.015
4	5.70 ± 0.01	76 ± 1	$0.519\,^{+0.050}_{-0.030}$

The deviations, when T_c is expressed in units of Λ_L via eqs. (27) and (28), can be explained by higher order terms in the renormalization group relation, without any violation of general scaling behaviour. Indeed, when T_c is given in units of $\sqrt{\sigma}$ [21] (σ is the string tension) no measurable deviations remain and

$$T_c - (0.519 \pm 0.050)\,\sqrt{\sigma} \simeq 208 \pm 20 \text{ MeV} \quad , \tag{36}$$

which is agreeing well with the value obtained in the SU(2) case[7].

Since for SU(3) we have a first order transition, we can calculate the latent heat - the difference in energy density between the two states at the transition. In fig. 6 we show $\Delta\varepsilon^G/T_c^4$, which is a dimensionless quantity and should scale.

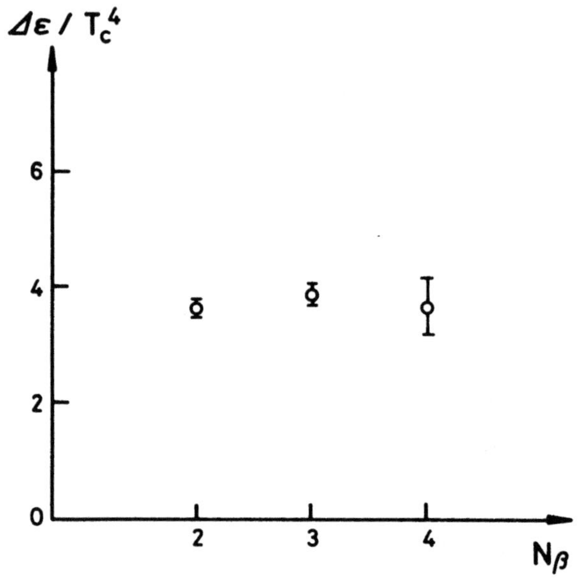

Figure 6 : The latent heat $\Delta\varepsilon^G/T_c^4$ as function of temperature lattice size N_β from ref. 10.

In fact, the result[10)] is independent of N_β and therefore of g_c (see table 1):

$$\Delta\varepsilon^G/T_c^4 = 3.75 \pm 0.25 \quad , \tag{37}$$

$$\Delta\varepsilon^G = 875 \pm 80 \text{ MeV/fm}^3 \quad . \tag{38}$$

The error in the last equation does not include the error in T_c .

IV. The Contribution of the Fermions in the Quenched Approximation

Evaluating eq. (25) for the fermion contribution ε^F to the energy density at $\xi = 1$ requires the derivatives $\partial K^F_{\sigma,\beta}/\partial\xi$. If we neglect the ξ-dependence of k_σ and k_β in eq. (20), we obtain

$$\left.\frac{\partial K^F_\sigma}{\partial\xi}\right|_{\xi=1} = -\frac{1}{4}\,k \quad , \quad \left.\frac{\partial K^F_\beta}{\partial\xi}\right|_{\xi=1} = \frac{3}{4}\,k \quad . \tag{39}$$

Also, at $\xi = 1$ the matrix Q from eq. (18) simplifies to

$$Q = \mathbb{1} - k\,M \quad , \tag{40}$$

where

$$M = \sum_{\mu=0}^{3} M_\mu \quad . \tag{41}$$

The inverse of Q is then readily computed via a hopping parameter expansion (HPE)[13,22,23]

$$Q^{-1} = \sum_{l=0}^{\infty} k^l M^l \quad . \tag{42}$$

Insertion of the last equation into eq. (25) leads then - in quenched approximation - to a HPE for the fermion energy density

$$\varepsilon^F a^4 = - \frac{3}{4N_\sigma^3 N_\beta} \sum_{l=0}^{\infty} k^{l+1} \times$$

$$\times \{ \langle \mathrm{Tr}\, M_o M^l \rangle - \frac{1}{3} \sum_{\nu=1}^{3} \langle \mathrm{Tr}\, M_\nu M^l \rangle \} \quad . \tag{43}$$

Note, that because of the approximation in eq. (39), on a symmetric lattice $(N_\sigma = N_\beta)$ ε^F is zero, i.e. no additional vacuum correction has to be introduced. As a consequence of the Kronecker deltas in eq. (19), only closed loops of length $l+1$ contribute to $\langle \mathrm{Tr}\, M_\mu M^l \rangle$. On an infinite lattice $l+1 = 2n$ and the first non-vanishing loop has length 4, it is proportional to a plaquette. For $N_\beta = 2,3$ however, the loop can be closed in temperature direction already with length $2,3$ because of the (anti)periodic boundary conditions. The loop is then proportional to the thermal Wilson loop $\langle L \rangle$ [14].

The size of the hopping parameter $k(a)$ has to be determined by a separate calculation, either by requiring that the pion mass is zero[23], or by determining the convergence radius of the HPE for $\langle \bar{\psi}\psi \rangle$ [8]. Taking Susskind fermions instead of Wilson fermions, where the bare quark mass is the parameter corresponding to the hopping parameter, involves a comparable problem: one must compute the results for finite quark mass and then extrapolate to zero quark mass[9].

In fig. 7 the fermion energy density ε^F/T^4 for one quark flavour and SU(2) is shown as a function of temperature. The energy density ε^F approaches the value of the free theory - the Stefan-Boltzmann limit on a lattice of the same size (indicated by a dashed line) - after a sudden jump at around the same temperature, where (see fig. 2) the gluon energy density ε^G has presumably a second order transition. That this indeed must happen - at least in the quenched approximation - is clear from the expansion in eq. (43), since the first non-vanishing term is proportional to $\langle L \rangle$, which stays an order parameter in the quenched approximation,

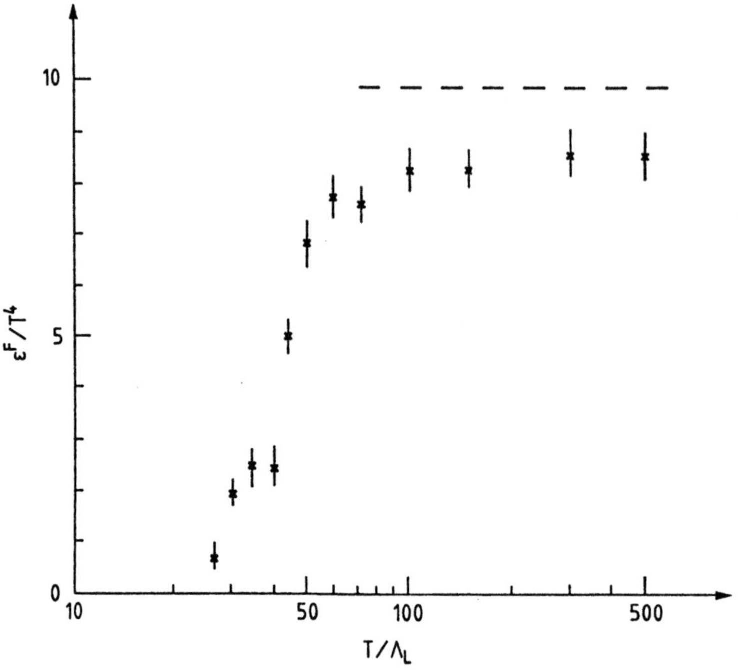

Figure 7 : The fermionic contribution ε^F to the energy density of SU(2) lattice QCD divided by the fourth power of temperature as a function of temperature, calculated on an $8^3 \times 3$ lattice with a 46th order hopping parameter expansion for massless quarks[8]. The dashed line is the Stefan-Boltzmann result on a lattice of the same size.

and since the fourth order term, which is proportional to $\bar{P}_\sigma - \bar{P}_\beta$, behaves at the transition essentially like ε^G. For a first order transition like in SU(3) the coincidence becomes even more obvious, as can be seen from fig. 8. Kogut et al.[11] estimate for the latent heat of the total system and four flavours

$$\Delta(\varepsilon^G + \varepsilon^F) = 1.50 \pm 0.5 \text{ GeV/fm}^3 \quad . \tag{44}$$

Finally let us comment on the problem of chiral symmetry restoration. This question is in the case of Wilson fermions particularly complex, since chiral symmetry is by construction broken on the lattice. Even a system of non-interacting massless fermions leads to a non-zero $\langle\bar{\psi}\psi\rangle_{SB}$. To study chiral symmetry restoration, one would therefore first have to show that $\langle\bar{\psi}\psi\rangle$, after subtraction of an adequate term, exhibits scaling behaviour and then check at what T it leads to a vanishing expectation value indicating chiral symmetry. A first step in the right direction consists in simply considering[13]

$$<\bar{\psi}\psi> - <\bar{\psi}\psi>_{SB} \tag{45}$$

as an order parameter of chiral symmetry for Wilson fermions.

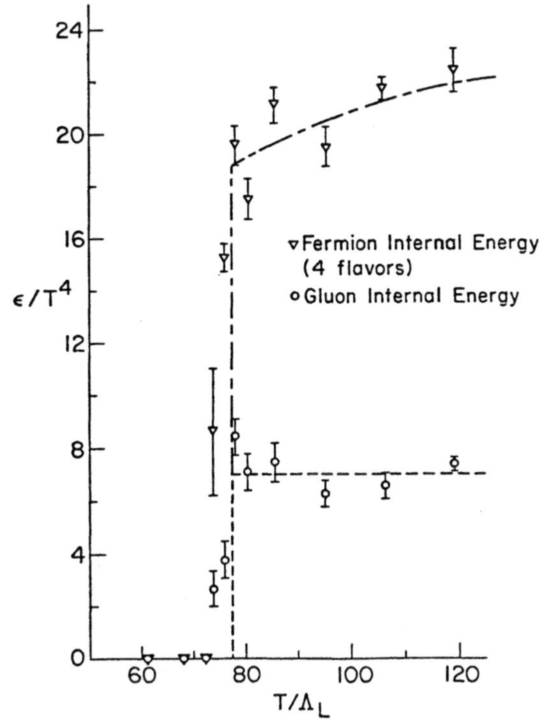

Figure 8 : The gluon and fermion energy densities ε^G/T^4 and ε^F/T^4 for 4 Susskind fermions versus temperature, calculated on an $8^3 \times 4$ lattice from ref. 11.

In fig. 9 this quantity is shown for SU(2) as function of temperature[8]. At the transition ($T_c \sim 40\ \Lambda_L$), expression (45) is still finite, suggesting $T_c < T_{ch}$ (the chiral transition temperature). The same result was found by Kogut et al.[9] for SU(2) Susskind fermions. However, the impression, that T_c and T_{ch} are different for SU(2) may be due to the second order nature of the transition and of course the approximations in the calculation. A clear picture is found for SU(3)[9]. As can be seen in fig. 10 both the order parameter $<L>$ for the deconfinement transition and the order parameter $<\bar{\psi}\psi>$ for Susskind fermions and the chiral transition change their behaviour drastically at the same critical coupling. Like in the case of ε^G and ε^F also the first order deconfinement and chiral transitions must coincide in the quenched approximation, because

$$<\bar{\psi}\psi> = <\mathrm{Tr}\ (\text{colour} + \text{spin})\ Q^{-1}> \tag{46}$$

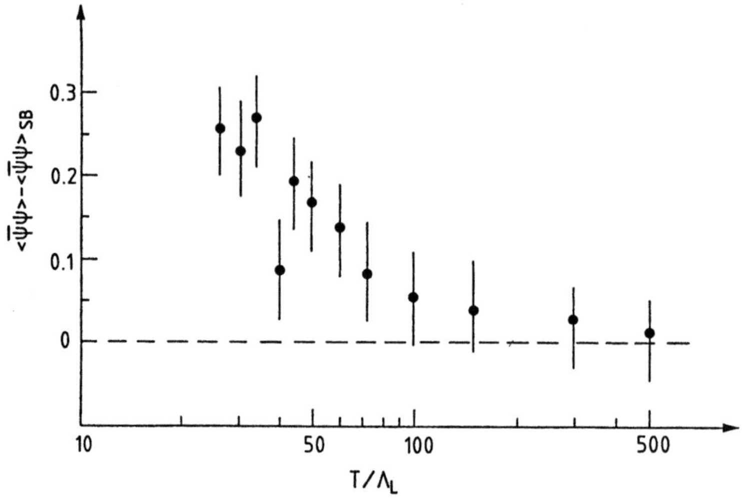

Figure 9 : The quantity $<\bar{\psi}\psi> - <\bar{\psi}\psi>_{SB}$ versus temperature for SU(2) and Wilson fermions from ref. 8, calculated on an $8^3 \times 3$ lattice.

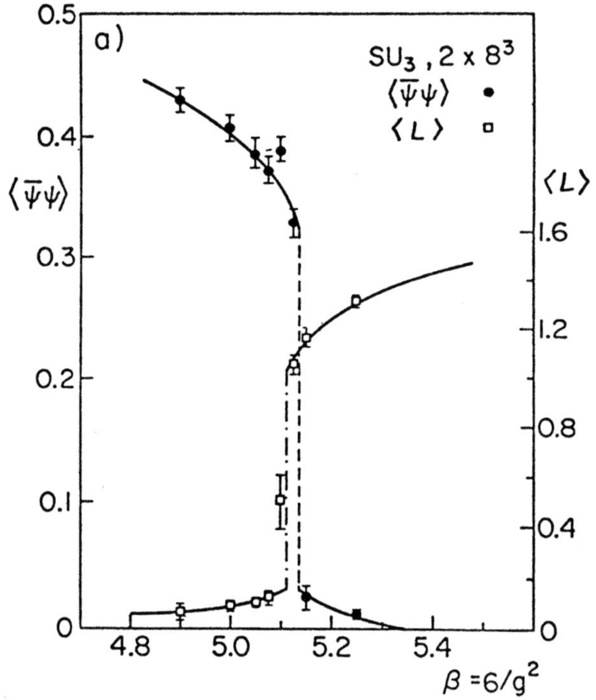

Figure 10: The quantity $<\bar{\psi}\psi>$ for Susskind fermions and the thermal Wilson loop $<L>$ for SU(3) versus $6/g^2$ from ref. 9, calculated on an $8^3 \times 2$ lattice.

and the corresponding hopping parameter expansion contains a term proportional to $\langle L \rangle$.

V. Summary and Outlook

In summarizing the results of Monte Carlo evaluations of lattice QCD in the quenched approximation and for chemical potential $\mu = 0$, one finds for SU(3)

i) a first order deconfinement transition at $T_c = 208 \pm 20$ MeV for both the pure gauge field and the fermion contributions

ii) the latent heat at the transition is

$$\Delta\varepsilon^G = 875 \pm 80 \text{ MeV/fm}^3$$

or

$$\Delta(\varepsilon^G + \varepsilon^F) \simeq 1.5 \pm 0.5 \text{ GeV/fm}^3 \quad \text{for} \quad N_f = 4$$

iii) the chiral transition temperature is the same as the deconfinement temperature.

For SU(2) the deconfinement transition is probably of second order, the critical temperature has about the same value as for SU(3).

The removal of the quenched approximation, i.e. the introduction of virtual quark loops into the calculation may change the above results considerably. First, exploratory calculations indicate[24,25], that with decreasing quark mass the first order transition is replaced by a second order transition, which may even eventually disappear, when the quark mass becomes zero. Also, because the thermal Wilson loop $\langle L \rangle$ is no longer an order parameter in the full theory, the conclusion from the HPE, that the chiral and deconfinement transitions should coincide, if they are of first order, is no longer true. Should these questions be settled, the next step would be the investigation of the phase diagram (fig. 1) for non-zero chemical potential. There is a lot of work still to be done.

References

1) E.V. Shuryak, Phys. Reports 61 (1980) 71;
D.J. Gross, R.D. Pisarski and L.G. Yaffe, Rev. Mod. Phys. 53 (1981) 43;
H. Satz, Proc. 5th High energy heavy ion study, Berkeley/Cal. 1981.

2) G. Baym, Proc. of the Bielefeld Workshop 1982, H. Satz and J. Jacob eds., World Scientific, Singapore, 1982.

3) M. Creutz, Phys. Rev. D21 (1980) 2308; Phys. Rev. Lett. 45 (1980) 313.

4) L. McLerran and B. Svetitsky, Phys. Lett. 98B (1981) 195;

J. Kuti, J. Polónyi and K. Szlachányi, Phys. Lett. 98B (1981) 199; J. Engels, F. Karsch, I. Montvay and H. Satz, Phys. Lett. 101B (1981) 89.

5) K. Kajantie, C. Montonen and E. Pietarinen, Z. Phys. C9 (1981) 253.

6) I. Montvay and E. Pietarinen, Phys. Lett. 110B (1982) 148; 115B (1982) 151.

7) J. Engels, F. Karsch, I. Montvay and H. Satz, Nucl. Phys. B205 [FS5] (1982) 545.

8) J. Engels and F. Karsch, Phys. Lett. 125B (1983) 481.

9) J. Kogut et al., Phys. Rev. Lett. 50 (1983) 393.

10) T. Çelik, J. Engels and H. Satz, Phys. Lett. 125B (1983) 411; 129B (1983) 323.

11) J. Kogut et al., Phys. Rev. Lett. 51 (1983) 869.

12) J. Kogut et al., Illinois preprint ILL-(TH)-83-10, April 1983.

13) C.B. Lang and H. Nicolai, Nucl. Phys. B200 [FS4] (1982) 135.

14) J. Engels, F. Karsch and H. Satz, Phys. Lett. 113B (1982) 398.

15) C. Bernard, Phys. Rev. D9 (1974) 3312.

16) K. Wilson, Phys. Rev. D10 (1974) 2445; in "New Phenomena in Subnuclear Physics", ed. A. Zichichi, Plenum Press, New York 1977 (Erice 1975)

17) A. Hasenfratz and P. Hasenfratz, Nucl. Phys. B193 (1981) 210.

18) T. Matthews and A. Salam, Nuovo Cim. 12 (1954) 563; 2 (1955) 120.

19) H. Hamber and G. Parisi, Phys. Rev. Lett. 47 (1981) 1792; E. Marinari, G. Parisi and C. Rebbi, Phys. Rev. Lett. 47 (1981) 1795; D. Weingarten, Phys. Lett. 109B (1982) 57.

20) F. Karsch, Nucl. Phys. B205 [FS5] (1982) 285.

21) F. Gutbrod, P. Hasenfratz, Z. Kunszt and I. Montvay, CERN preprint TH 3591, May 1983.

22) A. Hasenfratz and P. Hasenfratz, Phys. Lett. 104B (1981) 489.

23) A. Hasenfratz, P. Hasenfratz, Z. Kunszt and C.B. Lang, Phys. Lett. 110B (1982) 289.

24) P. Hasenfratz, F. Karsch and I.O. Stamatescu, CERN preprint TH 3636 (1983).

25) T. Çelik, J. Engels and H. Satz, Bielefeld preprint BI-TP 83/15, August 1983 (Phys. Lett. in press).

THE QUARK-QUARK INTERACTION IN THE NONRELATIVISTIC AND IN THE LATTICE APPROXIMATION

F. Palumbo
INFN - Laboratori Nazionali di Frascati, P.O. Box 13 - 00044 Frascati, Italy

Abstract

The Galilean Hamiltonian for QCD is presented and illustrated. In a small quantization box such Hamiltonian exhibits a confining potential which, however, is a finite size effect, becoming purely Coulombic for infinite quantization volume. This does not necessarily imply absence of confinement in Galilean QCD, and a possible mechanism for confinement is discussed, suggested by the exact solution of the infrared sector in Galilean QED.

Finally the derivation of classical Galilean gauge theories and their quantization are sketched.

1. - Introduction

Nonrelativistic approximations of the so far well understood relativistic theories provide an accurate description of low-energy phenomena, in particular of bound states. A nonrelativistic approximation to QCD would therefore seem useful in the study of confinement, unless this is a purely relativistic feature of the theory. In fact not all of the low-energy properties of relativistic theories must necessarily be retained by their nonrelativistic approximations, and radiative corrections in QED are usually given as an example. This example is of particular relevance, because just the divergences which are related to radiative corrections in QED are thought to be responsible for quark confinement in QCD. But radiative corrections are related to gauge invariance. Therefore the question: Is it possible to maintain gauge invariance in the nonrelativistic approximation? A positive answer would imply the possibility of formulating a nonrelativistic version of QED which should give rise to radiative corrections.

Before going on, however, we must specify the meaning of the world nonrelativistic. Here it will be used as a synonym for Galilean. When the velocity of light c becomes infinite, the Poincaré group contracts into the Galilean group. We therefore require that the limit $c \to \infty$ of gauge theories be gauge-invariant Galilean theories. We will see that such theories exist and that in Galilean QED there are indeed radiative corrections which are the $c \to \infty$ limit of the radiative corrections of relativistic QED.

The fact that Galilean QED retains the infrared features of relativistic QED gives some confidence that the same could be true for QCD.

Galilean QCD has been formulated and it has been shown that the q-q interaction is not confining in the usual sense. Since this is a nonperturbative result, it is natural to compare it to lattice calculations which give indication of a linear potential. I will therefore present Galilean QCD from this point of view, which is the reason for the title of my talk, although I will not discuss lattice calculations.

In Galilean QCD the q-q interaction depends on the size of the quantization volume L^3. Calculations must be done at finite L and the limit $L\to\infty$ must be taken at the end. Now while for infinite volume the q-q interaction is purely Coulombic, for small volume it has a confining term (a finite size effect) which is of the order of magnitude of that resulting from lattice calculations.

The absence of a confining potential does not necessarily imply, however, absence of confinement, and a possible mechanism for confinement will be discussed.

I will first show and illustrate the Hamiltonian in Galilean QCD. From this Hamiltonian I will derive the effective q-q interaction to be compared to that obtained in lattice calculations. After that I will present Galilean QED whose infrared sector has been solved exactly and I will compared it to relativistic QED. The solution of the infrared sector in QED suggests the mentioned mechanism for confinement in the absence of a confining potential.

Finally I will sketch the derivation of classical Galilean gauge theories and their quantization.

2. - Galilean QCD and the effective q-q interaction in a small quantization volume.

I will present the Galilean QCD Hamiltonian[1)] for two quarks in the fundamental representation of SU(2). This Hamiltonian is separated into two terms

$$H = H_o + V, \tag{1}$$

the potential having the form

$$\begin{aligned} V = \frac{1}{8}\alpha_s \frac{1}{|x|}\, \sigma_a(1) \Big\{ \delta^{ab} - \left[1-\cos g\, L^{-3/2}\, \vec{q}\cdot\vec{x}\right] \left[\delta^{ab} - \frac{v^a v^b}{v^2}\right] \\ + \varepsilon^{abc}\frac{v^c}{v} \sin g\, L^{-3/2}\, \vec{q}\cdot\vec{x} \Big\}\, \sigma_b(2)\,. \end{aligned} \tag{2}$$

In Eq. (2) $\vec{x}$ is the separation distance between the quarks, σ_a the Pauli matrices in color space, $\vec{q}$ and v^a two variables originating from the gluon field, g the coupling constant, $\alpha_s=g^2/4\pi$, L^3 the volume of the quantization box, and we have put $\hbar=c=1$. For $L\to\infty$ V becomes purely Coulombic, but as we will see, the prescription is to calculate at finite L and take the limit $L\to\infty$ in the final results.

H_o is given by

$$H_o = -\frac{1}{2m_q}(\Delta_1+\Delta_2) + 2\,m_q + \frac{1}{2}p^2 + \frac{1}{2}\omega^2 q^2 + \frac{1}{2}\frac{1}{q^2}(\vec{v}\wedge\vec{\pi})^2 + \tag{3}$$
$$+\,\omega\, q_k \frac{v^a}{v}\,\frac{1}{\sqrt{2\,m_q}}\, i\,(\sigma_a(1)\;\nabla_{1k} + \sigma_a(2)\Delta_{2k})$$

where m_q is the quark mass, p_k and π^a are the momenta conjugate to q_k and v^a respectively and

$$\omega^2 = \frac{g^2}{2\,m_q L^3}. \tag{4}$$

The above equations hold for quantization with periodic boundary conditions (p.b.c.), which seem appropriate to QCD. I have no time to discuss this point for which I refer to the literature[2].

The physical states are subject to the constraint

$$\left[(\vec{v}\wedge\vec{\pi})_a + \frac{1}{2}\,\sigma_a(1) + \frac{1}{2}\,\sigma_a\,(2)\right]\Psi = 0. \tag{5}$$

Note that H and therefore Ψ do not depend on v but only on $\frac{v^a}{v}$.
Evaluation of cross sections at finite L is not trivial, and we will show in the next section such an evaluation in the abelian case as an illustration and in order to suggest a possible mechanism for confinement.

In order to compare to lattice calculations, however, we must keep L finite. If we take two quarks in color singlet state, according to Eq. (5) their wave function Ψ does not depend on v^a/v, so that we can define an effective Hamiltonian as the average of H over v^a/v

$$H_{eff} = -\frac{1}{2\,m_q}(\Delta_1+\Delta_2) + 2\,m_q + \frac{1}{2}\,\omega^2 q^2 - \frac{3}{8}\,\alpha_s\frac{1}{|x|} + \frac{1}{4}\,\alpha_s\frac{1}{|x|}(1-\cos g\,L^{-3/2}\,\vec{q}\cdot\vec{x}), \tag{6}$$

and the effective potential energy is

$$W = -\frac{3}{8}\,\alpha_s\frac{1}{|x|} + \frac{1}{2}p^2 + \frac{1}{2}\,\omega^2 q^2 - \frac{3}{2}\,\omega + \frac{1}{4}\,\alpha_s\frac{1}{|x|}(1-\cos g\,L^{-3/2}\,\vec{q}\cdot\vec{x}). \tag{7}$$

This potential energy still depends on $\vec{q}$. At fixed $\vec{x}$ the lowest eigenvalue of W as an operator acting on $\vec{q}$ is the effective q-q potential energy. The evaluation of this eigenvalue has already been done for finite quark mass[3], but in order to compare to lattice calculations we need it for infinite quark mass. Denoting by W_o this eigenvalue we have[4]

$$W_o = -\frac{3}{8}\,\alpha_s\frac{1}{|x|} + \sigma\,|x|, \tag{9}$$

$$\sigma = \frac{1}{2}\sqrt{\pi}\,\alpha_s\,\frac{1}{L^2}\sqrt{\frac{L}{|x|}}\;. \tag{10}$$

The present results comes from an asymptotic expansion valid for large values of the parameter

$$K = \frac{1}{2\pi}\left(\frac{L}{x}\right)^3. \tag{11}$$

For $L \sim 1$ fm, $\alpha_s \sim 1$, $x = 0.2$ L, $\sigma \sim 400$ MeV/fm.

Thus we see that although the true potential is not confining in the usual sense ($\sigma \longrightarrow 0$ for $L \longrightarrow \infty$), for small L it appears to have a confining term. This term is spurious and such a spurious contribution to the string tension should also be present in lattice calculations with p.b.c., due to the contribution of Galilean configurations of the gauge field.

3. - Galilean QED

The e^+e^- Hamiltonian in Galilean QED is[2)]

$$H_o = -\frac{1}{2m_e}(\Delta_1+\Delta_2) - \frac{1}{2}\alpha\frac{1}{|x|} + \frac{1}{2}p^2 + \frac{1}{2}\omega^2 q^2 + \omega q_k \frac{1}{\sqrt{2m_e}} i(\nabla_{1k} - \nabla_{2k}), \qquad (12)$$

with

$$\omega^2 \simeq \frac{2e^2}{m_e L^3}. \qquad (13)$$

Eq. (12) holds for p.b.c.. Although phenomenology definitely requires that quantization be done by requiring that the fields vanish on the surface of the quantization box in QED, I will consider here Galilean QED with p.b.c. for comparison with Galilean QCD and with relativistic QED.

If we let $L \longrightarrow \infty$ in Eq. (12) the coupling between the "photon" and the electrons disappears. But we know that we must perform this limit only on the final results.

In order to solve the infrared sector[2)] it is convenient to introduce creation and destruction operators

$$a_k^+ = (2\omega)^{-1/2}(p_k + i\omega q_k), \qquad (14)$$

and rewrite H as

$$H = -\frac{1}{2m_e}(\Delta_1 + \Delta_2) + \frac{3}{2}\omega + \omega a_k^+ a_k + \frac{1}{2}\sqrt{\omega}\, i(a_k^+ - a_k) I_k - \frac{1}{2}\alpha\frac{1}{|x|}, \qquad (15)$$

where

$$I_k = -\frac{i}{\sqrt{m_e}}(\nabla_{1k} - \nabla_{2k}). \qquad (16)$$

If we denote the purely electronic states by $|i\rangle$, $|f\rangle$, ... the eigenstates of H can be written

$$|i, n_1, n_2, n_3\rangle = \prod_{k=1}^{3}\frac{1}{\sqrt{n_k!}}\left[a_k^+ + \frac{i}{\sqrt{2\omega}} I_k(i)\right]^{n_k} e^{\frac{i}{\sqrt{2\omega}}(a_k^+ + a_k) I_k(i)} |i\rangle, \qquad (17)$$

where $I_k(i)$ is the eigenstate of the operator I_k in the state $|i\rangle$. Let $|i\rangle$ be the initial state and $|f\rangle$ the final state, $\Delta I_k = I_k(f) - I_k(i)$. Assuming a proper choice of axes, $\Delta I_k = \delta_{k3}\Delta I$, and the S-matrix elements have the expression

$$|\langle i,0,0,0,|S|\,f,n_1,n_2,n_3\rangle|^2 = \delta_{on_1}\,\delta_{on_2}\,|\langle i\,|S\,(n_3\,\omega)|\,f\rangle|^2 \cdot \frac{1}{n_3!}\left[\frac{(\Delta I)^2}{2}\right]^{n_3} e^{-\frac{(\Delta I)^2}{2}}, \tag{18}$$

which is the probability for a classical source to radiate n_3 photons of energy ω.

The limit $L \rightarrow \infty$ can now be performed in Eq. (18) showing that

$$\lim_{L\rightarrow\infty} |\langle i,0,0,0\,|\,S\,|\,f,n_1,n_2,n_3\rangle|^2 = 0. \tag{19}$$

In an actual experiment, however, there will be a finite energy resolution ΔE in a final state. The theoretical cross section to be compared with experiment is

$$\sigma \sim \lim_{L\rightarrow\infty} \sum_{n_3=0}^{\infty} |\langle i,0,0,0\,|S|\,f,n_1,n_2,n_3\rangle|^2 \theta(\Delta E - n_3\,\omega) = |\langle i\,|\,S\left[\frac{1}{2}(\;I)^2\right]|f\rangle|^2 \theta\left[\Delta E - \frac{1}{2}(\Delta I)^2\right]. \tag{20}$$

Now we must compare this formula to the corresponding one in relativistic QED. The Bloch-Nordsieck theorem with p.b.c. gives[5)]

$$\sigma \sim \left[\frac{\Delta E - \frac{1}{2}(\Delta I)^2}{E}\right]^{\beta} \theta\left[\Delta E - \frac{1}{2}(\Delta I)^2\right], \tag{21}$$

where E is some experimental energy and β vanishes for infinite light velocity so that the Galilean limit is reproduced.

I have derived here Eq. (20) for two reasons. First to show that it agrees with the limit of the relativistic formula for the radiative corrections, which gives confidence in Galilean gauge theories. The second reason is that it suggests a new mechanism for confinement.

The radiative correction of Eq. (20) arises from the overlapping of the soft photon state relative to the initial state to the soft photon state relative to the final state of the electrons, the photon state being a function of the current. This is because $\vec{q}$ is coupled only to the current in the electronic Hamiltonian.

In Galilean QCD the gluon appears not only with the quark color current, but also in the quark-quark potential, so that the gluon state will depend on the structure of the q-q wavefunction. Confinement would result if the gluon state relative to a bound state were orthogonal to the gluon state relative to a free state of quarks in the $L\rightarrow \infty$ limit. In such a case, in fact, the disintegration cross-section would vanish.

4. Classical gauge theories and their quantization

In this final part I will show how classical Galilean gauge theories have been formulated[6] and quantized[1]. I will confine myself to pure gauge theories, without coupling to matter fields, for the sake of brevity. I will maintain $\hbar=1$ but I have obviously to reintroduce the light velocity c.

The relativistic gauge field Lagrangian density is

$$\mathcal{L}_G = \frac{1}{2}(\mathcal{F}^a_{oi})^2 - \frac{1}{4}(\mathcal{F}^a_{ij})^2 , \tag{22}$$

where

$$\begin{aligned} \mathcal{F}^a_{oi} &= \frac{1}{c}\ \partial_t \mathcal{A}^a_i - \partial_i \mathcal{A}^a_o + \frac{g}{c}\ f^{abc} \mathcal{A}^b_o \mathcal{A}^c_j \\ \mathcal{F}^a_{ij} &= \partial_i \mathcal{A}^a_j - \partial_j \mathcal{A}^a_i + \frac{g}{c}\ f^{abc} \mathcal{A}^b_i \mathcal{A}^c_j , \end{aligned} \tag{23}$$

f^{abc} being the structure constants of the color group.

Now if we let $c \to \infty$ other things remaining constant, $\mathcal{F}^a_{ij} \to \partial_i \mathcal{A}^a_j - \partial_j \mathcal{A}^a_i, \mathcal{A}^a_{oi} \to -\partial_i \mathcal{A}^a_o$ and we loose gauge invariance, without reaching Galilean invariance (see below). We therefore rescale the gauge fields according to

$$\begin{aligned} \mathcal{A}^a_o &= -V^a \\ \mathcal{A}^a_i &= c\ A^a_i , \end{aligned} \tag{24}$$

so that

$$\begin{aligned} \mathcal{F}^a_{oi} &\overset{\text{def}}{=} F^a_{oi} = \partial_t A^a_i + \partial_i V^a - g\ f^{abc}\ V^b\ A^c_i \\ \mathcal{F}^a_{ij} &\overset{\text{def}}{=} F^a_{ij} = c(\partial_i A^a_j - \partial_j A^a_i + g\ f^{abc}\ A^b_i A^c_j) , \end{aligned} \tag{25}$$

$$\mathcal{L}_G = +\frac{1}{2}(F^a_{oi})^2 - \frac{1}{4}\ c^2\ (F^a_{ij})^2 . \tag{26}$$

Now gauge invariance is maintained and Galilean invariance is acquired in the $c \to \infty$ limit

$$\mathcal{L}_G \xrightarrow[c\to\infty]{} \frac{1}{2}\ (F^a_{oi})^2 , \tag{27}$$

with the constraint

$$F^a_{ij} = 0 . \tag{28}$$

In fact, under Galilei transformations of parameters v_k

$$\begin{aligned} &A^a_i \to A^a_i ,\quad V^a \to V^a + v_k\ A_k \\ &F^a_{ij} \to F^a_{ij} ,\quad F^a_{oi} \to F^a_{oi} + v_k\ F^a_{ki} , \end{aligned} \tag{29}$$

so that the constraint (28) is Galilei-invariant and under such constraint the Lagrangian (27) is Galilei invariant.

In order to quantize it is convenient to rewrite $\mathcal{L}_G$ in first order formulation, including the constraint by means of Lagrange multipliers Λ^a_{ij}

$$\mathcal{L}_G = F^a_{oi}\, \partial_t\, A^a_i - \frac{1}{2} F^a_{oi}\, F^a_{oi} - V^a\, \mathcal{D}^{ab}_i\, F^b_{oi} - \Lambda^a_{ij}\, F^a_{ij}\,, \tag{30}$$

where

$$\mathcal{D}^{ab}_i = \partial_i\, \delta^{ab} - g\, f^{abc}\, A^c_i \tag{31}$$

The dynamical varaibles are A^a_i with canonical momenta

$$F^a_{oi} = \frac{\partial \mathcal{L}_G}{\partial \partial_t A^a_i}\,, \tag{32}$$

while V^a are Lagrange multipliers for the constraints

$$\phi^a = \mathcal{D}^{ab}_k\, F^b_{ok}\,, \tag{33}$$

which are common to the relativistic case.

The Hamiltonian is

$$H = \frac{1}{2} F^a_{oi}\, F^a_{oi}\,, \tag{34}$$

apart from a linear superposition of first class constraints.

We apply Dirac's theory of canonical quantization of constrained systems, and find that the only secondary constraints are

$$\chi^a_i = \varepsilon_{ijk}\, \mathcal{D}^{ab}_j\, F^b_{ok} = 0\,. \tag{35}$$

Among the constraints F^a_{ij}, ϕ^a, χ^a_i, only the ϕ^a are first calss, i.e. they commute among themselves, with the other constraints and with the Hamiltonian. According to Dirac's theory first class constraints can either be supplemented by gauge fixing, or can be used as conditions to be satisfied by physical states. We will use an intermediate procedure.

We will expand fields and constraints in Fourier series, and will single out the zero-momentum component of ϕ^a

$$\phi^a_o = g\, L^{-3/2} \int d^3x\, \phi^a = g\, L^{-3/2}\, Q^a_G\,, \tag{36}$$

for its physical meaning, which is the gauge field color charge Q^a_G, and we will use it as constraint on physical states. We will supplement the other Fourier components of ϕ^a by the gauge fixing

$$A^a_{3\vec{n}} = 0, \; n_3 \neq 0$$

$$A^a_{2\vec{n}} = 0, \; n_3 = 0, \; n_2 \neq 0 \tag{37}$$

$$A^a_{1\vec{n}} = 0, \; n_3 = n_2 = 0, \; n_1 \neq 0 \, .$$

The solution to Eqs. (28), (33), (35) and (37) for the color group SU(2) is

$$A^a_i = \frac{1}{L^{3/2}} q_i \frac{v^a}{v}$$

$$E^a_i = \frac{1}{L^{3/2}} \left(p_i \frac{v^a}{v} + \frac{q_i \ell^a_\perp}{q^2}\right) \tag{38}$$

with

$$\ell^a_\perp = (\pi^a - v^a \frac{1}{v^2} v^b \pi^b) \, v, \tag{39}$$

and q_k, p_k, v^a, π^b satisfying canonical poisson brackets.

Introducing Eqs. (38) into Eq. (39) we obtaine the pure gauge part of the Hamiltonian (3).

References

1) F. Palumbo, to be published.
2) F. Palumbo, Phys. Letters B132, 165 (1983).
3) F. Palumbo, in "Mesons, Isobars, Quarks and Nuclear Excitations" Erice 6-18 Aprile 1983.
4) G. De Franceschi and F. Palumbo, in preparation.
5) F. Palumbo and G. Pancheri, in preparation.
6) F. Palumbo, Nuclear Phys. B182, 261 (1981).

QUARK AND GLUON CONDENSATES IN HADRONS AND IN QUARK INTERACTIONS

and

THE ROLE OF THE QUARK-ANTIQUARK PAIR IN THE STRUCTURE OF NUCLEI

J.M.Namysłowski

Institute of Theoretical Physics, Warsaw University, PL-00-681 Warsaw

Abstract

Some methods of the perturbative and non-perturbative QCD are reviewed. The QCD sum rules are presented on the example of the ρ-meson; and the extension of the QCD sum rules from mesons to baryons, and to the meson-baryon coupling constant are given. There are proposed some modifications of the short distance interactions; and the role of the quark-antiquark pair in the structure of nuclei is discussed in some detail.

We introduce the perturbative and the non-perturbative QCD methods, and write the operator product expansion near the light cone. On the example of the ρ-meson we show the technique of Shifman-Vainstein-Zakharov, and the extension of the QCD sum rules from the meson to the baryon sector, and to the evaluation of the meson-baryon coupling constant. We use the numerical results of Ioffe, and Reinders, Rubinstein and Yazaki.

Using the basic idea of Shifman, Vainstein and Zakharov, we propose a suggestion how to enrich the short distance quark interactions by the explicit inclusion of the quark and gluon condensates. We also consider possible dynamical improvements of the standard QCD sum rules. For that we show the Brodsky-Lepage evolution equation method as a mean of coping with the QCD dynamics in the reduced degrees of freedom. We note a particular 3-quark irreducible force, and reviewing some work of Frankfurt and Strikman we show how the quark-antiquark pair influences the structure of : nucleon, deuteron, and iron. This force is a "core" of the two-nucleon correlation in nuclei, and accounts for the cumulative effect in the semi-inclusive processes.

1. Perturbative and non-perturbative QCD

Quantum chromodynamics, with its asymptotic freedom as a bonus, and the confinement problem as a real challange, has two distinct sets of methods to work at different ranges of momenta. They are: the perturbatove QCD rules (in practice very close to the QED rules), and the non-perturbative methods, among which the QCD sum rules, and various bag models are the most prominent ones. We focus our attention on the non-perturbative QCD, and review the basic ingredients of the QCD sum rule technique, with the view of extending it to modify the perturbative quark interactions.

In the non-perturbative QCD the vacuum contains a lot of structure. It is the physical vacuum, and the quark and gluon condensates are the first measures of its complicated character. These condensates show up crucially in the QCD sum rules (remaining the universal parameters in all channels), and in the long distance modifications of the known (short distance)quark interactions. This approach was initiated by Shifman, Vainstein and Zakharov[1], and it is the approach to hadrons (as the composite systems of quarks) from the "short distance side" up to larger distances. Including some power correction terms, with coefficients being appropriate quark and gluon condensates, one extends the perturbative QCD methods down to the region of the masses of Δ, N, ρ , etc, as well as to the evaluation of the hadronic coupling constants. Similarly, the one-gluon exchange force can be modified, to include the simplest signal of the physical vacuum - the gluon condensate. In all these cases we start out from the known QCD perturbative diagrams, operating at short distances, and work up our way to larger distances by including new "diagrams".

To set up the notation let us quote the vocabulary: the QCD Lagrangian, the running coupling constant α_s and its relation to the ratio of the cross sections of $e^+e^- \rightarrow$ hadrons to the cross section of $e^+e^- \rightarrow \mu^+\mu^-$. They are

$$L_{QCD} = \bar{\Psi}(i\gamma_\mu D^\mu - m)\psi - \frac{1}{4} G^a_{\mu\nu} G^{\mu\nu}_a$$

with $G^a_{\mu\nu} \equiv \partial_\mu A^a_\nu - \partial_\nu A^a_\mu + g f^{abc} A^b_\mu A^c_\nu$, $D^\mu \equiv \partial^\mu - ig\frac{1}{2}\lambda^a A^\mu_a$, λ^a obeying the SU(3) algebra $[\lambda^a, \lambda^b] = 2if^{abc}\lambda^c$, ψ denoting schematically the spinor quark fields of different colour and flavour, m an appropriate mass matrix, and g the coupling constant, entering both in $G^a_{\mu\nu}$ and D^μ. The renormalization group method (in a "one-loop" approximation) expresses the asymptotic freedom property of QCD in the following formula for the running coupling constant

$$\alpha_s(Q^2) = 4\pi/[(11 - \tfrac{2}{3}n_f)\ln Q^2/\Lambda^2] ,$$

where n_f is the number of flavours, and Λ is the QCD scale parameter (fixed by experiment), now in the range (100 ÷ 200) MeV. The physical meaning of α_s follows from the ratio of cross sections

$$R \equiv \frac{\sigma_{e^+e^- \rightarrow \text{hadrons}}}{\sigma_{e^+e^- \rightarrow \mu^+\mu^-}} = 3\sum_f e_f^2 \left[1 + \frac{\alpha_s(Q^2)}{\pi} + O(\alpha_s^2)\right],$$

where $Q^2 = +q^2$ is such, that q is the momentum of the virtual photon connecting the electromagnetic vertex $e^+e^-\gamma$ with the hadronic vertex. The cross section $\sigma_{e^+e^- \rightarrow \text{hadrons}}$ can be obtained by taking a cut through diagrams illustrated in Fig. 1a, which contain a typical vacuum polarization part

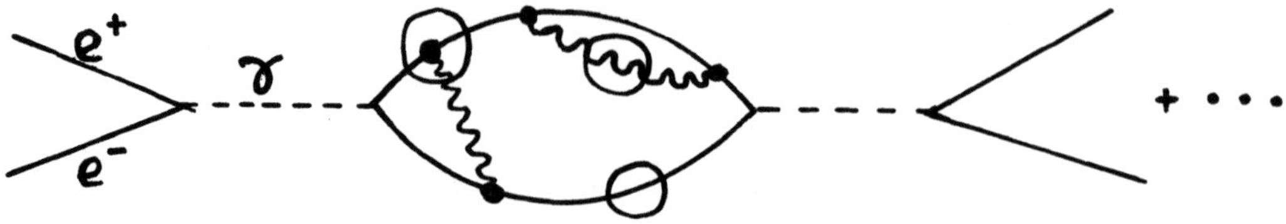

Fig. 1a

In Fig.1a the wavy lines represent gluons, and the full dots the running coupling constant α_s. The circles around the vertices and lines indicate the renormalized vertices and propagators. They represent an infinite set of "elementary" diagrams, which are indicated in Fig. 1b

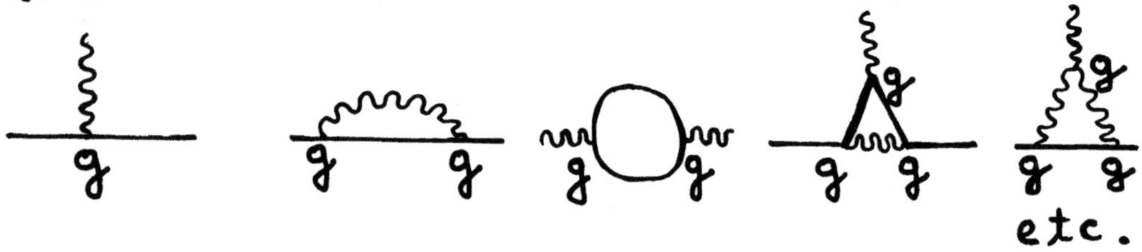

Fig. 1b

From now on we shall not indicate the circles around vertices and lines, as in Fig. 1a, but only dots at vertices, which remind us that a graph  is proportional to α_s.

The above summary of the perturbative QCD is needed later on, and now we turn to the non-perturbative QCD phenomena. The scale of the large distance QCD is roughly the inverse of the typical hadron mass, μ^{-1}. At that distance confinement is at work, and there appears the spontaneous breakdown of the chiral symmetry (the massless QCD has the chiral symmetry). One of the most evident examples of that is the large mass splitting between π, ρ, and A_1, which from the point of view of the perturbative QCD should be degenerate.

The non-perturbative phenomenon is the complex structure of the physical vacuum. It manifests itself among other things, in the non-vanishing value of the vacuum expection value $\langle 0|\bar{\psi}\psi|0\rangle \neq 0$. That is responsible for the mass splitting $m_\rho \neq m_{A_1}$, and it leads to a massless π (if the current quark masses are set equal to zero). To detect the structure of the physical vacuum one has to polarize it by different currents. Similarly to the diagram for the cross section $e^+e^- \rightarrow$ hadrons in Fig.1a, we shall be considering diagrams in Fig.2a

Fig.2a

where the dashed line with a cross represents a current, However, to enrich our formalism with the non-perturbative QCD phenomena we shall add to Fig.2a a new set of "diagrams" shown in Fig.2b

Fig.2b

where a line with a bra and ket ending depicture the presence of a non-zero vacuum expectation value of either the quark, or the gluon field. The hadrons observed in the real life give us signals about the structure of the physical vacuum in terms of the following non-zero vacuum expectation values

$$\langle 0|\bar{\psi}\psi|0\rangle = -(240\,\mathrm{MeV})^3, \quad \langle 0|\frac{\alpha_s}{\pi} G^a_{\mu\nu} G_a^{\mu\nu}|0\rangle = (340\,\mathrm{MeV})^4 \tag{1}$$

In the next sections we shall briefly review the QCD sum rule method of Shifman-Vainstein-Zakharov[1], its application by Reinders, Rubinstein and Yazaki[2], and by Ioffe[3], and finally we indicate its extention to make a modification of the perturbative QCD quark interactions by the inclusion of the gluon condensate.

2. QCD sum rules

There are two complementary techniques applicable to the quantitative treatment of the vacuum polarization. Both of them are used in the QCD sum rules. The first technique is based on the application of the operator product expansion near the light cone, (an extension of the Wilson expansion near the tip of the light cone). The second technique is based on the dispersion relation, and its Borel improvement. We shall explain them in turn, taking as an example the ρ meson. In the actual applications[1,2,3] of the QCD sum rules very many resonances are studied, and one of the successes of this approach is, that in all cases the same values for the quark and gluon condensates are used (as stated in Eq.(1)). It should be so since Eq.(1) characterizes (partially) the physical vacuum, which is common to all cases. Then, the agreement of the QCD sum rules in very many cases justifies the correctness of this approach to the nonperturbative QCD hadron physics.

For the ρ-meson we take the current

$$j_\mu^{(\rho)}(x) = \frac{1}{2}\left(\bar{u}(x)\gamma_\mu u(x) - \bar{d}(x)\gamma_\mu d(x)\right),$$

where u(x) and d(x) denote the u and d valence quark fields, present in the ρ-meson. This current can polarize the physical vacuum, and the appropriate vacuum polarization "operator" Π is defined by

$$\Pi(q^2)(q_\mu q_\nu - q^2 g_{\mu\nu}) = i\int d^4x\, e^{iqx} \langle 0|T(j_\mu^{(\rho)}(x)\, j_\nu^{(\rho)}(0))|0\rangle, \qquad (2)$$

where q denotes the momentum of the current, which flows in and out in Fig.`s 2a,b . The Π defined by Eq.(2) is not really an operator, because it is expressed through the Fourier transform of the matrix element. The actual operator product expansion near the light cone (x^2=0) can be written for $T(j_\mu^{(\rho)}(x)\, j_\nu^{(\rho)}(0))$, or, in a more handy way, for its Fourier transform

$$i\int d^4x\, e^{iqx}\, T(j_\mu^{(\rho)}(x)\, j_\nu^{(\rho)}(0)) = C_{1\mu\nu}(q^2)1 + \sum_n C_{n\mu\nu}(q^2)\, O_n, \qquad (3)$$

where C_n are coefficients, and O_n are operators, classified according to their dimensions. The list of the relevant (lowest dimension) operators is following

O_n	dimension
$m\bar{\Psi}\Psi$	4
$G^a_{\mu\nu}G^{\mu\nu}_a$	4
$\bar{\Psi}\Gamma\Psi\bar{\Psi}\Gamma\Psi$	6 ,

where Γ stands for one of sixteen 4x4 matrices used in the spinor calculus. The coefficients $C_n(q^2)$ have the characteristics of behaving like $(Q^2)^{-n}$, with $Q^2 = -q^2$. (The light cone region (x^2=0) in the x-space, corresponds to large Q^2 in the momentum space[4].) All coefficients C_n can be evaluated using the rules of the perturbative QCD, and the most modern technique of getting them is based on the use of the fixed point gauge ($A_\mu x^\mu = 0$). It is given in the form of algorithms in ref.5. The operator product expansion near the light cone can be proven in perturbation theory[6], while its use in the domain of the non-perturbative QCD has to be assumed, and verified[7] a posteriori.

The operator product expansion is a good mean to show the difference between the physical vacuum and the perturbative vacuum. If we take the perturbative vacuum expectation value of both sides of Eq.(3), then only the term with the 1 operator on the r.h.s. of Eq.(3)survives. The coefficient C_1 is just the standard perturbative QCD result. However, if we take the operators in Eq.(3)between the state $|0\rangle$(denoting the physical vacuum), then because of the appearence of the quark and the gluon condensates (as well as non-zero values of operators of higher dimensions) we get several terms. Every such term is a product of the perturbative coefficient C_n, and the non-perturbative matrix element. In the chosen case of the ρ-meson we get from Eq.(3) the following expression for the vacuum polarization operator

$$\Pi(Q^2) = -\frac{1}{8\pi^2}\left[\left(1+\frac{\alpha_s}{\pi}\right)\ln\frac{Q^2}{\mu^2} + 3\,\frac{m_u^2+m_d^2}{Q^2}\right] + \\ + \frac{1}{2Q^4}\left[\langle 0|m_u\bar{u}u + m_d\bar{d}d|0\rangle\right] + \frac{1}{24Q^4}\langle 0|\frac{\alpha_s}{\pi}G^a_{\mu\nu}G_a^{\mu\nu}|0\rangle + \cdots, \tag{4}$$

where in the first line on the r.h.s. of Eq.(4)there is the expression for C_1, while in the second line there are written the new terms, which are characteristic to the non-perturbative QCD. Eq.(4) can be illustrated "diagramatically" in Fig. 3

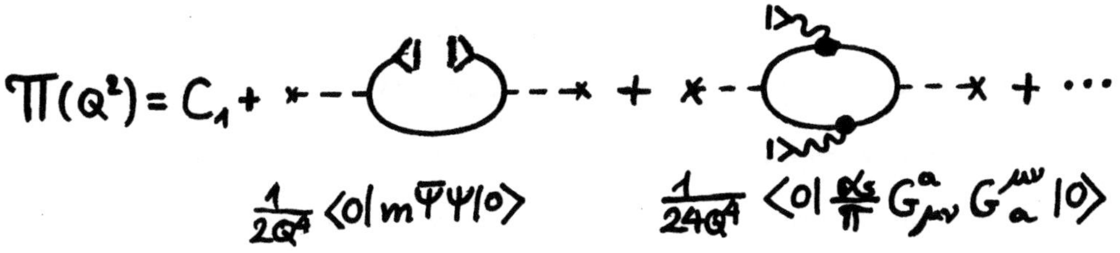

Fig.3

where the lines with the bra and ket endings are not the genuine lines (they are neither propagators nor the wave functions), but they only represent "diagramatically" the presence of the appropriate condensates. These faked lines are however useful in evaluating the coefficients C_n (we first consider some momenta attached to these lines, make an expansion, and then go to the zero limit with the momenta attached to these faked lines). The factorization of the perturbative and the non-perturbative quantities in the physical vacuum matrix elements of the operator product expansion is its great virtue, which enables us gradually to account for the complicated structure of the vacuum, by taking higher terms in Q^{-2n}. This enables us to work up our way from the small distance region to the larger one, by collecting more terms in Q^{-2n}, which are more relevant for the lower values of Q^2.

The second main tool used in the QCD sum rules is the dispersion relation, and the resonance (hadron in general) dominance of this relation. The vacuum polarization "operator" $\Pi(Q^2)$ has the following dispersion representation

$$\Pi(Q^2) = \frac{1}{\pi}\int \frac{\mathrm{Im}\,\Pi(s)}{s + Q^2}\, ds, \tag{5}$$

with disregarded subtractions, since they are inessential in the Borel improved version of Eq.(5). The spectral function Im $\Pi(s)$

is obtained, as usual, by cutting the diagram for $\Pi(Q^2)$ in the middle, and puting the cut particles (hadrons) on their mass shell. Moreover, the cut channel, which carries the quantum numbers of the current, is assumed to be dominated by the appropriate resonance (or sum of resonances), with the quantum numbers of that current. (In the actual applications of the QCD sum rules in ref.`s 1,2,3 one includes also the contribution from the continuum (which for simplicity is disregarded in our presentation). Graphically, we illustrate the evaluation of Im $\Pi(s)$ in Fig.4

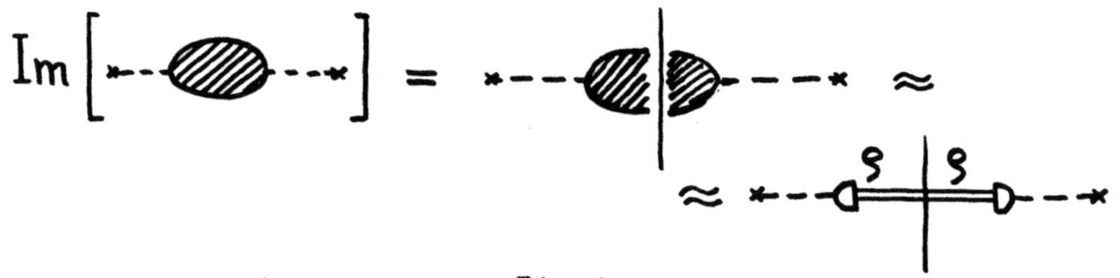

Fig.4

and algebraically we write

$$\mathrm{Im}\,\Pi(s) = \frac{12\pi^2 m_\rho^2}{g_\rho^2}\,\delta(s-m_\rho^2),$$

where m_ρ, and $g_\rho^2/4\pi$ are the mass, and the coupling constant of the ρ-meson ($\Gamma(\rho\to e^+e^-) = \frac{1}{3}\alpha^2 m_\rho 4\pi/g_\rho^2$).

To enhance the contribution to the dispersion integral of the resonance pole, over the contribution of the continuum part, one can differentiate the dispersion relation, or better take a limiting procedure L_M of it, where

$$\lim_{\substack{Q^2\to\infty \\ n\to\infty}} \Bigg|_{\frac{Q^2}{n}=M^2,\,\text{fixed}} \frac{1}{(n-1)!}(Q^2)^n\left(-\frac{d}{dQ^2}\right)^n \equiv L_M . \qquad (6)$$

The L_M operation replaces the previous variable Q^2 by the new one M^2, and the simple (relevant) examples are

$$L_M\left(\frac{1}{Q^2}\right)^k = \frac{1}{(k-1)!}\left(\frac{1}{M^2}\right)^k$$

$$L_M\left(\frac{1}{s+Q^2}\right) = \frac{1}{M^2}\exp(-s/M^2).$$

Therefore, instead of the dispersion relation in Eq.(5) we get

$$L_M\left(\Pi(Q^2)\right) = \frac{1}{\pi M^2}\int \mathrm{Im}\,\Pi(s)\exp(-s/M^2)\,ds, \qquad (7)$$

from which we see, that if M^2 is around m_ρ^2 then the contribution of the ρ- meson is enhanced over the contributions from the continuum part, which are suppressed by the exponential factor $\exp(-S/M^2)$.

The QCD sum rule is obtained if the r.h.s. of Eq.(7) is set equal to the Borel improved version of the operator product expansion of $\Pi(Q^2)$, given in Eq.(4). The integrated function $\mathrm{Im}\,\Pi_{(s)}$ can be expressed in terms of the ratio R of cross sections, and for the case of the ρ - meson channel we have

$$\int R\Big|_{\substack{e^+e^-\to \text{hadrons}/\,e^+e^-\to\mu^+\mu^- \\ S=1,\ I=1}} \exp(-s/M^2)\,ds =$$

$$= \frac{3}{2}M^2\left[1+\frac{\alpha_s(M^2)}{\pi}+\frac{4\pi^2}{M^4}\langle 0|m_u\bar{u}u+m_d\bar{d}d|0\rangle+\frac{\pi^2}{3M^4}\langle 0|\frac{\alpha_s}{\pi}G^a_{\mu\nu}G_a^{\mu\nu}|0\rangle+\cdots\right]. \qquad (8)$$

The left hand side of the sum rule in Eq.(8) is referred to as the "phenomenological" side, while the r.h.s. as the "theoretical" side. The QCD sum rule means, that there exists such a domain of M^2, that both sides (obtained from completely different sources) are equal. Of course the domain of M^2 is limited, since both sides of the sum rule are only approximate expressions for the (Borel transformation of the) function Π. In the actual numerical calculations, performed in ref.2, they get

$$m_\rho = 770 \text{ MeV} \qquad g_\rho^2/4\pi = 2.42$$

$$(\text{exp.} 776 \pm 3) \qquad (\text{exp.} 2.36 \pm .18),$$

taking for the quark and gluon condensates their values stated in Eq.(1).

3. Baryons and baryon-meson coupling constants

The QCD sum rule technique can be extended from the quark-aniquark to the 3-quark (baryon) sector. To do that one has to build a current out of three valence quark fields. For the Δ resonance the construction is unique, while for the nucleon there are several options. B.I. Ioffe in ref.8 has given arguments to sellect the most appropriate current. He puts two criteria for the sellection of the optimal current: 1) the continuum contribution should be minimal, and 2) the disregarded terms in the operator product expansion (corresponding to operators with higher dimensions), should be small in comparison with the retained terms (in the domain of the characteristic values of the Borel mass parameter M^2). Taking Ioffe`s reasoning we write for nucleon the following current

$$\eta(x) = (u^{aT}(x)\, C\gamma_\mu u^b(x))\gamma_5\gamma^\mu d^c(x)\, \varepsilon_{abc},$$

where two u-quark and one d-quark fields correspond to the valence quark fields in proton, and C is the standard charge conjugation matrix.

The appropriate vacuum polarization "operator" in the nucleon channel is defined by

$$\Pi^{3q}(q^2) = i\int d^4x\, e^{iqx}\langle 0|T(\eta(x)\bar\eta(0))|0\rangle,$$

and the operator product expansion of Π^{3q} can be schematically written as

$$\Pi^{3q}(q^2) = C_1(q^2)1 + C_m(q^2)\langle 0|m\bar\Psi\Psi|0\rangle + C_G(q^2)\langle 0|\frac{\alpha_s}{\pi}G^a_{\mu\nu}G^{\mu\nu}_a|0\rangle + \dots, \tag{9}$$

where the implicite summation over different quark flavours (denoted here by Ψ) has to be understood. The pure perturbative term, corresponding to $C_1 1$ in Eq.(9), is drawn in Fig.5a

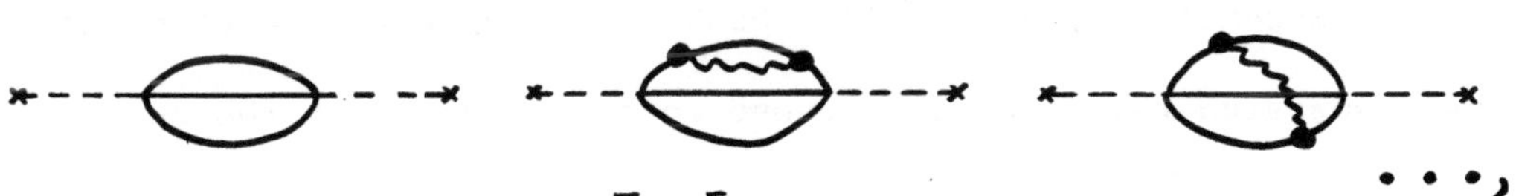

Fig.5a

while the mixed (perturbative and non-perturbative) "diagrams" are shown in Fig.5b

$C_m \langle 0| m\bar{q}q |0\rangle$ $\qquad$ $C_G \langle 0| \frac{\alpha_s}{\pi} G^a_{\mu\nu} G^{\mu\nu}_a |0\rangle$

Fig.5b

In Fig.'s 5a and 5b there are "strange" vertices, corresponding to the "coupling" of the current to three quark fields. They should be understood only symbolically, and they arrise from a static approximation, in which we put all three quark fields at the same point x. In reality the valence quarks move, therefore there should be a structure at the current-3quark "vertex". In the future one should include the motion of the valence quarks in the scheme of the QCD sum rules, which amounts to the inclusion of the non-static properties of hadrons in terms of their formfactors. The motion of the valence quarks is taken care of in the scheme of the evolution equations, and a step toward this direction is taken in the next section, where we modify the quark interactions by the presence of the gluon condensates. (The quark interactions are kernels in the integrodifferential evolution equations.)

The discontinuity of Π^{3q}, which is the spectral function in the dispersion relation, is assumed to be dominated by the one-nucleon state $|N(p)\rangle$ of momentum p. We write it as follows

$$\mathrm{Im}\,\Pi^{3q}(p^2) \approx \langle 0|\eta|N(p)\rangle\langle N(p)|\bar{\eta}|0\rangle ,$$

and graphically we illustrate it in Fig.6

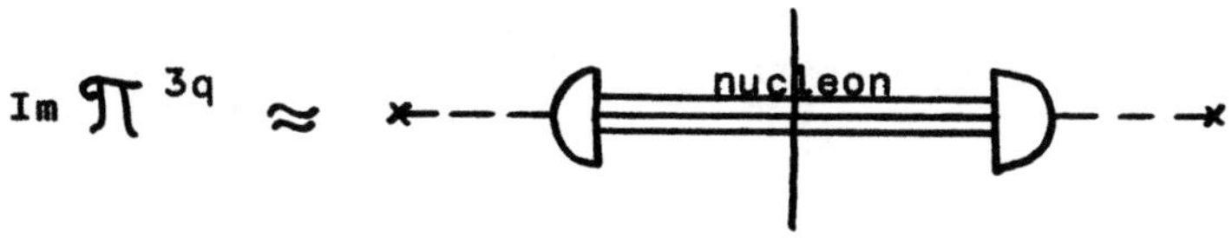

Fig.6

The matrix element $\langle 0|\eta|N(p)\rangle$ we write as $\lambda v^{(r)}(p)$, where λ is a constant, and $v^{(r)}(p)$ is the spinor describing the nucleon with momentum p and polarization r $(\bar{v}(p)v(p)=2M_N)$.

The QCD sum rule is obtained similarly as in the meson case, though the algebra is now more complicated than before. The Δ resonance channel is treated similarly as the N channel, and taking the quark and gluon condensates as stated in Eq.(1), B.I.Ioffe found[9,10] the following values for the nucleon and Δ masses

M_N = 985 MeV	M_Δ = 1.41 GeV
(Erratum .90 GeV)	(Erratum 1.30 GeV).

Yet another problem - the πNN coupling constant - can be attacked with the QCD sum rule technique. The corresponding physical quantity we depict in Fig.7

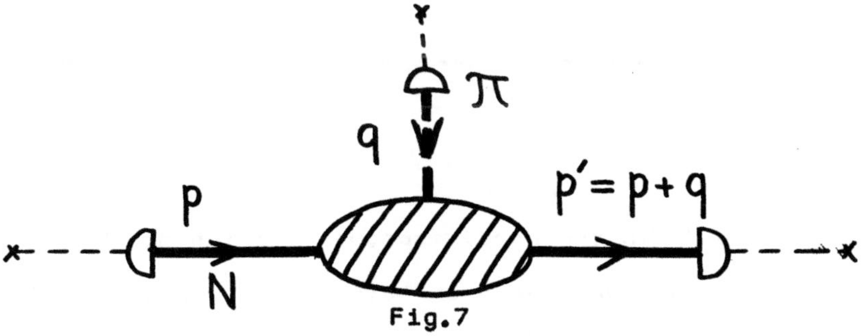

Fig.7

and write for it the following mathematical expression

$$A(p,p'=p+q,q) = \int dx\,dy\, e^{ip'x-iqy} \langle 0|T(\eta(x) j^{(\pi)}(y)\, \eta(0))|0\rangle,$$

where $j^{(\pi)}(y)$ is the current operator for the π- meson, given by the formula

$$j^{\pi}(y) = \bar{u}(y) i\gamma_5 u(y) - \bar{d}(y) i\gamma_5 d(y).$$

The quantity $A(pp'q)$ as indicated in Fig.7, can be also represented (in a phenomenological way) by the following expression

$$A(p,p'=p+q,q) = \lambda^2 \frac{M_N}{(p^2-M_N^2)^2} i\gamma_5 g_{\pi NN} \frac{1}{q^2-m_\pi^2} \frac{f_\pi}{\sqrt{2}} \frac{m_\pi^2}{m_q}, \quad (10)$$

where $g_{\pi NN}$ is the πNN coupling constant, and f_π=133 MeV ($f_\pi/\sqrt{2}$ = 95MeV, as usual). Note that Eq.(10)is the definition of $g_{\pi NN}$ (for our purpose) in terms of A.

The leading mixed "diagrams", which constribute significantly in the QCD sum rule evaluation of $g_{\pi NN}$, are shown in Fig.8

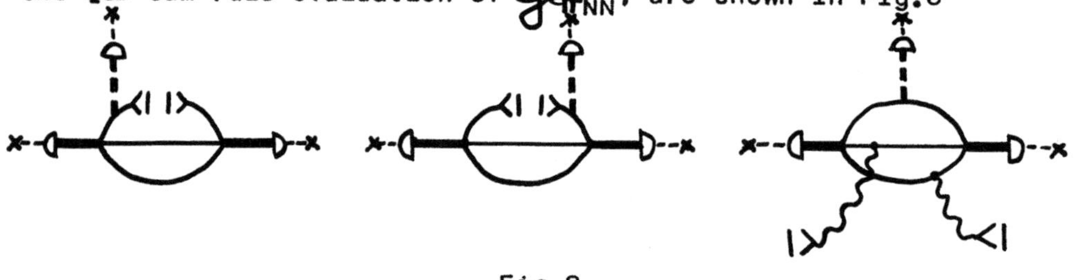

Fig.8

If the mass variable M^2 in the Borel improved version of the dispersion relation for the function A is taken to be equal 1 GeV^2, then the numerical work, curried out in ref.2, leads to

$$g_{\pi NN} = 12.5 \qquad (\text{exp. } 13.5).$$

The authors of that calculation comment about this result, that the error in the evaluation of $g_{\pi NN}$ is at least 20%, due to uncertainties in the values for the quark and gluon condensates. However, the stability of the result for $g_{\pi NN}$, with the variation of the mass parameter M^2 (in the Borel improved version) seems to be more questionable, than the 20% error mentioned above. The actual formula for $g_{\pi NN}$, which can be derived from the QCD sum rule, is

$$g_{\pi NN} = 2(2\pi)^2 \sqrt{2}\, \frac{f_\pi}{m_N} \frac{1}{1 + \frac{4}{3}\frac{a}{M^6} - \frac{b}{M^4}},$$

where $a \equiv -(2\pi)^2 \langle 0|\bar{\psi}\psi|0\rangle$, $b \equiv \pi^2 \langle 0|\frac{\alpha_s}{\pi} G^a_{\mu\nu} G^{\mu\nu}_a|0\rangle$. ($\chi^2$ is taken to be equal $1.2 \times 10^{-3}\,\mathrm{GeV}^{-6}$). Changing the value of M^2 about (30 ÷ 50)% one gets about 300% variation of $g_{\pi NN}$. The actual numbers are

$$g_{\pi NN} = \begin{cases} 6.81 \\ 12.5 \\ 19.13 \end{cases} \quad \text{for} \quad M^2 = \begin{cases} 0.8\ \mathrm{GeV}^2 \\ 1.0\ \mathrm{GeV}^2 \\ 1.3\ \mathrm{GeV}^2 \end{cases},$$

and they come from Prof.M.Kremer`s talk at the Mainz Workshop (June 1983). The lack of stability of $g_{\pi NN}$ formula remains a puzzle, which possibly can be healed by a special interplay of the, so far omitted, higher dimension operators in the operator product expansion. That requires some more thorough work, and deserves full attention.

In connection with the above stability problem let us also mention the criticism of the QCD sum rule technique by people[11] playing with potential models. They claim, that the Shifman-Vainstein-Zakharov method underestimates the value of the anolog of the gluon condensate, because in their view this method is equivalent to the Rayleigh-Ritz variational calculus. There is however a second criticism made by Voloshin[12] of the idea of enforcing the potential picture to the QCD sum rule method. Voloshin shows[12], that the potential approach is inapplicable to describe the non-perturbative effects. He finds terms in the operator product expansion result, which can not be generated as iterations in the potential approach. Also the time scale argument used in the potential approach is in conflict with the time scale connected with the non-perturbative gluon fluctions. Therefore, the potential model arguments apply to a different type of physical phenomena, and the QCD sum rule technique is free from this challenge.

4. Quark interactions with non-perturbative modifications

Hadrons are composite systems of quarks, and obey appropriate bound state equations. The most natural framework to make a sensible Fock state decomposition of hadronic states is the light-cone quantization. It is described in full detail in several papers of Brodsky, Lepage and collaborators[13,14,15] /the most relevant reference, connected with the present discussion, is ref. 15/. The hadronic wave functions are the equal $x_+ = t + z$ wave functions, and the appropriate bound state equation is the Weinberg /infinite momentum, or equivalently light-cone/ integral equation. For nucleon, the three quark valence wave function $\Psi(x_1 x_2 x_3 q_k Q_k)$, where $x_i = p_{i+}/P_+$, $P = \sum_{i=1}^{3} p_i$, $\sum_{i=1}^{3} x_i = 1$, p_i are the individual quark's momenta, $q_k = (x_j p_i - x_i p_j)(x_i + x_j)^{-1}$, $Q_k = (x_i + x_j)p_k - x_k(p_i + p_j)$, obeys the following bound state equation

$$\psi(x_i q_{V_k} Q_k) =$$
$$= G_0 \int [dy] (16\pi^3)^{-2} d^2 q'_{V_m} d^2 Q'_m \, V(x_i q_{V_k} Q_k y_l q'_{V_m} Q'_m) \psi(y_l q'_{V_m} Q'_m), \quad /11/$$

where $[dy] \equiv dy_1 dy_2 dy_3 \delta(1 - y_1 - y_2 - y_3)$

$$G_0 \equiv D_0^{-1}$$
$$D_0 \equiv M^2 - q^2_{V_k}(x_i + x_j)(x_i x_j)^{-1} - Q^2_k (x_i + x_j)^{-1} x_k^{-1} - m^2 \sum_{l=1}^{3} x_l^{-1}$$

M is the nucleon mass, and m is the quark mass.

The kernel V of the Weinberg equation is well known at short distances, and there the leading forces are shown in Fig. 9 /only the representative cases are indicated/

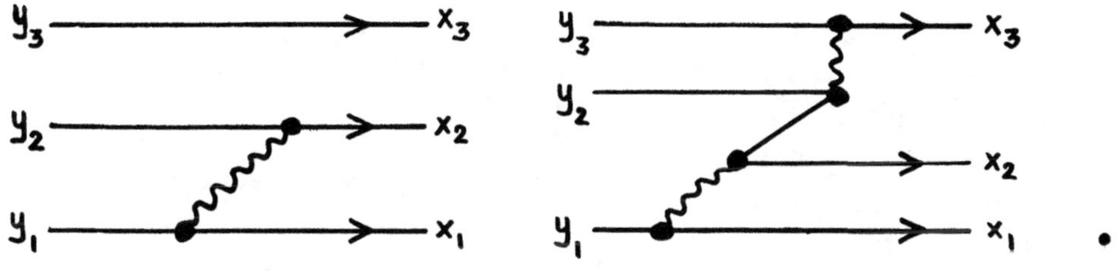

Fig. 9

The large distance extension of these forces is not known so far /except for phenomenological ansatz/, and to use Eq./11/ Brodsky and Lepage[13] developed a collinear projection technique to change Eq./11/ into an evolution equation for a distribution amplitude. The collinear projection operator $P_{\tilde{Q}}$, corresponding to a scale $\tilde{Q}$, is defined as follows

$$P_{\tilde{Q}} \equiv (16\pi^2)^2 \delta(q^2_{V_i}) \delta(Q^2_i) \Theta(\tilde{Q}^2 + D_0),$$

and the distribution amplitude is a collinear average of the bound state wave function in the perpendicular components of the quark`s momenta. However, the dependence on the longitudinal quark`s momenta /i.e. on x_1, x_2, x_3/ is fully retained in the distribution amplitude. It is advantagous to make a partial wave projection[16] of the bound state wave function before making the collinear projection. If we perform the relative angular momentum decomposition, corresponding to the rotation about the angle φ_k between q_k and Q_k in their plane, then the partial wave distribution amplitude ϕ_{l_i} is defined as follows

$$\phi_{l_i}(x\tilde{Q}) \equiv \left[d_F(\tilde{Q}^2)\right]^{-3/2} \int dq_k^2\, dQ_k^2\, \Theta(\tilde{Q}^2 + D_0)\, \psi_{l_i}(x q_k Q_k),$$

where d_F is the quark wave function renormalization, related to the quark field anomalous dimension γ by $\gamma = -\tilde{Q}^2 d \ln d_F^{-1} / d\tilde{Q}^2$. The distribution amplitudes obey the integro-differential evolution equations. For $l_i = 0$ we have the Brodsky-Lepage[13] equation, while for $l_i \neq 0$ we get inhomogeneous[16] evolution equations. They are, respectively,

$$x_1 x_2 x_3 \left(\frac{\partial}{\partial \xi} + \frac{3}{2}\frac{C_F}{\beta}\right)\tilde{\phi}_0 = \frac{C_B}{\beta}\int [dy]\, V^{1\,gluon}\, \tilde{\phi}_0(y)$$

$$x_1 x_2 x_3 \left(\frac{\partial}{\partial \xi} + \frac{3}{2}\frac{C_F}{\beta}\right)\tilde{\phi}_{l_i \neq 0} = \frac{C_B}{\beta}\int [dy]\, V^{zigzag}_{l_i \neq 0, 0}\, \tilde{\phi}_0(y), \quad (12)$$

where $\tilde{\phi}_{l_i} \equiv (x_1 x_2 x_3)^{-1} \phi_{l_i}$, $\xi \equiv \ln\ln \tilde{Q}^2/\Lambda^2$, $C_F = 2C_B = 4/3$

$$\beta \equiv 11 - \tfrac{2}{3} n_f \ (= 9 \text{ for } n_f = 3),\quad \Delta\tilde{\phi}_0(y\tilde{Q}) \equiv \tilde{\phi}_0(y\tilde{Q}) - \tilde{\phi}_0(x\tilde{Q})$$

$$V^{1\,gluon} \equiv 2\sum_{i=1}^{3}\left[x_i y_j x_k \left(\frac{\Delta}{y_i - x_i} + \frac{\delta_{\lambda_i \bar{\lambda}_j}}{x_i + x_j}\right)\Theta(y_i - x_i) + \right.$$

$$\left. + (i \leftrightarrow j)\right]\delta(x_k - y_k),$$

$$V^{zigzag}_{l_i\neq 0,\,0} \equiv c_B \alpha_s (x_i x_j)^{1/2} (x_k - y_k)^{-2} \Theta(x_k - y_k) \Big[\Theta(y_i - x_i)\Theta(y_i - 1 + x_k) y_i^{-1} \delta_{2\lambda_i l_i}^{+} + \Theta(y_j - x_j)\Theta(y_j - 1 + x_k) y_j^{-1} (2\pi)^{-1} \int_0^{2\pi} d\varphi_i \exp(-i l_i \varphi_i - 2\lambda_j \varphi_j) \Big].$$

Note, that in the evolution equations the kernels are fully known, since they are entirely given by the short distance /perturbative QCD/ quark interactions. This is the basic reason to go from the bound state wave equation to the evolution equation.

To solve evolution equations one has to specify input, which gives the shape of the distribution amplitude at some low value of $\tilde{Q}^2$. Then, the perturbative QCD kernels of the evolution equations enable us to see how that input "evolves" with the increase of the scale $\tilde{Q}^2$. In the input to the evolution equations there are included /in a phenomenological way/ some informations from the non-perturbative QCD. We want to enrich these informations by incorporating the quark and gluon condensates in a similar way as they are included in the QCD sum rules. We shall illustrate that in the following paragraph. Besides that it is also helpful to find some constraints on the non-perturbative input itself. This has not been done so far, but a way to approach that would be to use the spectral representation of the full Green function. The residuum of the bound state pole of the Green function is determined by the wave function. Thus, writing on one hand the operator product expansion for the product of field operators in the Green function, and on the other hand a dispersion relation /with the Borel improvement/ one can get a relation similar to the QCD sum rule. That leads us to a constraint, which should be respected by the input to the evolution equations. /Some other proposals to include the QCD non-perturbative constraints in the hadron wave functions are given in ref. 17./

Our non-prturbative modifications of the short distance quark interactions follow from the investigation of the higher order irreducible forces /containing a loop integral/, in the same way as Shifman-Vainstein-Zakharov studied the vacuum polarization loops. The mixed /perturbative and non-perturbative/ type diagrams, which give the non-perturbative contributions to the quark interactions are shown in Fig. 10 /also only the representative cases are indicated/

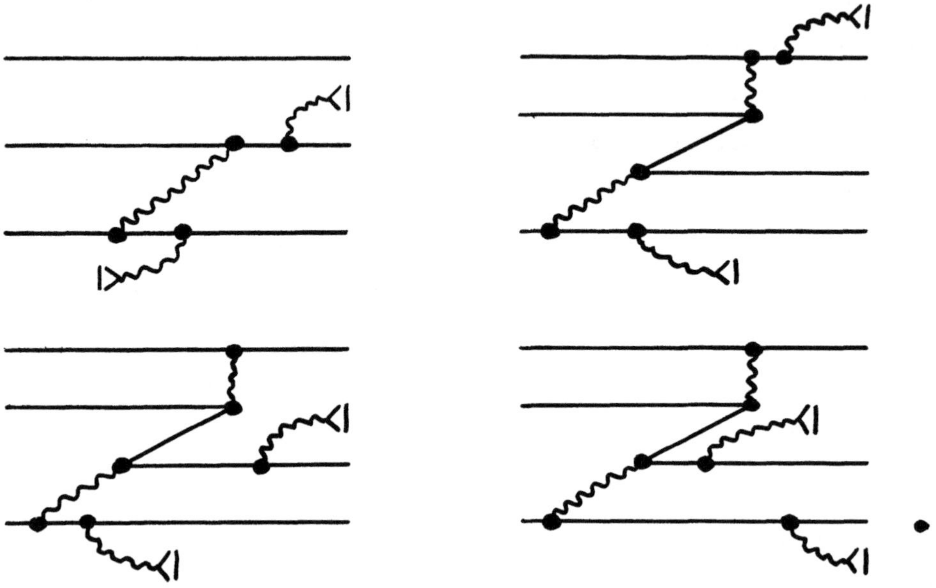

Fig. 10

When counting powers of the running coupling constant α_s one should keep in mind, that the gluon condensate, as written in Eq./1/, contains already one power of α_s. Therefore, the mixed type diagrams in Fig. 10 are partners of the diagrams in Fig. 9 at the same α_s-order. On the other hand, counting the powers of Q^{-2}, the mixed type diagrams in Fig. 10 are higher order than the diagrams in Fig. 9. This is the usual situation if one includes corrections corresponding to the higher twists.

The higher twist corrections arrise in many physical situations, when we use the operator product expansion and take matrix elements between different states, not necessairly the vacuum state. More generally /i.e. even without invoking the operator product expansion we get the higher power Q^{-2} terms, if we try to separate out the hard interaction part /perturbative QCD/ from the soft one /non-perturbative part/. This type of approach is presented in ref.`s 18, and 19. The ref.18 is of particular interest in connection with the presented approach, since these authors make also use of the advantages of the infinite momentum /light-cone/ scheme.

Our final comment concerns the improvement of the QCD sum rules themselves by the inclusion of the non-static properties of quarks which form a hadron. The current-valence-quark vertex should contain a nontrivial structure, to account for a hadron form factor. We have to allow the individual quarks to be at different points, and to move. The inclusion of the motion in all degrees of freedom may be a complicated task, but to allow for the dynamics in the $x_i = p_{i+}/P_+$ variables is a doable program, which can follow the techniques used in the evolution equations. In terms of diagrams, one has to include more mixed type diagrams, than these which are present in the standrd QCD sum rules. The extra mixed type diagrams have the same order of the Q^{-2} power, but their perturbative part contains loops /infinitely many/ with moving quarks. These classes of diagrams sum up to make the hadron form factors.

5. The role of quark pairs in the structure of nuclei

The $q\bar{q}$ pairs play crucial role in the non-perturbative structure of the physical vacuum, and in the spontaneous breakdown of the chiral symmetry. The Nambu-Jona-Lasinio model is one example of the mechanism of these phenomena . In this example there emerge collective states, and the quark condensate shows up. Such states influence crucially the hadron mass spectrum, as was seen explicitly in the QCD sum rules. Also, the hadron coupling constants, and the quark interactions are influenced by the appearence of the quark, and the gluon condensates. Therefore, the structure of nuclei, starting from the simplest one (hydrogen) is sensitive to the collective $q\bar{q}$ states. These are, in a **sense, obvious** effects (but also very complex one, on the level of nuclei). We shall not tackle this theme, but rather facous our attention on just one $q\bar{q}$ pair, and notice its rôle in the nucleon, deuteron, and iron.

In the nucleon, the $q\bar{q}$ pair is the crucial part of the "zigzag" force, illustrated in Fig.9. It is responsible for producing the driving force in the inhomogeneous evolution equation (Eq./12/). The effect of such force is to break the SU(6) symmetry, in contrast to the one-gluon exchange force, which reteins this symmetry. To show that we take, as the input to the evolution equations, the SU(6) symmetric part of the eigenfunction of the Brodsky-Lepage[13] equation. This solution corresponds to the leading (lowest) anomalous dimension, and it is

$$\phi_0 = C\, x_1 x_2 x_3 \exp\left(-\frac{2}{3\beta}\xi\right)(\chi_1\chi + \chi_2\chi'),$$

where

$$\chi_1 \equiv 2uud - udu - duu \qquad \chi_2 \equiv udu - duu$$
$$\chi \equiv 2\uparrow\uparrow\downarrow - \uparrow\downarrow\uparrow - \downarrow\uparrow\uparrow \qquad \chi' \equiv \uparrow\downarrow\uparrow - \downarrow\uparrow\uparrow .$$

The one-gluon exchange force reproduces ϕ_0 with an appropriate constant. However, the zigzag force when acting on ϕ_0 gives terms of the type

$$\sqrt{x_1 x_2}\, x_3^2 \exp\left(-\xi\left(1+\frac{2}{3\beta}\right)\right)(\chi_1\chi'' + \chi_2\chi'),$$

where $\chi'' \equiv -(\uparrow\downarrow\uparrow + \downarrow\uparrow\uparrow)$. The terms of such structure break the SU(6) symmetry. The presence, or the absence of the SU(6) symmetry can be detected experimentally in the measurments of the ratio of the d-quark distribution to the u-quark distribution in proton.

The SU(6) symmetry requires $1/2$ for this ratio, while the experimental points significantly deviate from $1/2$. The zigzag force brings this ratio down to the value .2 for x=1, corresponding to the kinematical situation, when one of the proton`s valence quarks takes almost the whole proton momentum. Another interesting effect of the zigzag force is its contribution to the leading terms $((1-x)^3)$ in deep inelastic structure function $F_2(x)$. In contrast to that the one-gluon exachange force contributes to the terms in $F_2(x)$, which are proportional to $(1-x)^5$, thus they are non-leading.

The zigzag force plays also a significant rôle in the structure of deuteron. Frankfurt and Strkman[20] have shown, that the zigzag force is responsible for the existence of the point-like structure in nucleons (in deuteron), and explains the cumulative effect in the semi-inclusive electron-deuteron scattering. Using that force they were able to show a dual picture of this processes, once viewed as the standard nuclear physics process, and otherwise as the QCD hard subprocess in the space of 6 valence quarks . The leading QCD diagram explains the two-nucleon correlation, with the zigzag force, joining quarks, which belong to different nucleons. To indicate the Frankfurt-Strikman reasoning let us introduce a variable $\hat{x} = Ax$, where A is the nucleus mass number. In the case of deuteron $\hat{x} \in [0,2]$.

We consider the semi-inclusive process

$$e + d \longrightarrow e + N + X ,$$

and measure the deep inelastic structure function of deuteron, with the extra constraint, that we detect a fast nucleon N in the laboratory frame. Such structure function is denoted $F_2{}^{d/N}$. In the standard nuclear physics picture this process looks as indicated in Fig.11

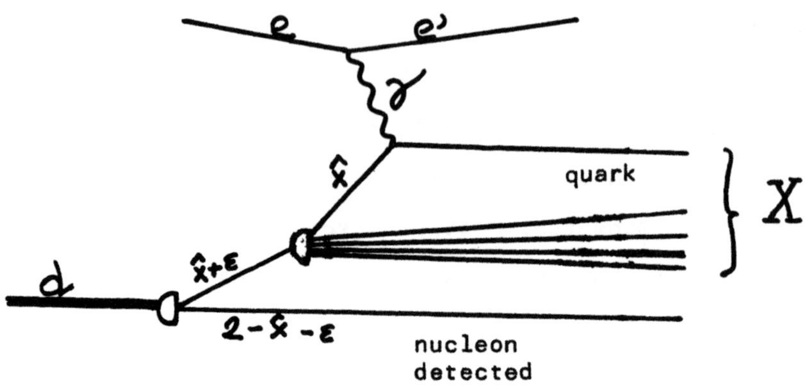

Fig.11

In this figure $\hat{x} = 2x$, where x is the usual ratio of the + component of the quark to the + component of deuteron. The nucleon detected in the laboratory system carries $2 - \hat{x} - \varepsilon$, where ε is a very small number. The value of $\hat{x}$ is $\hat{x} \approx \frac{1}{3} = 2 \cdot \frac{1}{6}$, where $^1/_6$ is the evarage quark`s $x$ in the 6-quark model of deuteron. The detected nucleon has $2 - \hat{x} - \varepsilon \approx 1.7$. The other nucleon (from which the inclusive hadron matter $X$ gets out) has $\hat{x} + \varepsilon$, and this subprocess is described by the nucleon`s structure function $F_2^N(\frac{\hat{x}}{\hat{x}+\varepsilon}) \sim$ $\sim (1 - \frac{\hat{x}}{\hat{x}+\varepsilon})^3$. The standard nuclear physics' description of this process is to account yet for the two-nucleon structure of deuteron (described by the nuclear density ρ_d, measured as a function of the spectator nucleon) and write the structure function for the whole process $F_2{}^d/N$ in the form

$$F_2^{d/N} \sim \rho_d(2-\hat{x}-\varepsilon)\, F_2^N\left(\frac{\hat{x}}{\hat{x}+\varepsilon}\right) \sim \rho_d \frac{\varepsilon^3}{(\hat{x}+\varepsilon)^3} \,.$$

The same semi-inclusive process can be dually interpreted in the 6-quark model of deuteron. Using perturbative QCD for hard scattering Frankfurt and Strikman draw the leading QCD diagram, with the zigzag force, in Fig.12

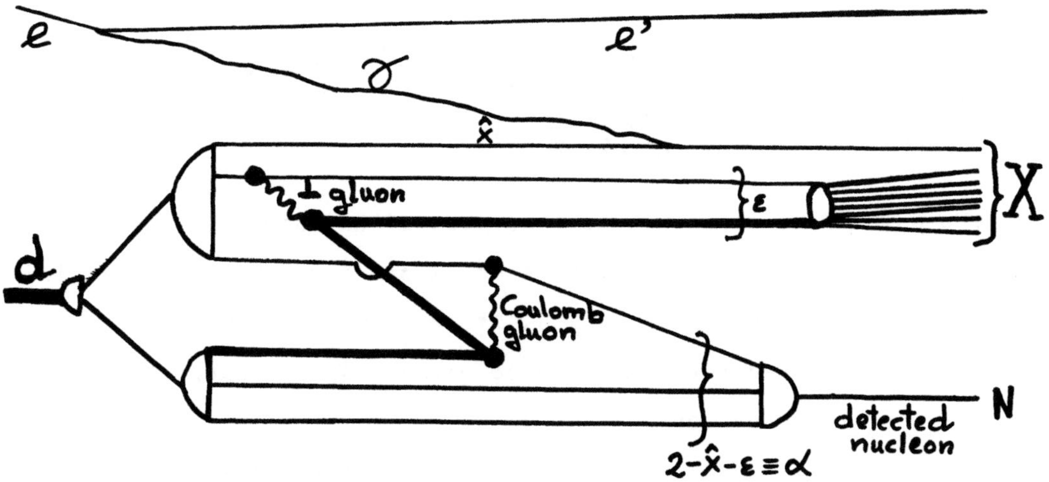

Fig.12

(The quarks "making" the letter Z are drawn with a thick line to be better seen.) The diagram in Fig.12 can be evaluated, using the rules[13] of QCD on the light-cone. Counting the p-denominators, which have terms behaving like ε^{-1}, and including the appropriate factors in the vertices (due to the spin of quarks) we get, that the whole process is proportional to ε^{3},

$$F_2^{d/N} \sim \varepsilon^3 ,$$

i.e. the same ε-behaviour as obtained in the dual (nuclear physics) description. Moreover, if denote by $\alpha \equiv 2 - \hat{x} - \varepsilon$ the variable of the spectator nucleon, then we get

$$\rho_d \sim (\hat{x} + \varepsilon)^3 \equiv (2-\alpha)^3 \sim |\Psi_{d/NN}|^2 / (2-\alpha) ,$$

where the last factor $(2-\alpha)^{-1}$ comes from the phase space. Therefore, we get for deuteron wave function the result

$$\Psi_{d/NN} \sim (2-\alpha)^2 ,$$

and the Hulthen type deuteron wave function (written in the light-cone notation) has that form, namely

$$\Psi^{\text{Hulthen}}_{d/NN} \sim \frac{1}{\left[\frac{k_\perp^2 + m_N^2}{\alpha(2-\alpha)}\right]^2} \sim (2-\alpha)^2 .$$

The point-like structure is observed, if we note that the detected quark, although it looks like a usual one from the point of view of deuteron (it has $\hat{x} \approx 2 \cdot \frac{1}{6}$), it looks very unusual as a member of the non-observed nucleon, because within ε it carries all momentum of that nucleon .

Thus, it is a point-like object in this nucleon. On the other hand the spectator nucleon is also unusual, because it is very fast (having $\alpha = 2 - \hat{x} - \varepsilon \approx 1.7$), and such nucleon corresponds to the cumulative effect, which can happen only in nuclei (not when we consider free nucleons). The two-nucleon correlation, which is easily seen in Fig.12, can happen in any nucleus, and seems to be the main reson for the appearence of the cumulative effect.

The above mechanism (produced by the zigzag force) can be also used as one of the mechanisms contributing to the EMC effect. In iron we observe more slower, and fewer faster quarks, than in a free nucleon. One can argue, that the existence of the extra processes like the one shown in Fig. 12 (which can only happen in a nucleus, but not in a free nucleon) adds up to the probability of finding slow quarks ($x \approx \frac{1}{6} \approx \frac{1}{2}\hat{x}$). In the same time the quarks being in nucleus can have $\hat{x}$ as high as A. However the whole deep inelastic structure function is normalized, thus the excess in the region of small x (or $\hat{x}$), and the alongation of the distribution for very large $\hat{x}$, is only possible, if there is a depletion in the region of the medium x . In the EMC iron experiment there is indeed a depletion for $x \approx .6$. Of course, one does not claim, that this accounts for the whole EMC effect, but it joins in the many available explanations, at present.

Acknowledgements

The author would like to acknowledge warm and stimulating hospitality extended to him in the Bonn, Regensburg, and Tübingen Universities, where most of this work was done. In particular a series of lectures in Regensburg, for a devoted audience, were very helpful in preparing this paper. Very many discussion with my collaborators Dr. E.A.Bartnik, and Dr.St.Głazek were crucial in finding the final understanding of this subject. I would also like to acknowledge very illuminating discussions with Professors K.Bleuler, S.J.Brodsky, A.Faessler, H.Petry , D.Schütte, W.Weise, and W.Werner.

References

1. M.A.Shifman, A.I.Vainstein and V.I.Zakharov, Nucl.Phys.B147, 385; 448; 519 (1979)
2. L.J.Reinders, H.R.Rubinstein and S.Yazaki, Nucl.Phys.B 186, 109 (1981); B 196, 125 (1982);B 213, 109 (1983); Phys.Letters 94B, 203 (1980); 95B, 103 (1980); 97B, 257 (1980); 103B, 63 (1981); 104 B, 305 (1981); 113B, 411 (1982); 120B, 209(1983)
3. B.L.Ioffe, Nucl.Phys.B188, 317(1981), Erratum B191, 591 (1981)
4. Y.Frishman, Ann. of Phys. 66 , 373 (1971)
5. A.V. Smilga, Sov.J.Nucl.Phys.35,2711(1982)
6. W.Zimmermann in Lectures on Elementary Particles and Quantum Field Theory, ed.S.Deser et al (MIT Press. Cambridge, Mass.1971)
7. V.A.Novikov, M.A.Shifman, A.I.Vainstein, and V.I.Zakharov, Nucl. Phys.B174, 378 (1980)
8. B.L.Ioffe - ITEP (Moscow) preprint 131/82 titled: "On the choice of quark currents in the QCD sum rules for baryon masses" .

See also Y.Chung, H.G.Dosch, M.Kremer and D.Schall in Zeit. f. Physik C15, 367 (1982)

9. B.L.Ioffe, Nucl.Phys.B188, 317 (1981)
10. B.I.Ioffe, Erratum Nucl.Phys.B191, 591 (1981)
11. J.S.Bell and R.A.Bertlmann, Nucl.Phys.B177, 218(1981); B187,285 (1981); B.Durand, J.B.Shitenton and L.Durand, Phys.Lett. 124B, 410 (1983)
12. M.B.Voloshin, Sov.J.Nucl.Phys.36,143 (1982); Nuclear Phys.B154, 365 (1979)
13. G.P.Lepage and S.J.Brodsky, Phys.Rev.D22, 2157 (1980)
14. S.J.Brodsky and G.P.Lepage (SLAC-PUB-2447) published in the 1979 Summer Institute on Particle Physics at SLAC, Stanford, California; Physica Scripta 23, 945 (1981); S.J.Brodsky, T.Huong and G.P.Lepage (SLAC-PUB-2868) published in the Proceedings of the KFK Summer School on Quarks and Nuclear Forces (1981), "Quarks and Nuclear Forces" ed.D.Fries and B.Zeitnitz, Spring field (1982)
15. G.P.Lepage, S.J.Brodsky, T.Huang and P.B.Mackenzie, Banf Summer Institute on Particle Phys.Banff, Alberta, Canada, August 1981
16. E.A.Bartnik and J.M.Namysłowski -"3-quark interaction - the driving force in the inhomogeneous evolution equations", Warsaw Univ.preprint IFT/14/83 , submitted to Phys.Rev.D.
17. V.L.Chernyak and A.R.Zhitnitsky , Nucl.Phys.B201, 492 (1982); V.L.Chernyak, A.R.Zhitnitsky and I.R. Zhitnitsky, Nucl.Phys.B204, 477 (1982)
18. R.L.Jaffe and M.Soldate, Phys.Rev.D26, 49 (1982); M.Soldate, "Pion and nucleon twist-4 structure functions in the limit $x \to 1$", SLAC-PUB-2998 (October 1982) (T); J.Gunion, P.Nason and R.Blankenbecler, Phys.Letters 117B, 353 (1982)
19. R.K.Ellis, W.Furmański and R.Petronzio, Nucl.Phys.B207, 1,(1982) B212, 29 (1983)
20. L.L.Frankfurt and M.I.Strikman, Phys.Rep.76,215 (1981); Nucl. Phys.B148, 107 (1979).

QUARK EFFECTS IN NUCLEAR PHYSICS

Gerald A. Miller
CERN, Geneva 23, Switzerland
and
Physics Department
University of Washington
Seattle, WA 98195, U.S.A.

I. INTRODUCTION

In this talk I discuss how nuclear properties may be treated in terms of interactions between quarks. In my view the need for this new approach to Nuclear Physics involving fundamental constituents arises from the evidence that the conventional picture of the nucleus as a collection of nucleons (even as modified by including explicit mesons and isobars) is breaking down. There are significant problems in many areas. For example, the electromagnetic properties of nuclei at medium and even low momentum transfer are not fully understood.[1-3] Our comprehension of the nuclear weak interaction has always been very dependent on knowing various short, long and medium range correlations of the nuclear wave function.[4] The high momentum transfer process of pion production in proton-nucleus collisions is very badly understood.[5] Even the age-old problem of deriving nuclear saturation properties from first principles continues to resist solution.[6] Despite the fact that Nuclear Physics is more than fifty years old there are many interesting problems that remain to be solved!

Let me mention at the beginning that it is not my intention to solve all of these problems during this talk. I merely want to note that there is an aspect common to each of the subjects of the previous paragraph: Computation of the necessary observables requires the knowledge of the two-nucleon wave function (either in the medium, or in the free space) at medium and short distances. However, the typical approach is to treat the constituent baryons as if they were point-like. This is at variance with the overwhelming evidence[7-9] that baryons have a considerable extent, characterized by a sphere of radius about one fermi. Indeed one needs the two-nucleon wave function, $\psi(r)$, at separation distances, r, of the order of the nucleon radius, R. It is therefore not surprising that conventional calculations can run into trouble.[10]

There are also sound fundamental reasons for exploring the quark degrees of freedom of the nucleus. It is widely believed that quantum chromodynamics or QCD[11] is the precise microscopic theory of the strong interaction. The basic interactions are the exchanges of gluons between quarks, not the exchanges of mesons between nucleons. Thus the motivation for describing nuclear properties in terms of quarks can be summarized by the simple statement: If you are having trouble with your calculations, use a better theory. QCD is the better theory.

But can one use QCD in Nuclear Physics? Despite great progress in its perturbative[12] and non-perturbative[13,14] aspects, the reliable relevant information from QCD is very sparse. Indeed there are only two basic pieces of information. Nonabelian gauge theories possess the property of asymptotic freedom. When quarks are close together they do not interact. Conversely, as is widely believed, the large separation distance interaction between quarks is both strong and attractive. This means that quarks, gluons, and any object carrying color are confined.

It would be extremely interesting for Nuclear Physics applications if more detailed information about the confinement process were available. Even so, the ideas of asymptotic freedom and confinement are very powerful and we shall try to carry them as far as possible during this talk.

Let us discuss the nucleon-nucleon interaction in light of the small and large separation distance properties of QCD. Consider first a single nucleon. It consists of three valence quarks, gluons, and quark anti-quark ($q\bar{q}$) pairs carrying various quantum numbers. It is worth noting that treating part of the $q\bar{q}$ sea as a pionic cloud leads to considerable improvements of phenomenological treatments of hadronic structure.[15,16] In principle any one of these constituents of one nucleon may interact with those of a second nucleon. However if one nucleon is to interact with another that is well separated from it, all terms save the exchanges of color singlets are prohibited by the confinement property. Since the pion is the lightest color singlet formed from a quark anti-quark pair one expects the exchange of a single pion to dominate the very long range nucleon-nucleon interaction. One also expects considerable contributions from the exchange of two pions.[17] Thus at large separation distances the conventional picture of mesonic exchanges is in complete accord with QCD. However for small separations between nucleons the quarks in one nucleon are close to the quarks in the other. In this case it is proper to treat the system as a set of confined quarks interacting via various perturbative terms. Hence one can expect that QCD will be most relevant in changing our ideas about the short distance aspects of nucleon-nucleon wave functions.

The aim of this talk is to use the above ideas as guidelines in constructing a model that enables one to assess the importance of quark effects in various nuclear reactions. The approach is a coordinate space division of the two nucleon wave function.[18-23] For separations, r, larger than some chosen distance, r_0, the conventional two-nucleon wave function obtainable from, e.g. one and two pion exchanges, is used. For small separations, defined by $r > r_0$, the two-nucleon system is treated as six quarks confined in a single spherical bag.[24] In treating the nucleons as completely overlapping or completely separated we are making a tremendous simplification of a very difficult dynamical problem.[25] However, this approach is consistent with the available information from QCD. Furthermore a significant number of experimental constraints can readily be implemented. The large distance part of the two-nucleon wave function is constructed in explicit agreement with experimental phase shifts as well as theoretical potentials. The principle of conservation of probability current yields powerful constraints on the normalization of the six-quark bag components of the wave functions, as we show below. We certainly believe that the present approach is adequate for our present purpose which is only to determine whether quark degrees of freedom are relevant for a given nuclear process.

The outline of the remainder of the talk is as follows. In Section II the general features of our procedure are described and specific constraints yielded by conservation of probability are derived. The applications of our approach to parity violation in proton-proton scattering and to the deuteron asymptotic D-state are presented in Sections III and IV. Our goal is to explain how the interactions between quarks influence those aspects of nuclear physics. We find that quarks are indeed very relevant. A brief summary of the talk and discussion of its possible implications is given in Section V.

II. THE MODEL

Consider a two nucleon wave function, Ψ_E, where E is the energy of the system. The two nucleons may be either bound or unbound, in the nuclear medium or in free space. Our method relies on a coordinate space division of the description. For relative distances greater than or equal to a parameter, r_0, use the wave function

$$\Psi_E = \psi_E^{NN}(\vec{r})\phi_1^N\phi_2^N \ , \ r > r_0. \tag{1}$$

Here $\psi_E^{NN}(\vec{r})$ is a conventional wave function obtained from solving the appropriate two-nucleon Schroedinger equation. The term $\phi_1^N\phi_2^N$ is to represent the product of the internal wave functions of each of the nucleons. In writing Eq. (1), angular momentum and isospin quantum numbers, as well as effects of antisymmetrization have been ignored to simplify the presentation. Note that Ψ_E depends on the quark coordinates even for $r \geq r_0$; however these play no role. (It is possible to generalize Eq. (1) so that the effects of nucleon isobars can be included.[22] One could have, e.g. an explicit delta-nucleon component of the physical two nucleon wave function. These effects are ignored here.)

If one accepts the arguments of the Paris group,[26] the quantity $\psi(\vec{r})$ may be determined from nucleon-nucleon scattering experiments and the theoretically constructed meson exchange potential, so long as r_0 is greater than about 0.8 fm. This is because the Paris potential which is constructed to fit the phase shifts requires no phenomenological modifications for separations greater than about 0.8 fm.

A totally different treatment of the short distance, $r \leq r_0$, behavior of Ψ is dictated by the considerations of QCD. We choose to treat the system as six quark constrained to move within a single spherical region of confinement. We have

$$\Psi_E = \Phi_E(\vec{r}_1,\vec{r}_2,\ldots,\vec{r}_6)\,,\quad r < r_0 \tag{2}$$

in which $\vec{r}_i$ is the coordinate of a single quark. One can define a six-body operator, $\vec{r}$, to represent the relative distance operator. For example,

$$\vec{r} = \frac{\vec{r}_1+\vec{r}_2+\vec{r}_3}{3} - \frac{\vec{r}_4+\vec{r}_5+\vec{r}_6}{3}\,. \tag{3}$$

Then we may use a simplified notation

$$\Phi_E(\vec{r}_1,\vec{r}_2,\ldots,\vec{r}_6) = \Phi_E(\vec{r},\vec{\xi}) \tag{4}$$

in which $\vec{\xi}$ represents the other, internal, variables. Again complications of angular momentum, isospin and antisymmetrization are ignored in order to clarify the presentation.

II.1 THE VALUE OF r_0

Before turning to the details of the model it is worthwhile to discuss the reasonable expectations regarding the value of the parameter r_0.

First we use purely geometrical considerations. An upper limit is obtained by saying that a single six quark bag is formed whenever two nucleons touch. If that were true r_0 would be equal to 2R where R is the nucleon radius. Then $r_0 \approx 2$ fm. But that would be absurd. It is certainly better to describe the two nucleons in the configuration of Fig. 1a as two nucleons instead of six quarks in a single bag. Indeed if two nucleons have small overlap, as in Fig. 1b one still expects

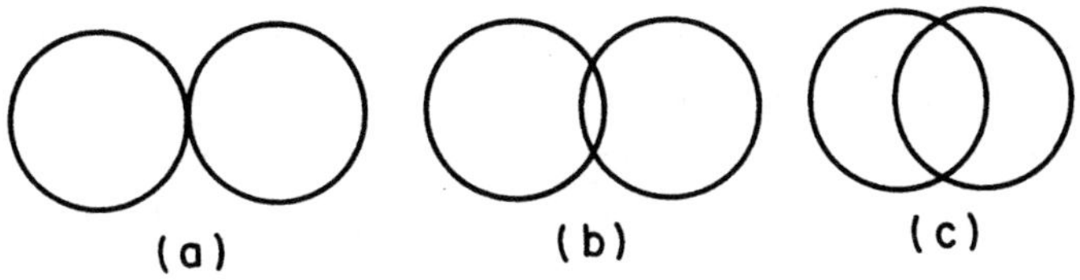

Figure 1. Two nucleon overlaps

the two nucleon description to be more valid. Indeed the ratio of the volume of the overlap region between the two spheres to the total volume is $\frac{3}{16}((r/R)^2)(1 - \frac{r}{8R})$, so that r must be about equal to R, as in Fig. 1c (the edge of one nucleon is at the center of the other), in order to get an overlap ratio that differs significantly from zero. For bags of radius of 0.8 fm to 1 fm, r_0 is between 0.8 fm and 1 fm. Such values are in accord with several of our previous studies.[19-23]

Another way of getting at r_0 is to consider the Paris potential.[26] As noted above, the theoretically determined part of the potential requires very little phenomenological modification for $r \geq 0.8$ fm. Thus r_0 should be in the neighborhood of 0.8 fm.

II.2 CONSTRAINING Φ

Since $\psi(r)$ is reasonably well-known, the main task is to determine Φ. Of course, this is very difficult to do in a reliable model-independent manner. However it does turn out that one can use the ideas of conservation of probability current,

$$\frac{\partial \rho}{\partial t} = -\vec{\nabla}\cdot\vec{j} \tag{5}$$

and the continuity of $\vec{j}$ across the boundary at $r = r_0$

$$\hat{r}\cdot\vec{j}\Big|_{r=r_0-\varepsilon} = \hat{r}\cdot\vec{j}\Big|_{r=r_0+\varepsilon}, \tag{6}$$

where ε is very small, to determine the size of the six quark components of the wave function. The motivation for Eq. (6) is that although our description of the wave function changes suddenly as r increases beyond r_0, the basic physics should not.

To make things quantitative, define a six-quark "probability", P_{6q} as

$$P_{6q} = \int d\xi \int d^3r\ \theta(r_0-r)\,|\Phi_E(\vec{r},\xi)|^2. \tag{7}$$

For bound states P_{6q} is actually a probability, but for scattering states it is not, since Ψ is not normalizable to unity. Nevertheless, we refer to the quantity defined by the right-hand side as the six quark probability. The interesting thing is that Eqs. (5) and (6) can be used to show that P_{6q} depends only on $\psi_E(\vec{r})$ and its radial derivative at r_0. Thus P_{6q} is as well determined as $\psi_E(\vec{r})$ at $r = r_0$. The answer is

$$P_{6q} = \frac{1}{2\mu}\, r_0^{\,2} \int \Big[\frac{\partial\psi_E}{\partial E}\,\frac{\partial\psi_E}{\partial r} - \psi_E\,\frac{\partial^2}{\partial E\partial r}\,\psi_E\Big]_{r=r_0}\, d\hat{r} \tag{8}$$

where μ is the reduced mass of the NN system.

The proof follows an old argument due to Wigner[27] who was working on the analogous problem of compound nucleus formation in nucleus-nucleus interactions. Consider a wave packet χ with $r < r_0$, and

$$\chi = \Phi_{E_1}\, e^{-iE_1t} + \Phi_{E_2}\, e^{-iE_2t}. \tag{9}$$

From χ one can form the probability, ρ, and current, $\vec{j}$, densities. However it is not necessary to explicitly construct these quantities. Instead consider the current conservation equation, Eq. (5), and take its integral over ξ and $\vec{r}$ with $r < r_0$. This gives

$$\int d\xi\ d^3r\ \theta(r_0-r)\,\frac{\partial\rho}{\partial t} = -\int d\xi\ d^3r\ \theta(r_0-r)\,\vec{\nabla}\cdot\vec{j}\ . \tag{10}$$

Except for the time derivative, the left-hand side resembles P_{6q}. The right-hand side may be simplified by the use of the divergence theorem which allows one to replace the volume integral by one over the surface. So the right-hand side of Eq. (10) depends on $\vec{r}\cdot\vec{j}$. Now use the equation of continuity of current, Eq. (6), to replace a very difficult expression involving six quarks by the one expressed in terms of two nucleons. This is the basic idea behind the result (Eq. (8)). Use of probability current conservation and continuity enables one to express an internal probability based on quark degrees of freedom in terms of the ordinary nucleonic wave functions evaluated at $r = r_0$.

All that remains is a bit of algebra. Since ρ and $\vec{j}$ are quadratic in χ, each contains four terms. Equation (5) holds for the term involving $e^{-i(E_1-E_2)t}$. (If it didn't it would not be valid at all.) Keep only that term in Eq. (10). This gives

$$-i \int d\xi\, d^3r\, \theta(r_0-r)(E_1-E_2)\Phi^*_{E_2}\Phi_{E_1}$$

$$= -\frac{\hbar}{2\mu} i \int \hat{r}\cdot(\psi^*_{E_2}\vec{\nabla}\psi_{E_1} - \psi_{E_1}\vec{\nabla}\psi^*_{E_2})\Big|_{r=r_0} d\hat{r}, \tag{11}$$

which incorporates the previously indicated simplifications of the right-hand side of Eq. (10). Now let E_2 approach E_1 and keep only terms of order (E_1-E_2) on both sides of Eq. (11). As a result the left-hand side is clearly proportional to P_{6q}. By dividing through by (E_1-E_2) one obtains immediately Eq. (8).

The utility of Eq. (8) is that one uses general principles to obtain significant and well-determined constraints on the six quark probability.

Equation (8) is to be employed in a situation in which the NN potential for $r \geq r_0$ is known, but the short distance interaction is unknown. This corresponds to the situation at present, in my view. However, if one has a phenomenological NN potential for all values of r one may derive an alternate expression for P_{6q} in terms of $\psi_E(\vec{r})$ at $r < r_0$. Using only the Schroedinger equation one may show that Eq. (8) is equivalent to

$$P_{6q} = \int d^3r\, |\psi_E(\vec{r})|^2\, \theta(r_0-r) \tag{12}$$

assuming that $\psi_E(\vec{r})$ is known for $r \leq r_0$. Expression (12) holds for negative as well as positive values of E. The result (12) is equivalent to Eq. (8) for the Reid soft core[28] and Paris potentials, even though the $\psi_E(\vec{r})$ of these potentials is not expected to be the correct short distance wave function. Equation (12) also may be derived immediately from the assumption that the complete 6 quark plus NN wave function obeys the same normalization condition as an ordinary NN wave function.

II.3 EXAMPLES OF P_{6q}

Although P_{6q} is to be calculated numerically from Eqs. (8) or (12) it is worthwhile to consider a few simple examples before presenting any numerical results. This is done in order to gain a physical understanding of P_{6q} and to better interpret the numerical results.

To simplify matters neglect spin effects, and let the positive energy scattering wave $\psi_E(\vec{r})$ be given in terms of a single partial wave as

$$\psi_E(\vec{r}) = \frac{u_L(kr)}{kr(2\pi)^{3/2}} P_L(\hat{k}\cdot\hat{r}) \tag{13}$$

with $u_L(kr)$ real and, where $k^2/M_N = E$. Then Eq. (8) can be rewritten as

$$P_{6q} = \frac{1}{4\pi^2 k^3}\left[\frac{\partial u_L}{\partial k}\frac{\partial u_L}{\partial r} - u_L \frac{\partial^2 u_L}{\partial k \partial r}\right]_{r=r_0} . \tag{14}$$

Note that P_{6q} has dimensions of volume. This is as it should be: the volume integral of the absolute square of a dimensionless scattering wave function has dimensions of volume. Were we to consider the consequences of Eq. (8) for a bound state wave function, the resulting six quark probability would be dimensionless. However for scattering problems P_{6q} is a volume.

Now consider the consequences of Eq. (14) for the model problem in which the scattering phase shift δ vanishes for all energy and in which the NN potential $V(r)$ also vanishes for $r \geq r_0$.

First examine s-wave scattering. With $L = 0$,

$$u_0(kr) = \sin kr. \tag{15}$$

The use of (15) in (14) leads to the result

$$P_{6q} = \frac{1}{4\pi^2}\frac{r_0}{k^2}\left(1 - \frac{\sin 2kr_0}{2kr_0}\right) \tag{16}$$

It is instructive to consider the high low and high energy limits of Eq. (16). We have

$$\lim_{E\to 0} P_{6q} = \frac{1}{6\pi^2} r_0^3 \tag{17}$$

and

$$\lim_{E\to\infty} P_{6q} = \frac{1}{4\pi^2}\frac{r_0}{2EM_N} . \tag{18}$$

At low energies the probability of forming the six quark compound system is proportional to allowed volume. The larger r_0, the easier it is to form a six quark bag. This is because at low energies and with no external interactions, the only parameter available is the value of the matching radius, r_0.

At high energies P_{6q} falls off inversely as $1/E$ so that at sufficiently high energies it is negligible. The interpretation of this is that if two nucleons approach each other with high relative velocities, the time available for forming a six-quark bag state is too small for the necessary significant interactions to take place.

Now consider again the same situation with $V(r) = 0$ for $r > r_0$ and $\delta(E) = 0$ for all energy. However let $L \neq 0$. Then

$$u_L(kr) = kr\, j_L(kr) \quad . \tag{19}$$

The use of Eq. (19) in Eq. (16) leads to an expression for P_{6q} that is somewhat similar, but more complicated than Eq. (16). However the low and high energy limits are instructive. At low energies the relevant effect is the centrifugal barrier which inhibits formation of the six quark compound state. The fact that

$$\lim_{E\to 0} kr_0\, j_L(kr_0) = \frac{(kr_0)^L}{(2L+1)!!} \tag{20}$$

leads to the result

$$\lim_{E\to 0} P_{6q} \propto r_0^3 (kr_0)^{2L} \quad . \tag{21}$$

Thus at low energies it is very difficult to form a six quark compound state of high angular momentum.

At high energies, we have

$$\lim_{E\to\infty} kr_0\, j_L(kr_0) = \sin(kr_0 - \frac{L\pi}{2}) \quad , \tag{22}$$

so that

$$\lim_{E\to\infty} P_{6q}\ (L\neq 0) = \frac{1}{4\pi^2}\,\frac{r_0}{2EM_N} \quad . \tag{23}$$

That Eq. (23) is of the same form as Eq. (18) should not be surprising because of the similarities between Eqs. (22) and (15). Formation of high angular momentum six quark compound states is also suppressed at sufficiently high energies. Of course one must go to quite high energies indeed before the limit of Eq. (22) is satisfied. This is because the corrections to Eq. (22) go like $1/(kr_0)$, which falls quite slowly with E.

As a final illustration let us examine a case in which the scattering phase shift does not vanish. Set $V(r) = 0$ for $r \geq r_0$ so that the interactions originate amongst the confined quarks. Again work with s-waves at very low energy. Then we have

$$u(kr_0) = \sin(kr_0 + \delta) \tag{24}$$

where in the effective range expansion

$$k \cot \delta = -\frac{1}{a} + \frac{1}{2} r_e k^2 , \tag{25}$$

and a and r_e are the scattering length and the effective range. We especially wish to consider the situation in which $|a| >> r_0$, which is to represent the actual situation in the $^1S_0$ channel. The use of Eqs. (24) and (25) in Eq. (14) gives (after some algebra) the result

$$P_{6q} \approx \frac{1}{4\pi^2} a^2 (2r_0 - r_e/2) . \tag{26}$$

Terms of order r_e/a or r_0/a or their square relative to the right-hand side of Eq. (26) have been neglected in obtaining Eq. (26), consistent with the idea that $|a| >> r_0$. For this model the effective range, r_e, must be less than r_0 since there are no interactions for $r > r_0$. Thus P_{6q} of Eq. (26) is positive definite.

The point of obtaining Eq. (26) is to compare it with the corresponding result of Eq. (17) which obtains when δ vanishes. P_{6q} of Eq. (26) is enhanced by a huge factor of $\sim 3(a^2/r_0^2)$ or about $3(-20 \text{ fm}/1 \text{ fm})^2 = 1200$ using realistic input numbers for the $^1S_0$ channel. Under the simplified conditions of this subsection the strong interaction responsible for generating a large value of $|a|$ can only take place if there is significant formation of the six quark compound state.

II.4 NUMERICAL RESULTS

We now compute the six quark probabilities for the proton-proton system. The Reid soft core potential[28] is used to generate the necessary exterior ($r \geq r_0$) wave functions. The generalization of the previous results, Eq. (14) for cases with non-zero spin is trivial. Simply define $P_{6q}(L,S,J)$ to be obtained from Eq. (14) using the appropriate wave functions $u_{LSJ}(kr)$. These results are used in calculations of parity violation in proton proton scattering[20] (see below) and in estimating quark contributions to the $pp \to d\pi^+$ cross section.[19]

Shown in Fig. 2 are results for P_{LSJ} as a function of energy for a variety of partial waves. The value of r_0 is 0.8 fm. The most dramatic feature is the large value of P_{000} at low energies. This is in contrast with the Born approximation results of the preceding section, but in accord with the idea expressed in Eq. (26) the significant attractive effects can give a large value of P_{LSJ}. In this case it is the attractive force, due mainly to two pion exchange effects, occurring for $r > r_0$ which brings the two nucleons together for a long enough time to form a six quark compound system, in our interpretation. At high energies, for which the potential for $r > r_0$ is not so significant, the P_{000} falls roughly as 1/E in accord with Eq. (18). Note also that at low energies the Paris potential gives essentially the same value of P_{000} as shown here.

The value of P_{000} seems very small. However the magnitude depends on the chosen normalization of the scattering wave function. To get a better idea about the size of P_{000} it is necessary to compare it with some relevant quantity having the dimensions of volume. Consider the volume integral of the square of a plane wave function $\frac{e^{i\vec{k}\cdot\vec{r}}}{(2\pi)^{3/2}}$, over a sphere of radius r_π, the pion Compton wave length, 1.4 fm. The resulting volume, V_π, is a rough estimate of the probability available in a region defined by the range of the strong interaction. Then form the ratio, α^2_{000} of P_{000}/V_π. We have

$$\alpha^2_{000} = P_{000}/(r^3_\pi/6\pi^2) \ . \qquad (27)$$

At 15 MeV $\alpha^2_{000} \cong 0.5$, so that the six-quark bag formation probabilities seem large enough to be relevant in nuclear reactions.

The results for the $^3P_0$, $^3P_1$ and $^1D_2$ partial waves are also shown in Fig. 2.

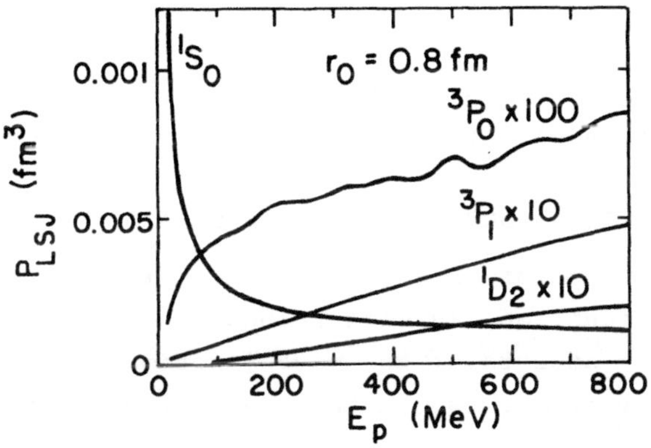

Figure 2. The six quark bag formation probabilities, P_{LSJ} in various partial waves, as a function of energy. The matching radius, r_0=0.8 fm.

The corresponding values of P_{LSJ} are suppressed relative to P_{000} in accord with the centrifugal barrier effect of Eq. (21). Interaction effects play a significant role also, as seen from the difference between $^3P_0$ and $^3P_1$. For energies less than or equal to 800 MeV, the expected high energy behavior of 1/E is not in evidence for waves with

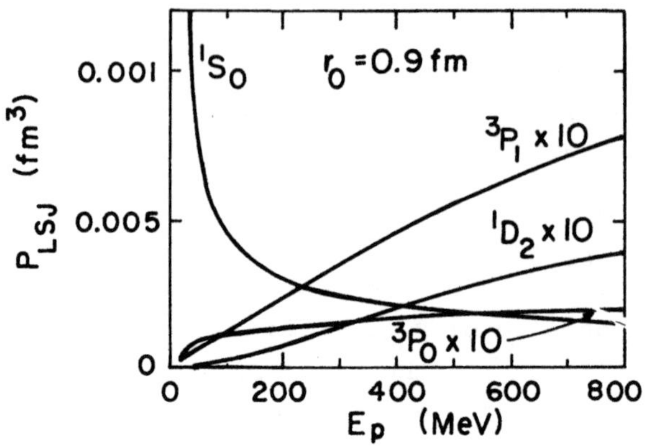

Figure 3. Same as Fig. 2 with r_0 = 0.9 fm.

non-zero angular momentum. This behavior does occur for higher energies.

Results for r_0 = 0.9 fm are shown in Fig. 3. The values of P_{LSJ} increase with r_0 as expected on the basis of Eq. (17). This is because the volume available for six-quark bag formation increases with r_0.

The discussion of the NN bound state is straightforward in our formalism. It is easier to use Eq. (12) instead of Eq. (8). We have

$$P_D = \int_0^{r_0} (u^2(r) + w^2(r))dr, \tag{28}$$

with u and w the s and D wave radial wave functions of the deuteron. For the Reid soft-core potential we obtain P_D = .029 for r_0 = 0.9 fm.

II.5 UNKNOWN FEATURES OF Φ

The preceding parts of this section indicate how the size of Φ, the six-quark wave function, is constrained. However nothing has been said regarding other aspects of the wave function. Unfortunately we really do not know very much about these wave functions.

One may use antisymmetrized products of six single quark confined wave functions, coupled to the correct angular momentum and isospin quantum numbers. However this information is not sufficient to uniquely specify the wave function. For example, even parity states can be of the [6] or [42] spatial symmetry.[29] Our procedure, which is discussed below in specific examples, is to make the simplest choice consistent with current information about six quark wave functions.

Another problem is determining the size of the six quark confinement region. It is certainly related to r_0, and increases if r_0 increases. However we do not compute the radius of the six quark bag. The choices we make are discussed below.

To some extent we can find out if the choice of a given configuration is biasing the results by making different choices and seeing how the results change. This enables us to make some assessments of the model dependence of our results.

At present the model is incomplete, and should be thought of as a framework which can be developed into a more extensive theory. One possibility is to explicitly construct six quark wave functions in accord with Eq. (6), but this has not yet been done.

This completes the discussion of the general features of the model. On to the applications!

III. PARITY VIOLATION IN PROTON-PROTON SCATTERING

The outline of this section is as follows. We discuss:

1. WHAT is measured and computed.
2. WHY we expect the process to involve quarks. (The weak interaction proceeds via W and Z boson exchange between quarks.)
3. HOW we compute quark effects.
4. RESULTS - W and Z boson exchange effects can give a reasonably good description of the low energy data.

III.1 WHAT IS MEASURED AND COMPUTED

Theorists can discuss parity violation in proton-proton scattering only because of the ability of experimentalists to prepare beams in states of (pretty) good helicity. The helicity quantum number is ± 1 according to whether the spin of the protons is polarized parallel or anti-parallel to the direction of the momentum of the beam. Two (total or angle-integrated) cross sections $\sigma_\pm$ corresponding to the two helicity states of the incident beam can therefore be defined. Under the parity transformation the pseudoscalar helicity operator, $\vec{\sigma}\cdot\hat{p}$ changes sign so that a parity-violating Hamiltonian leads to an inequality between σ_+ and σ_- .

To obtain a numerical evaluation of the amount of parity violation define an asymmetry parameter, A, by the relation

$$A = \frac{\sigma_+ - \sigma_-}{\sigma_+ + \sigma_-} . \tag{29}$$

If parity is conserved the quantity A vanishes. To compute $\sigma_{\pm}$, and therefore A, it is necessary to know the parity conserving, F, and parity violating, f_W scattering amplitudes. Since parity violation is a perturbative effect, A is of the order of the ratio f_W/F. Using standard expectations regarding the weak and strong coupling constants, we expect that A is of the order of 10^{-6}. A is expected to be a very small quantity.

The experimental values of A are even smaller. A Los Alamos experiment[31] at 15 MeV gives $A = 1.7 \pm 0.8 \times 10^{-7}$, whereas the SIN measurement[32] at 45 MeV gives $A = 2.3 \pm 0.8 \times 10^{-7}$.

It is our intention of understanding these results by using a model in which the parity violating amplitude is generated by W and Z boson exchanges between quarks. The parity conserving amplitude, F, is obtained directly from phase shifts and the Coulomb interaction.

II.2 WHY ARE QUARK EFFECTS RELEVANT IN PROTON-PROTON PARITY VIOLATION?

The first answer to this question is based on principle. It is well known that the parity violating electroweak force is caused by the exchange of gauge bosons, W and Z between quarks. The relevant fermions are quarks, not nucleons. Of course, analogous statements can be made about any nuclear interaction.

There is another more practical reason for believing quark interactions to be particularly relevant to understanding the weak force between protons. It stems from the conventional boson exchange picture of the parity violating nucleon-nucleon interaction[33,34] shown in Fig. 4. For example, in pion exchange S is of the form $\vec{\sigma}\cdot\vec{q}$ where $\vec{\sigma}$ is the nucleon spin operator and $\vec{q}$ is the momentum of the exchanged pion. The vertex W is of the form f_W^{π}. The potential is proportional to the product of the two factors. The momentum space dependence on $\vec{\sigma}\cdot\vec{q}$ is trans-

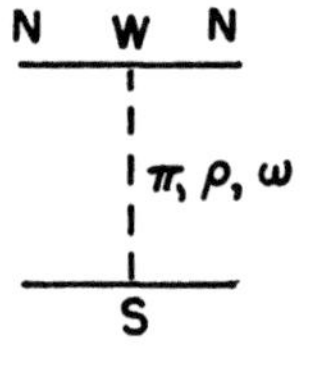

Figure 4. Boson exchange model of the parity-violating nuclear potential. S(W) means that the boson-nucleon interaction is strong (weak).

formed into a coordinate space term proportional to $\vec{r}$ that violates parity. It is interesting to note that f_w^π is both the largest and best known of the weak boson-nucleon coupling constants,[35] so that in general one expects to have a very significant long range parity violating force.

But the proton-proton has NO such one pion exchange parity violating force.[36] Conservation of electric charge dictates that only a neutral pion can be exchanged between two protons. The π^0 is its own antiparticle; it is even under the charge conjugation transformation. Thus a parity violating π^0 nucleon vertex would also violate CP. Although CP violating effects have been observed, they are about a thousand times weaker than the small effects we are discussing and may be neglected. Thus there is no long-ranged parity violating interaction in the proton-proton system. This is a crucial advantage for proton-proton scattering: the longest ranged, most important parity violating one-boson exchange term is not present.

There are parity violating ρ and ω exchanges. In our picture these take place only for $r > r_0$, and are therefore significantly suppressed. Parity violating exchanges of two uncorrelated pions can occur, in principle, but are expected to be small.[33,34]

The result of the above considerations is that, for the proton-proton system, the dominant parity violating interactions are expected to occur within the six-quark compound system.

III.3 HOW WE COMPUTE PARITY VIOLATING EFFECTS IN QUARK-QUARK INTERACTIONS

The starting point is the Weinberg-Salam[37] theory of the electroweak interaction as shown in Fig. 5a. Note that both charged and neutral currents are included. The latter are of special importance for us since the six quark system that represents the two proton system contains four up quarks.

This amazingly successful theory has been described in many places[38] so that we need only show a few examples. Consider the weak interaction between two up quarks. The Hamiltonian density, $\mathcal{H}_{ws}^{eff}$, is of the form

$$\mathcal{H}_{ws}^{eff} = \frac{G}{\sqrt{2}} .20\ [\bar{u}\gamma_\mu u][\bar{u}\gamma^\mu\gamma_5 u]. \qquad (30a)$$

One has an interaction of the VA form. The coupling constant, G, has the value $1.03\times10^{-5}/M^2$. The local form of the interaction (and the small value of G) results from the fact that the Z boson propagator has been replaced by a delta function divided by the square of the Z boson mass. The coefficient .20 is determined from experimental measurements of the Weinberg angle.

However the above interaction which is taken directly from the Weinberg-Salam model for leptons is not sufficiently accurate for application to hadronic and nuclear weak interactions. Whenever two quarks are close enough to interact weakly they will also interact strongly. Examples are shown in Fig. 5b-f. The strong interaction (gluon) corrections can be made to even higher order than shown using

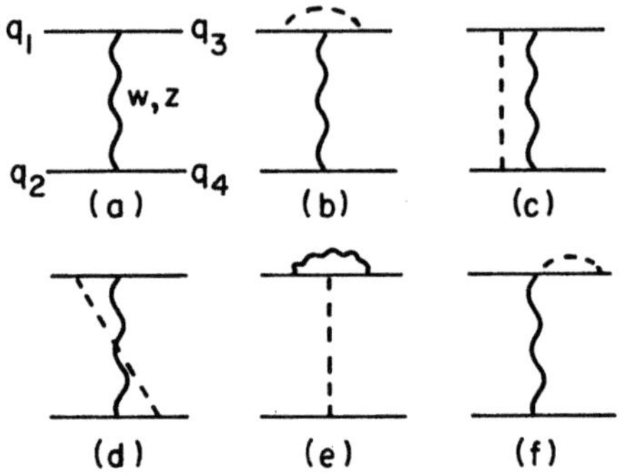

Figure 5. Weak quark-quark interaction. Wiggly line W,Z; dashed line, gluon.

renormalization group methods,[39] and are very important in understanding the $\Delta I = 1/2$ rule in $\Delta S = 1$ processes.[40] Several groups[41-43] have made the necessary calculations for the systems of light quarks of relevance here. The conclusions of different groups are consistent.

The resulting effective Hamiltonian is summarized in Ref. 20. Here we present the effects of gluon exchanges (enhancements) corresponding to the term of Eq. (30a). For interactions between two u quarks we have

$$\mathcal{H}^{eff}_{enh} = \frac{G}{\sqrt{2}}\ .40\ [\bar{u}\gamma_\mu u][\bar{u}\gamma^\mu\gamma_5 u] \tag{30b}$$

and the effect of including strong interaction effects is to double the strength of the effective weak interaction Hamiltonian.

The weak scattering amplitude, f_w, is computed from matrix elements in which the six-quark part of a proton-proton state of a given parity is converted via H^{eff}_{enh} into a six-quark state of the opposite parity. We need to compute such matrix elements as:

$$\langle \Phi(^3P_0)|H^{eff}_{enh}|\Phi(^1S_0)\rangle \tag{31}$$

The weak interaction is specified in Eq. (30b) and Ref. (20), so we turn to a description of the wave functions. For the $^1S_0$ state we have

$$|\Phi(^1S_0)\rangle = e^{i\delta_{^1S_0}} [C_0|(S_{1/2})^6\rangle + C_1|(P_{1/2})^2(S_{1/2})^4\rangle + C_2|(P_{3/2})^2(S_{1/2})^4\rangle] \qquad (32)$$

where $\delta_{^1S_0}$ is the $^1S_0$ scattering phase shift. The normalization is constrained by the probability conservation considerations of Section II according to

$$C_0^2 + C_1^2 + C_2^2 = P_{000}. \qquad (33)$$

Each ket represents an antisymmetrized wave function of six quarks confined in an MIT bag of radius r_0, and is normalized to unity. The size of C_1 and C_2 represents the amount of [42] symmetry and is constrained[29] by

$$\frac{C_1^2 + C_2^2}{C_0^2} = 8. \qquad (34)$$

Other six-quark wave functions are chosen for simplicity. For example

$$|\Phi(^3P_0)\rangle = e^{i\delta_{^3P_0}} |S_{1/2}P_{1/2}, (S_{1/2})^4\rangle. \qquad (35)$$

It turns out that the results are not very sensitive to the choices of the configurations $|\Phi\rangle$. For example, varying the ratio C_1/C_2 does not change the results very much so long as Eq. (34) is satisfied.

Having described the theory in some detail, we can now turn to a discussion of the results.

III.4 NUMERICAL RESULTS

The computed values of A are shown as a function of lab energy in Fig. 6 for two values of r_0. One can see that quite a reasonable description of the data can be achieved. It is tempting, at first glance, to try to associate the interesting peak at about 100 MeV with some specific aspects of quark effects. However, the peak occurs solely as a result of effects associated with kinematics and the energy dependence of the strong interaction, as shown by Brown et al.[44] At low energies the dominant transition is from an s wave to a p wave, so that centrifugal barrier effects suppress the value of A at very low energies.

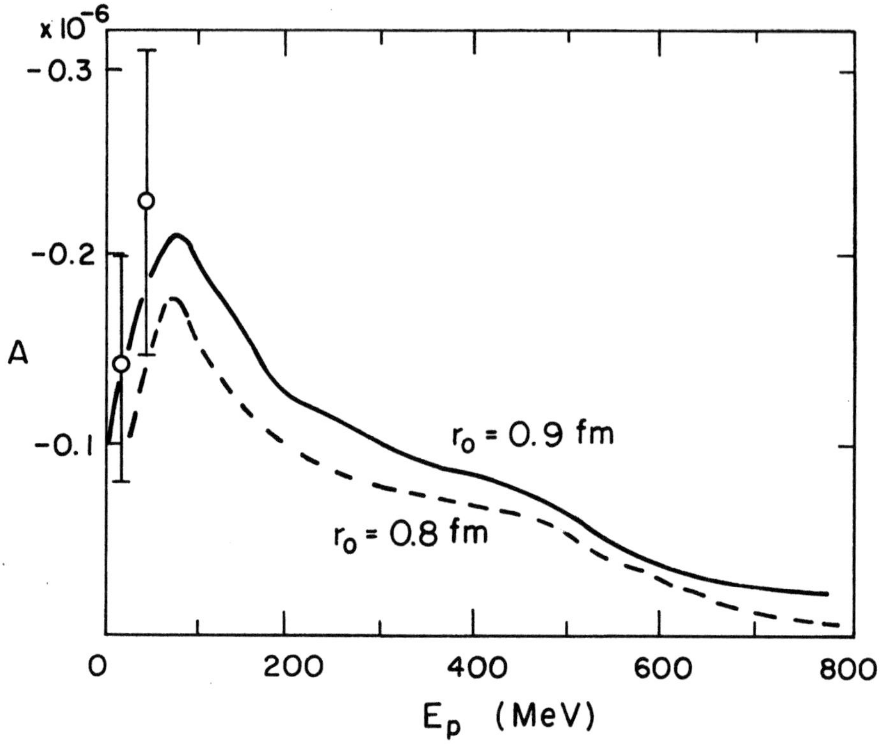

Figure 6. A as a function of r_0 and laboratory kinetic energy.

A starts from 0 to E=0 MeV and its magnitude can only rise as the energy increases. However, the rise cannot continue as the energy approaches 200 MeV. This is because A arises from an interference between the parity conserving and parity violating amplitudes. Both the $^1S_0$ and $^3P_0$ phase shifts vanish near 200 MeV, and the corresponding value of A is therefore reduced. A is prevented from passing through zero by the Coulomb amplitude which does not vanish. (The Coulomb effect is neglected in Ref. 44).

As stated above the radius of the six-quark bag is chosen to be r_0. A dimensional analysis of the matrix element of H^{eff}_{enh} shows that it behaves as R_6^{-3}, where R_6 is the radius of the six-quark bag. Even if R_6 differs with r_0, one can regain the results of Fig. 6 by varying the value of r_0.

Although our theory provides an adequate description of the data, it is not the only one to do so. According to Desplanques[45] the conventional boson exchange picture is also successful. That this would be so is not apparent from the early work of Henley et al. who obtained values of A of the wrong sign. This came about as a result of the lack of knowledge of the relative sign of the weak ρ and ω nucleon weak coupling constants. The necessary vertex functions have been analyzed in great detail by Donoghue et al.[46] and new results have been obtained. It is these newer coupling constants that lead to the results described in Ref. 45. However the coupling constants obtained in Ref. 45 do have error bars and, within the error bars are found values not so very different than those employed by Henley et al.

It seems that more information in the form of the energy dependence of A, or even the parity violating angular distributions will be needed to completely distinguish the two approaches.

IV. QUARKS AND THE DEUTERON ASYMPTOTIC D STATE

Before discussing this topic in any detail at all, it is necessary to consider the question: Why worry about quarks in treating an asymptotic property of the most loosely bound nucleus?

All quantum mechanical observables are matrix elements which involve an integral over the entire volume of the system. Thus if the experimental precision is sufficiently high, one can learn something about the medium and even short range contributions, even if these are not the dominant ones. Indeed Ericson and Rosa-Clot[47] and de Kam[48] have argued that calculations of the asymptotic ratio of D to S state wave functions, η, can be used to place limits on the size of the nucleon bag. They say that the nucleon bag must have a radius less than or equal to about 0.6 fm. This is rather small from the point of view of bag models, e.g., Refs. 8,9 and 16, which prefer a value in the range of 0.8 to 1.1 fm. In this section I argue that tensor forces due to pions and gluons exchanged between quarks, effects not included in Refs. 47 and 48, give additional contributions. The result is that medium and even large sized bags are not ruled out.

IV.1 POTENTIAL MODEL COMPUTATIONS

In order to see how this occurs, let's forget briefly about quarks and follow the arguments of Ericson and Rosa-Clot. As you all know, the deuteron comes in S and D states, with corresponding radial wave functions u(r) and w(r). At large separations we have

$$\lim_{r\to\infty} u(r) = A\, e^{-\alpha r}$$
$$\lim_{r\to\infty} w(r) = \eta A\, e^{-\alpha r} \qquad (36)$$

where $\alpha^2 = M_N\, 2.2$ MeV. The quantity η is the one I wish to discuss.

As an asymptotic quantity it is an observable, and very accurate experiments[49] have been made to determine it. The result is that η^{EXPT} = 0.0271 ± 0.0004.

Ericson and Rosa-Clot, using ordinary potential theory, have derived an exact expression for:

$$\eta^{P} = \frac{M}{N\sqrt{2\alpha}} \int_0^\infty r \, \mathcal{J}_2(r) V_{20}(r) u(r) dr \tag{37}$$

where $V_{20}(r)$ is the tensor force that connects the S and D states, and the function $\mathcal{J}_2(r)$ is the L=2 solution of the homogeneous Schroedinger equation that is regular at the origin; and, N is the well-determined asymptotic S state normalization. One should not be surprised by Eq. (37). It is quite analogous to electric multipole theory, with the term $V_{20}(r)u(r)$ playing the role of the source term and $\mathcal{J}_2(r)$ corresponding to the r^3 expression.

The integral in Eq. (37) is over all values of r so that medium and short range contributions are included. However there is no doubt at all that the integral is dominated by long range contributions. As argued in Ref. 47, there are several reasons for this. First of all, the tensor force is dominated by the nucleon-nucleon force of longest range, namely one pion exchange (OPEP). The function u(r) is very small for small values of r, in any existing model. The regular solution, $\mathcal{J}_2(r)$, is also inhibited for small r due to the combined repulsive effects of the centrifugal and OPEP (diagonal L=2) forces.

Using an OPEP obtained from pointlike nucleons and pions to compute V_{20}, Ericson and Rosa-Clot obtained the value

$$\eta^{OPEP} = 0.0276. \tag{38}$$

Corrections due to two pion exchange and other short-range effects lead to a theoretical value

$$\eta^{TH} = 0.0267 \tag{39}$$

which is in agreement with the experimental value and shows also that OPEP plays the dominant role in any computation of η.

IV.2 FINITE NUCLEON SIZE

However the treatment of the nucleon as an object without extent was crucial in obtaining these numerical values. It is the attempts to include the nucleon's finite size that bring up the subject of the role of quarks. The conventional manner of including the nucleon finite size in computing OPEP, employed by Refs. 47 and 48, is to use a form factor which has the effect of modifying the pion's propagator. In momentum space we have

$$\frac{1}{k^2+m_\pi^2} \rightarrow \frac{1}{k^2+m_\pi^2}\left(\frac{3j_1(kR)}{kr}\right)^2 / \gamma^2 \tag{40}$$

where R is the bag radius and the Cloudy Bag Model form factor[15,16] is employed. The factor $\gamma = \frac{3j_1(im_\pi R)}{im_\pi R}$ and is included so that the left and right hand sides of (40) are the same when evaluated at the pion pole, $k = im_\pi$. If this is done the propagator modification (40) introduces no change of the pion-nucleon coupling constant. The effect of the finite nucleon size is incorporated by using non-zero values of R. When R vanishes the left and right hand sides of (40) are identical.

R provides the largest uncertainty in calculations of η as shown in Refs. 47 and 48. As indicated schematically in Fig. 7, inclusion of the finite nucleon size severely reduces the magnitude of the OPEP tensor force for $r < 2R$. For the large values of r the finite size has <u>no</u> effect. This is analogous to conventional electromagnetic theory: When two extended sources of charge do not overlap the sources behave as if they were pointlike. However, the changes in V_{20} introduced for values of r corresponding to regions in which the nucleons overlap lead to big decreases in the calculated value of η. Indeed the result of Ref. 48 is that $R \lesssim 0.6$ fm.

It is important to realize that the form factor calculations described above are not complete. If the nucleon bag has a finite size it is because quarks are moving within. When two nucleons overlap, the two-baryon wave function must be antisymmetrized in the quark variables. The interactions must also be assumed to take place between the

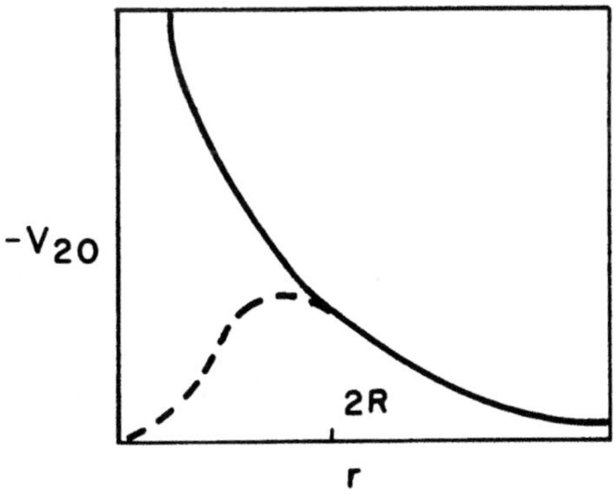

Figure 7. Schematic representation of $-V_{20}$. The solid curve is for pointlike nucleons, and the dashed line is for finite sized nucleons.

quarks, not between nucleons.

To be specific assume, as is done in chiral bag models, that pion-quark interactions give rise to the OPEP between nucleons. Let the nucleons overlap. As shown in Fig. 8 there are two relevant terms. The one of Fig. 8a is included in form factor calculations. This can be seen most easily in coordinate space. Each vertex acquires a factor that is the quark density, the Fourier transform of which is just the form factor. The term of Fig. 8b arises solely from the antisymmetrization of the quark wave function. It is the Fock term, and it is not included in typical form factor calculations.

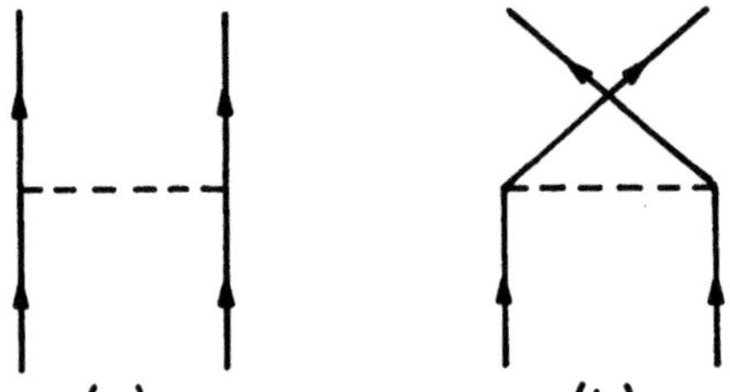

Figure 8. One pion exchange between quarks, a) The direct term. b) The exchange term caused by antisymmetrization of the quark wave function. This term is absent in typical OPEP calculations using form factors.

Another term that can contribute when two nucleons overlap is the exchange of a single gluon between quarks, which has a tensor component.

IV.4 QUARK EFFECTS

So the important question is: Can quark antisymmetrization and gluon exchange effects compensate for the loss in η encumbered by the use of finite-sized nucleons? To answer this question the hybrid model of the previous sections is applied.

For $r > r_0$ we use the OPEP force between point nucleons. For $r < r_0$ the interactions are the OPEP and one gluon exchange potential (OGEP) between the six quarks confined in a single spherical region. In order to simplify these first calculations, the nonrelativistic, constituent quark model is used.

In the present model which includes both quark and nucleonic contributions, we have

$$\eta = \eta^N + \eta^Q \tag{41}$$

in which the superscript denotes the nucleonic (N) or quark (Q) terms.

Since the nucleon wave function is used only for $r > r_0$ we have

$$\eta^N = \frac{M}{N\sqrt{2\alpha}} \int_{r_0}^{\infty} r\, \mathcal{J}_2(r) V_{20}(r) u(r) dr \tag{42}$$

so that, in this computation, cutting off the integral for $r < r_0$ plays the role of a form factor. Since one is concerned with the relatively small changes brought about by using this "form factor", it is convenient to define a quantity $\Delta\eta$ with

$$\Delta\eta = \frac{M}{N\sqrt{2\alpha}} \int_{0}^{r_0} r\, \mathcal{J}_2(r) V_{20}(r) u(r) dr. \tag{43}$$

$\Delta\eta$ represents the contribution lost due to the cutoff procedure. This is to be compensated eventually by quark effects. In computations of Eq. (43) we take u(r) from the Paris potential, $V_{20}(r)$ and $\mathcal{J}_2(r)$ are computed with the point OPEP as in Ref. 48.

In this model the short distance parts of the S(u) and D($\mathcal{J}_2(r)$) state wave functions are treated as collections of six confined quarks. Tensor interactions between the quarks cause the transition between the S and D states. The resulting expression for the quark contributions is

$$\eta^Q = \frac{M}{N\sqrt{2\alpha}} \sqrt{P_{6q}}\sqrt{\tilde{P}_D}\, \langle D_{6q}| V_{20}^{\pi} + V_{20}^{G} |S_{6q}\rangle \; . \tag{44}$$

Here $|S_{6q}\rangle$ is the antisymmetrized state formed by placing six quarks in the lowest harmonic oscillator S state. The non-relativistic quark model[50,51] is employed. The six quarks are coupled to the correct spin (1) and isospin (0) of the deuteron. The harmonic oscillator parameter b (the wave function of a single quark is proportional to $e^{-\frac{1}{2}r^2/b^2}$) is given by $b = r_0/\sqrt{3}$ which is obtained from the requirement that the average distance between quarks in the state $|S_{6q}\rangle$ be r_0. (The distances between quarks should not be too large since we wish to use a single confinement region when all the quarks are close together.) The state $|D_{6q}\rangle$ is generated by the action of a Galilean invariant operator on the state $|S_{6q}\rangle$:

$$|D_{6q}\rangle = \frac{1}{N_D} \sum_{i<j} r_{ij}^2\, Y^2(\hat{r}_{ij}) |S_{6q}\rangle \tag{45}$$

in which magnetic quantum numbers and a C-G coefficient are eliminated for simplicity of presentation. The factor N_D insures that the state $|D_{6q}\rangle$ is normalized to unity as is the state $|S_{6q}\rangle$. The use of Eq. (45) to generate $|D_{6q}\rangle$ means that no spurious excitation of the center-of-mass wave function is included. Our employment of $|S_{6q}\rangle$ and $|D_{6q}\rangle$ corresponds to reasonable choices, which should allow us to make realistic but rough estimates of quark effects. Of course, further developments in the physics of confinement could cause our results to be altered. The terms P_{6q} and $\tilde{P}_D$ are the "probabilities" for the six quark state to exist in the deuteron. These are determined by conservation of the probability current, Eqs. (8) and (11). For example, in the Paris potential the total probability to have an S state is 0.94. This is unchanged if the total probability is divided into two parts: a nucleonic one, P_N, and a six quark P_{6q} term so that

$$0.94 = P_N + P_{6q}. \tag{46}$$

In this hybrid treatment u(r) is unmodified for $r \geq r_0$ so that

$$P_N = \int_{r_0}^{\infty} u^2(r)\,dr, \tag{47}$$

which via Eq. (46) implies

$$P_{6q} = 0.94 - P_N = \int_0^{r_0} u^2(r)\,dr. \tag{48}$$

In a similar fashion $\tilde{P}_D$ is given by

$$\tilde{P}_D = \int_0^{r_0} r^2 \mathcal{J}_2^2(r)\,dr \tag{49}$$

where $\mathcal{J}_2(r)$ must be used since it is that wave function that appears in Eq. (37).

It is worthwhile to comment on Eq. (49). Whereas u(r) is a normalizable bound state wave function so that probability is a well-defined concept, $\mathcal{J}_2(r)$ is not normalizable[47] even though it is a regular solution (at negative energy) of the Schroedinger equation. However, solutions of Schroedinger's equation satisfy the constraints of conservation of probability current. Thus, it is only the knowledge of the probability current into a spherical region bounded by $r = r_0$ which is required to determine the "probability", $\tilde{P}_D$, contained within $r \leq r_0$. Furthermore, any interactions that lead to the same values of $\mathcal{J}_2(r)$ and $\mathcal{J}_2'(r)$ at $r = r_0$ must have the same value of $\tilde{P}_D$ [Eq. (49)]. This is the same as the treatment of scattering wave functions which has been discussed above. One form of the result is that $\tilde{P}_D$ is simply the integral of $\mathcal{J}_2^2(r)$ over the inner region ($r \leq r_0$). This is analogous to Eq. (48), but now $\tilde{P}_D$ has units of volume, as shown in Eq. (49).

To evaluate $\tilde{P}_D$, $\mathcal{J}_2(r)$ is taken from the Reid soft-core potential since short range attractive effects of the spin orbit potential are relevant. The results for $\tilde{P}_D$ would be the same if the Paris potential were employed.

The S and D six quark states are connected by OPEP and OGEP forces. To obtain OGEP, the force of Isgur and Karl[51] is used. The OPEP between quarks is of the conventional non-relativistic form, and the π-quark coupling constant is constrained so that it leads to the correct pion-nucleon coupling constant.

The matrix elements appearing in Eq. (44) are given in Ref. 21.

The above considerations describe the model completely. The only unspecified parameter is r_0. This is the quantity of greatest interest because it is related to the nucleon bag size. It is our intention to vary r_0 to determine its allowed values. Having said this we turn to the results.

IV.5 NUMERICAL RESULTS

The important question is whether or not the loss, $\Delta\eta$, due to "form factor" effects can be compensated for by η^Q. This question can be settled independently of the starting value η^P and independently of the current experimental value of η. One can simply compare $\Delta\eta$ with η^Q for each value of r_0. The results, shown in Fig. 9, are expressed as in terms of percentage of the value $0.0271 \equiv \eta^{exp}$. First consider the larger graph. The solid curve shows $\Delta\eta/\eta^{exp}$ as a function of r_0. For example, with $r_0 = 1.2$ fm about 10% of the calculated values of η^{OPEP} is lost. It is this loss which is to be compensated by quark interaction effects. Indeed, as the long-dashed curve (η^Q/η^{exp}) shows, much of the loss due to the r_0 cut-off is regained from π and gluon tensor interactions between quarks.

The short-dashed curve gives η^Q/η^{exp} neglecting the effects of gluons. Except for large values of r_0, for which the π mass becomes significant in reducing the radial integral, the gluon exchange term is small. This is in contrast with the Δ-N splitting (caused by the spin-spin interaction) in which the gluonic effect is two to three times larger than the pion effect. For the tensor force to be effective here the space and spin wave functions of two quarks must be symmetric. Hence the color isospin wave function must be antisymmetric. For six quarks and three colors, there are more pairs with a symmetric color wave function, hence antisymmetric isospin. This favors the $\vec{\tau}_i \cdot \vec{\tau}_j$ of OPEP interaction over the $\vec{\lambda}_i \cdot \vec{\lambda}_j$ term of OGEP.

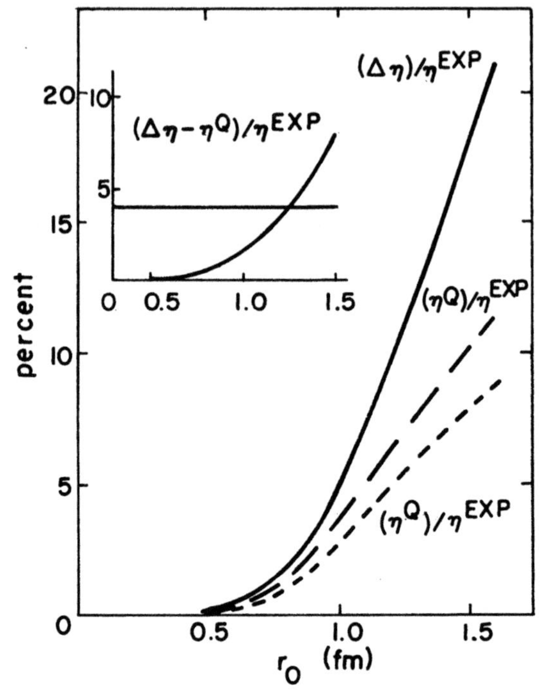

Figure 9. Changes in contributions to η/η^{exp}. $\Delta\eta$ is the amount lost due to the coordinate space cut-off. η^{Q} is the amount gained by pion and gluon exchanges between quarks (long-dashed curve). If gluon exchange is omitted, η^{Q}/η^{exp} given by the short-dashed curve. The percentage loss in computations of η, $(\Delta\eta-\eta^{Q})/\eta^{exp}$, is shown in the upper curve inset in the figure.

The entire effect of using a model with nucleon and explicit quark degrees of freedom can be examined by considering the difference $(\Delta\eta - \eta^{Q})/\eta^{exp}$, which is the contribution that is lost. This is shown in the figure insert in Fig. 9. One can have $r_0 > 1.5$ fm and still have a loss of η less than 10 per cent. Thus, one obtains weaker limits on r_0 than are implied in Refs. 47 and 48. To obtain a largest acceptable value of r_0, it is useful to define a maximum allowed value of the loss in η. We take this to be four per cent. There are three reasons focusing this instead of the quoted experimental error. 1) The Paris potential has a four per cent error with respect to η^{exp}. 2) Data from as late as 1979 and 1982 [see Ref. 47] have an error of about four or five per cent. 3) The theoretical value for the OPEP contribution is 0.02762. If we start with a value of 0.02762 a four per cent reduction would leave the number comparable with the current experiments. Thus, we take 4% as an allowable error. This is shown as a horizontal line in the inset. The intersection of this line with the solid curve defines an upper limit on r_0 with

$$r_0 \leq 1.25 \text{ fm}. \tag{50}$$

Thus for value of r_0 less than 1.25 fm the coordinate space cut-off has no significant consequences for η.

IV.6 LIMIT ON NUCLEON SIZE

It is tempting to convert Eq. (50) into a limit on the radius of a nucleon bag, R. This is difficult since we employ harmonic oscillator wave functions and since we deal with a six quark system. We nevertheless attempt an estimate of an upper limit on R. First we convert the limit on r_0 to one for a six quark bag of radius R_6. This is done by requiring that the root mean square radius of the harmonic oscillator wave function be the same as that of a quark wave function confined in a MIT bag of radius R_6. This gives $b^2 = R_6^2/3$ which when combined with $b^2 = r_0^2/3$ gives

$$r_0 = R_6. \tag{51}$$

Finally, one must relate R_6 to R, this can be done using the non-linear boundary condition (NLBC) which says $R_6 = (2)^{1/4}R$. We then have using Eq. (50)

$$R \leq 1.05 \text{ fm}. \tag{52}$$

Thus, even fairly large bags are not ruled out. However, caution must be exercised regarding the use of (52), since it is not at all clear that the NLBC can be used here.

IV.7 SUMMARY OF SECTION IV

Large bags are not ruled out in computations of η. It is necessary, however, to include the explicit effects of tensor interactions between quarks, and to use antisymmetrized quark wave functions. If this is done, results very similar to those obtained from pointlike OPEP are obtained. This is not too surprising because our calculations rely heavily on probability current conservation and, the idea that physical effects are continuous whether described in terms of explicit quarks or in terms of two baryons. For these reasons, we believe that the qualitative statements at the beginning of this paragraph will hold even after more detailed and accurate treatments of quark effects become available.

For nucleonic bag radii of about 1 fm the quark contribution to η

is about six per cent. This is rather large, considering that we are examining an asymptotic property of the most loosely bound nucleus. This result may further indicate that quark effects could be very significant in more tightly bound nuclei.

V. SUMMARY

In this talk a phenomenological approach which enables the size of quark effects in various nuclear processes is discussed. The principle of conservation of probability provides significant constraints on six quark wave functions. Using this approach, it is found that the low-energy proton-proton weak interaction can be explained in terms of W and Z boson exchanges between quarks. That the value of the asymptotic ratio of D to S state wave functions is influenced (at the 5% level) by quark effects, is another result of our approach. There was not time available to discuss the applications of our treatment to the $pp \rightarrow d\pi^+$ process.[19] This is a high momentum transfer (and therefore short distance) process. It is found that quark-quark interactions can contribute about half of the transition matrix element.

We have not discovered a nuclear effect that can be uniquely explained by quark-quark interactions. However it does seem that quark physics is very relevant for nuclear physics.

ACKNOWLEDGMENTS

This talk is based on work done in collaboration with P.A.M. Guichon, E.M. Henley and L.S. Kisslinger. Without their efforts this talk would not exist. I would also like to thank T.E.O. Ericson and A.W. Thomas for many useful discussions. This work is supported in part by the U.S. Department of Energy.

REFERENCES AND FOOTNOTES

1. J.P. Mathiot, Orsay preprint on Electrodisintegration of the Deuteron, 1983.

2. I. Sick, p. 236 in Few Body Systems and Nuclear Forces II, edited by H. Zingl, M. Haftel and H. Zankel, Springer-Verlag Berlin, 1978.

3. J.V. Noble, Phys. Rev. Lett. 46, 412 (1981).

4. A.B. MacDonald, at 1982 LAMPF Users meeting.

5. See for example D.F. Measday and G.A. Miller, Ann. Rev. Nuc. Part. Sci. 29, 121 (1979).

6. B.D. Day, Phys. Rev. Lett. 47, 226 (1981).

7. R. Hofstadter, Ann. Rev. Nuc. Part. Sci. 7, 231 (1957).

8. A. Chodos et al., Phys. Rev. D 9, 3471 (1974);
A. Chodos et al., Phys. Rev. D 10, 2599 (1974);
T. DeGrand et al., Phys. Rev. D 12, 2060 (1975).

9. R. Friedberg and T.D. Lee, Phys. Rev. D 15, 1694 (1977);
D 16, 1096 (1977); D 18, 2623 (1978);
T.D. Lee, Particle Physics and Introduction to Field Theory (Harwood Academic, New York, 1981);
R. Goldflam and L. Wilets, Phys. Rev. D 25, 1951 (1982).

10. R.L. Jaffe, 1982 Summer Institute on Theoretical Physics, Seattle, has stated that in assessing quark effects in nuclei one should use considerations based on energy; in this case the 300 MeV associated with the perturbative quark-gluon exchange mechanism. This is a large energy, by nuclear Standards, but not prohibitively huge. Indeed, the Δ hole excitations discussed by many have this excitation energy.

11. F. Wilczek, Ann. Rev. Nuc. Part. Sci. 32, 177 (1982);
P. Soding and G. Wolf, Ann. Rev. Nucl. Part. Sci. 31, 231 (1981).

12. L.S. Brown and S.D. Ellis, Phys. Rev. D 24, 2383 (1981).

13. K.G. Wilson, Phys. Rev. D 10, 2445 (1975);
M. Creutz, Phys. Rev. D 21, 2308 (1980).
F. Fucito, G. Martinelli, C. Omero, G. Parisi, R. Petronzio and F. Rapuano, Nucl. Phys. B210, 407 (1982).

14. M.A. Shifman, A.I. Vainshtein, and V.I. Zakharov, Nucl. Phys. B147, 385, 448, 519 (1979).

15. G.E. Brown and M. Rho, Phys. Lett. 82B, 177 (1979);
G.E. Brown, M. Rho and V. Vento, Phys. Lett. 94B, 383 (1979).

16. G.A. Miller, A.W. Thomas and S. Théberge, Phys. Lett. 91B, 192 (1980); S. Théberge, A.W. Thomas and G.A. Miller, Phys. Rev. D 22, 2838 (1980); D 23, 2106 (E) 1981;
A.W. Thomas, S. Théberge and G.A. Miller, ibid. 24, 216 (1981).
A.W. Thomas, Chiral Symmetry and the Bag Model, CERN preprint TH. 3368 (1982) and TRIUMF TRI-PP-82-29 (1982), to appear in Advances in Nuclear Physics, Vol. 13, eds. J. Negele and E. Vogt (Plenum, New York, 1982).
G.A. Miller, S. Théberge and A.W. Thomas, Comm. Nucl. Part. Phys. 10, 101 (1981).

17. M.H. Partovi and E.L. Lomon, Phys. Rev. D 2, 1999 (1970);
W.N. Cottingham, M. Lacombe, B. Loiseau, J.-M. Richard and R. Vinh Mau, Phys. Rev. D 8, 800 (1973).

18. L.S. Kisslinger, Phys. Lett. 112B, 307 (1982).

19. G.A. Miller and L.S. Kisslinger, Phys. Rev. C 27, 1669 (1983).

20. L.S. Kisslinger and G.A. Miller, Phys. Rev. C 27, 1602 (1983).

21. P.A.M. Guichon and G.A. Miller, "Quarks and the Deuteron Asymptotic D State", 1983 CERN Preprint TH 3649, submitted to Phys. Lett. B.

22. E.M. Henley, L.S. Kisslinger and G.A. Miller, Phys. Rev. C 28, 1277 (1983).

23. T.E.O. Ericson and G.A. Miller, "Charge Dependence of Nuclear Forces", 1983 CERN preprint TH 3615, to be published in Phys. Lett. B.

24. V.A. Matveev and P. Sorba, Nuovo Cimento Letters 20, 435 (1977).

25. C. de Tar, Phys. Rev. D 17, 323 (1978); 19, 1028 (F); 1451 (1979); M. Harvey, Nucl. Phys. A352, 326 (1981); H. Toki, Z. Phys. A294, 178 (1980); M. Oka and K. Yazaki, Prog. Theor. Phys. 66, 556; 572 (1981); D. Robson, Nucl. Phys. A308, 381 (1978); C.S. Warke and R. Shanker, Phys. Rev. C21, 2643 (1980); D.A. Liberman, Phys. Rev. D 16, 1542 (1977).

26. M. Lacombe, B. Loiseau, J.-M. Richard, R. Vinh Mau, J. Coté, P. Pires and R. de Tourreil, Phys. Rev. C 21, 861 (1980).

27. E.P. Wigner, Phys. Rev. 70, 15 (1946).

28. R.V. Reid, Jr., Ann. Phys. (N.Y.) 50,411 (1968).

29. A. Faessler et al., Phys. Lett. 112B, 201 (1982);
A. Faessler and F. Fernandez, Phys. Lett. 124B, 145 (1983).

30. P.J. Mulders and A.W. Thomas, CERN preprint TH. 3443 (1982); to appear in Journal of Physics G.

31. D.E. Nagle et al., in High Energy Physics with Polarized Beams and Targets, edited by G.H. Thomas, AIP Conference Proceedings, (AIP, NY, 1978), p. 224.

32. R. Balzer, R. Henneck, Ch. Jacquemart, J. Lang, M. Simonius, W. Haeberli, Ch. Weddigen, W. Reichart and S. Jaccard, Phys. Rev. Lett. 44, 699 (1980).

33. E.M. Henley, Ann. Rev. Nuc. Sci. 19, 367 (1969).

34. B. Desplanques and J. Missimer, Nucl. Phys. A300, 278 (1978); M. Box, A. Gabrić, K. Lassey, and B.H.J. McKellar, J. Phys. G2, L107 (1976).

35. E.G. Adelberger, C.D. Hogle, H.E. Swanson, R.D. Von Lintig, Phys. Rev. Lett. 46, 695 (1981); W.C. Haxton, Phys. Rev. Lett. 46, 698 (1981).

36. G. Barton, Nuovo Cimento 19, 512 (1961).

37. S. Weinberg, Phys. Rev. Lett. 19, 1264 (1967); A. Salam, "Elementary Particle Theory", ed. N. Svartholm, p. 367.

38. E.D. Commins and P.H. Bucksbaum, Ann. Rev. Nuc. Sci. 30, 1 (1980).

39. M.K. Gaillard and B.N. Lee, Phys. Rev. Lett. 33, 1081 (1974); G. Altarelli et al., Phys. Lett. 52B, 351 (1974).

40. F.J. Gilman and M.B. Wise, Phys. Rev. D 20, 2392 (1979); T. Karino et al., Prog. Theor. Phys. 65, 693 (1981).

41. R.D.C. Miller and B.H.J. McKellar, J. Phys. G8, 21 (1982); Los Alamos preprint (1982).

42. D. Preston and T. Goldman, Los Alamos preprint (1982).

43. A. Barroso and D. Tadić, Nucl. Phys. B.

44. V.R. Brown, E.M. Henley and F.R. Krejs, Phys. Rev. Lett. 30, 770 (1973); Phys. Rev. C9, 935 (1974).

45. B. Desplanques, J.F. Donoghue, and B.R. Holstein, Ann. Phys. (NY) 124, 449 (1980); V.M. Dubovik, V.S. Zamiralov, and S.V. Zenkin, Nucl. Phys. B182, 52 (1981); B. Desplanques, 1982 Grenoble preprint, SP 82-144.

46. B. Desplanques, J.F. Donoghue and B.R. Holstein, Ann. Phys. (N.Y.), 124, 449 (1980).

47. T.E.O. Ericson and M. Rosa-Clot, Phys. Lett. 110B, 193 (1982) and Nucl. Phys. A405, 497 (1983).

48. J. de Kam, Z. Phys. A310, 217 (1983).

49. I. Borbely, W. Gruebler, V. König, P.A. Schielzbach and B. Jenny, Phys. Lett.109B, 262 (1982). See also appendix Ref. 47.

50. A. De Rújula, H. Georgi and S. Glashow, Phys. Rev. D 12, 147 (1975).

51. N. Isgur and G. Karl, Phys. Rev. D 18, 4187 (1978).

52. Y. Simonov, at Few Body X, 1983.

CHIRAL SYMMETRY BREAKING AND THE PION

V. Bernard, R. Brockmann, W. Weise and E. Werner
Institute of Theoretical Physics
University of Regensburg
D - 8400 Regensburg, W. Germany

1. Introduction

The special properties of the pion (in particular: its small mass) are frequently interpreted as being related to chiral symmetry and its spontaneous breaking. Approximate chiral symmetry manifests itself in various ways in nuclear and low energy particle physics (through PCAC, the Goldberger-Treiman relation, the low energy theorems in pion-nucleon scattering etc.). Chiral bag models [1-3] use this concept to surround the quark core with a (massless) pion field, the Goldstone pion, coupled to the bag boundary such that the axial current becomes continuous at the boundary.

However, the pion in this model is a pointlike, structureless object. There is no obvious relationship to an underlying $q\bar{q}$-structure in this picture. The only reference to intrinsic pion properties is in terms of one (phenomenological) parameter, the pion decay constant $f_\pi = 93$ MeV.

The motivation of the present work is to investigate the connection between the Goldstone Boson nature and the quark-antiquark structure of the pion. We shall adapt the Nambu and Jona-Lasinio (NJL) chiral symmetry breaking scheme [4] to hadrons of finite size. This will be based on an effective Hamiltonian [5,6] which combines the underlying chiral symmetry of QCD for massless quarks with the phenomenology of confinement at hadronic length scales and asymptotic freedom at short distances.

The result is the pion as a (highly collective) superposition of $q\bar{q}$-states with a strongly reduced $q\bar{q}$ valence quark content. A test of the intrinsic pion wave function is provided by a calculation of the $\pi^\pm \to \mu^\pm\nu$ decay. Finally, a few related topics will be discussed, such as the role of the pion cloud in the nucleon charge and axial form factors.

2. Chiral Symmetry and Bag Model Phenomenology

As a starting point, consider the fundamental QCD Lagrangian L_{QCD} in the flavour SU(2) sector. For massless up and down quarks, $m_u = m_d = 0$, L_{QCD} is invariant under the chiral SU(2) x SU(2) transformation of the quark fields:

$$\psi \rightarrow \exp\,[i\gamma_5\,\vec{\tau}\cdot\vec{\theta}]\psi,\ \bar{\psi} \rightarrow \bar{\psi}\,\exp[i\,\gamma_5\,\vec{\tau}\cdot\vec{\theta}]. \tag{1}$$

Equivalently, the quark axial current

$$A^{\mu}_{\lambda}(x) = \bar{\psi}(x)\gamma^{\mu}\gamma_5\,\frac{\tau_\lambda}{2}\,\psi(x) \quad , \tag{2}$$

is a conserved quantity, i.e. $\partial_\mu A^{\mu} = 0$.

On the other hand, the non-linear self-couplings of the gluonic part of L_{QCD} are supposed to create confinement for colored objects such as individual quarks. The act of confinement necessarily implies a breaking of chiral symmetry at the quark level (dynamical symmetry breaking). This can be illustrated as follows: assume that the confining forces generate a scalar confining potential (or local mass) $M(r)$ for individual quarks inside the hadron. The single particle Dirac equation is

$$[i\,\gamma_\mu\partial^{\mu} - M(r)]\,\psi = 0 \tag{3}$$

Consequently, the divergence of the quark axial current becomes

$$\partial_\mu[\bar{\psi}\,\gamma^{\mu}\,\gamma_5\,\frac{\tau_\lambda}{2}\,\psi] = M(r)\bar{\psi}\,i\gamma_5\,\tau_\lambda\,\psi \equiv J_{5\lambda} \tag{4}$$

In bag models, the confining potential is replaced by a sharp boundary condition, in which case $J_{5\lambda}$ becomes proportional to a δ-function at the bag radius.

The non-vanishing divergence of the axial current measures the degree to which chiral symmetry is broken, at the quark level, in terms of a pseudoscalar-isovector source function $J_{5\lambda}$. In chiral bag models [2,3], this $J_{5\lambda}$ is interpreted as the (surface peaked) source function of the Goldstone pion, introduced such as to restore chiral symmetry. As mentioned before, the pion in such models is an elementary field, without any reference to an underlying $q\bar{q}$ substructure.

Chiral bag models have established successful phenomenology in several respects: they provide a simple model of pion-nucleon coupling; they reproduce basic chiral symmetry and current algebra results such as the Goldberger-Treiman relation; and they permit to obtain a certain amount of consistency with nuclear physics and the long range properties of the nucleon-nucleon interaction, by allowing for quark core radii considerably smaller than in the original MIT bag, and by generating the appropriate one-pion exchange potential between separated bags.

An apparently unrelated picture of the pion is that of a single $q\bar{q}$ pair in the lowest orbit of a bag or confining potential (we refer to this as the bag model pion). This bag model pion usually comes out much too heavy, except for the question of (ambiguous) center-of-mass corrections.

3. Chiral Effective Lagrangian

In trying to establish a connection between the Goldstone pion and the $q\bar{q}$ pion, we wish to construct a suitable effective Lagrangian which combines features expected from QCD with the underlying chiral symmetry .

We start again from the chiral invariant QCD Lagrangian L_{QCD} in the flavour SU(2) sector with current quark masses $m_u = m_d = 0$. We assume that the gluonic degrees of freedom can formally be eliminated from L_{QCD} without breaking its chiral SU(2) x SU(2) symmetry. This leads to the replacement of L_{QCD} by a chiral invariant effective Lagrangian in which an effective quark-quark interaction is the remnant of the non-linear gluon dynamics.

The assumed "minimal" effective action is:

$$S = \int d^4x\ \bar{\psi}(x)\ i\gamma_\mu \partial^\mu \psi(x) + \int d^4x_1 \ldots d^4x_4\ K(x_1 \ldots x_4)$$
$$[\bar{\psi}(x_1)\psi(x_2)\bar{\psi}(x_3)\psi(x_4) + \bar{\psi}(x_1)i\gamma_5\vec{\tau}\psi(x_2)\cdot\bar{\psi}(x_3)i\gamma_5\vec{\tau}\psi(x_4)]. \tag{5}$$

The interaction kernel K cannot be calculated from first principles; however, it must be consistent with certain features derived from or suggested by QCD:

(a) The kernel K cannot only depend on space-time distances between quarks. This is a natural consequence of the elimination of the gluonic degrees of freedom: since the gluonic fields have their own dynamics, their elimination manifests itself in the form of a distinguished reference frame. Therefore, if the original gluon field was centered at some space time point x_o, the quark-quark interaction should depend on the distance of each quark relative to x_o.

(b) Due to the asymptotic freedom properties of QCD the interaction strength must be weak at small distances so that quarks move essentially freely in the deep interior of hadrons.

(c) At distances comparable to hadron dimensions a rapid transition from weak to strong coupling sets in. This is suggested by recent Monte Carlo calculations on the lattice.

4. Model Hamiltonian for bound Quark-Antiquark Pairs

We now introduce a model Hamiltonian which has the properties discussed above and allows to demonstrate explicitly the connection between the $q\bar{q}$-pion and the Goldstone pion. The following simplifying assumptions will be made:

(i) The interaction part of the Hamiltonian is assumed to be static.

(ii) The static kernel is represented in the form

$$K = d(\vec{r}_1 \ldots \vec{r}_4;\ \vec{x}_o)\ \prod_{i=1}^{4} G(|\vec{r}_i - \vec{x}_o|)^{1/4}\ , \tag{6}$$

in such a way that K is invariant with respect to simultaneous translations of any one of the variables $\vec{r}_i$ and $\vec{x}_o$.

Here $G(|\vec{r} - \vec{x}_o|)$ is a coupling strength which depends on the distance between the position of each quark and the supposed center x_o of the multigluon field. Furthermore $d(\vec{r}_1 \ldots \vec{r}_4; \vec{x}_o)$ is a distribution function. Localizing the kernel in the variables $\vec{r}_i - \vec{r}_j$ (where i = 1,2; j = 3,4) and identifying the center-of-mass of the interacting quarks with the center x_o, one obtains the following Hamiltonian $H = H_o + H_{int}$ (with $x_o = 0$):

$$H_o = \int d^3r \; \overline{\psi}(\vec{r})\vec{\gamma}\cdot\vec{p}\psi(\vec{r})$$

$$H_{int} = - \int d^3r d^3r' \; G(r)\delta^3(\vec{r} + \vec{r}')[\overline{\psi}(\vec{r})\Gamma\psi(\vec{r})\overline{\psi}(\vec{r}')\Gamma\psi(\vec{r}') + \overline{\psi}(\vec{r})\Gamma\psi(\vec{r}')\overline{\psi}(\vec{r}')\Gamma\psi(\vec{r})], \tag{7}$$

where $\Gamma = 1, i\gamma_5\vec{\tau}$. We wish to apply it to the description of a bound $q\bar{q}$ pair with zero total momentum, with the center of mass localized at the origin. That is, in Eq. (7), $\vec{p} = -\frac{i}{2}\vec{\nabla}$, where the gradient is taken with respect to $\vec{r}$, the position of each individual quark with respect to the center.

The interaction H_{int} contains as an essential ingredient the position dependent coupling strength G(r). Following the discussion in section 3, G(r) should become small as $r \to 0$, while at values of r approaching hadronic dimensions G(r) should increase rapidly.

The model is a natural extension of the Nambu-Jona-Lasinio (NJL) model of chiral symmetry breaking [4)] to a situation where the coupling strength becomes position dependent.

5. Self-consistent Confining Mean Field

In order to obtain the selfconsistent confining potential experienced by each quark we add and subtract a term $M(r)\overline{\psi}(\vec{r})\psi(\vec{r})$ in the Hamiltonian (7). In analogy with the NJL model, M(r) is obtained by linearisation:

$$M(r) = \frac{5}{4} G(r) \; tr\langle\psi(\vec{r})\overline{\psi}(\vec{r})\rangle , \tag{8}$$

where use has been made of the reflection property $\psi(\vec{r})\overline{\psi}(\vec{r}) = \psi(-\vec{r})\overline{\psi}(-\vec{r})$. In the factor 5/4 in eq. (8), 1/4 comes from the exchange term of H_{int} eq. (7), after Fierz rearrangement.

The selfconsistency of the procedure requires that the quark fields $\psi(\vec{r})$ have to be evaluated using the Dirac equation

$$[\vec{\alpha}\cdot\vec{p} + \beta M(r)]\psi(\vec{r}) = E\psi(\vec{r}) \; . \tag{9}$$

(The role of nonzero current quark masses will be discussed later.) The quark fields can be expanded in terms of particle and antiparticle solutions of eq. (9):

$$\psi(\vec{r},t) = \sum_\alpha [u_\alpha(\vec{r})e^{-iE_\alpha t} b_\alpha + v_\alpha(\vec{r}) \; e^{iE_\alpha t} d^+_\alpha] \tag{10}$$

where b_α and d^+_α are quark annihilation and antiquark creation operators; $\alpha = (n\ell jm)$ denotes quark orbit quantum numbers. One obtains

$$\langle\psi(\vec{r})\bar{\psi}(\vec{r})\rangle = \frac{n_f}{2}\sum_\alpha [u_\alpha(\vec{r})\bar{u}_\alpha(\vec{r}) - v_\alpha(\vec{r})\bar{v}_\alpha(\vec{r})] \tag{11}$$

where n_f is the number of flavours.

In order to make the sum over α meaningful one needs a physically motivated cutoff procedure. For this purpose we choose a finite model space to obtain solutions of eq. (8). Subsequent conclusions about the Goldstone boson nature of the pion are independent of the size of this space. With the conventions for the upper and lower components g and f, respectively, of u_α and v_α defined in ref. [5,6], eq. (8) becomes

$$M(r) = \frac{5}{4} n_f\, G(r) \sum_{n\ell j} \frac{2j+1}{4\pi r^2} [g^2_{n\ell j}(r) - f^2_{n\ell j}(r)]. \tag{12}$$

Since G(r) is not well known from first principles, we turn the argument around and start with a confining potential M(r) roughly compatible with the systematics of hadron sizes and spectroscopy and calculate G(r) selfconsistently from eq. (12) once a model space is specified.

Potentials of the form $M(r) = cr^n$, $n = 2,3$ have been used quite successfully to describe various properties of the quark core in nucleons, like electromagnetic [7] and axial [8] form factors.

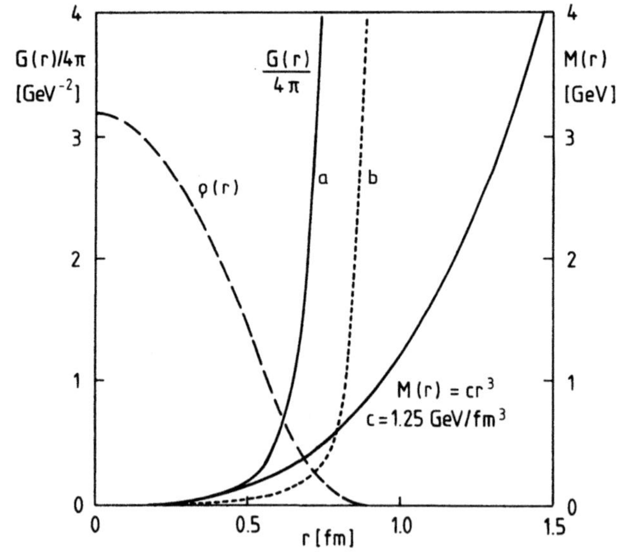

Figure 1: Coupling strength G(r) which generates a confining potential $M(r)=cr^3$ self-consistently, as determined by the mass equation, eqs. (8,12) Curves a and b show results for different model spaces: all {1s 1/2, 1p 3/2, 1p 1/2} quark and antiquark orbits (curve a); large space with all {1s,1p,1d,2p,1f} orbits (curve b). Also shown for orientation is the 1s 1/2 quark density distribution.

A typical result is presented in fig. 1 for n = 3. The coupling strength G(r) which generates $M(\vec{r})$ selfconsistently is shown for two different model spaces. One sees clearly the rapid transition from weak to strong coupling which takes place in the hadron surface.

6. Pion Mass and Wave Function

The effective Hamiltonian developed in the preceeding sections will now be used in the $q\bar{q}$-channel carrying pion quantum numbers. We construct properly coupled eigenfunctions of the Hamiltonian

$$H^{(o)} = -\frac{i}{2}(\vec{\alpha}_1-\vec{\alpha}_2)\cdot\vec{\nabla} + (\beta_1+\beta_2)M(r) \tag{13}$$

for $q\bar{q}$-pairs bound in the confining potential M(r). Note that this effectively introduces confinement in the relative coordinate of the pair. We define basis states $|(q\bar{q})_{\alpha\beta}\rangle = b^+_\alpha d^+_\beta|0\rangle$ for quark and antiquark in orbitals α and β. The corresponding r-space amplitude coupled to total angular momentum J, with center of mass fixed at the origin, is

$$\chi^J_{\alpha\beta}(\vec{r}) = [u_\alpha(\vec{r})\bar{v}_\beta(-\vec{r})]^J \tag{14}$$

and satisfies

$$\mathrm{tr}[\chi^+_{\alpha\beta}H^{(o)}\chi_{\alpha\beta}] = (E_\alpha + E_\beta)\mathrm{tr}[\chi^+_{\alpha\beta}\chi_{\alpha\beta}] \; . \tag{15}$$

The lowest states coupled to $J^\pi = 0^-$ are $(1s^1/2)^2$, $(1p^3/2)^2$, $(1p^1/2)^2$, the lowest energy being about 600 MeV. The mechanism which brings the energy down to zero is provided by the residual interaction in the pseudoscalar-isovector channel. It is useful to introduce the amplitude

$$D_{\alpha\beta}(r) = \langle(q\bar{q})_{\alpha\beta}|\bar{\psi}(\vec{r})i\gamma_5\tau_3\psi(-\vec{r})|0\rangle \tag{16}$$

where we consider for convenience neutral $q\bar{q}$-pairs. It is decisive to restrict the space $\{\alpha,\beta\}$ to the model space used for determining M(r). The sum of the direct and exchange matrix elements of H_{int}, eq. (7), is:

$$W_{\alpha\beta,\gamma\delta} = -\frac{5}{2}\int d^3r D_{\alpha\beta}(r)G(r)D_{\gamma\delta}(r) \, . \tag{17}$$

The eigenvalue equation for the pionic mode can be obtained from the Bethe-Salpeter equation for the amplitude $\phi(x_1,x_2)$ for finding a $q\bar{q}$-pair at the space-time positions x_1 and x_2, respectively, within a pion. (It is related to χ, eq. (14), by $\phi = i\,[\mathrm{tr}\,\gamma_5\chi]$). Since our effective interaction is not retarded the Bethe-Salpeter equation reduces to a Salpeter equation. Expanding the intermediate $q\bar{q}$-propagator in terms of the bound state pairs $(\alpha\beta)$ one arrives at the equation

$$(E_N^2 - E_{\alpha\beta}^2)C^N_{\alpha\beta} = 2E_{\alpha\beta}\sum_{\gamma\delta} W_{\alpha\beta,\gamma\delta}C^N_{\gamma\delta} \; , \text{ where}$$

$$E_{\alpha\beta} = E_\alpha + E_\beta, \qquad C^N_{\alpha\beta} = X^N_{\alpha\beta} + Y^N_{\alpha\beta} \; , \tag{18}$$

$$X^N_{\alpha\beta} = \langle o|d_\beta b_\alpha|\pi_N\rangle \;, \qquad Y^N_{\alpha\beta} = \langle o|b^+_\alpha d^+_\beta|\pi_N\rangle \;.$$

The modes $|\pi_N\rangle$ can be expressed by the amplitudes X and Y: the creation operator Q^+_N for the mode $|\pi_N\rangle$ defined by $|\pi_N\rangle = Q^+_N|o\rangle$ is

$$Q^+_N = \sum_{\alpha\beta} [X^N_{\alpha\beta} b^+_\alpha d^+_\beta - Y^N_{\alpha\beta} d_\beta b_\alpha] \;. \tag{19}$$

Eq. (18) is identical to the well-known RPA equations in many-body physics. The Goldstone boson nature of the pion has been investigated in a model space consisting of all 1s and 1p quark and antiquark orbitals. We find the Goldstone mode at E = O independent of the value of the potential parameter c in $M(r) = cr$.

We now discuss the effect of finite, but small quark masses m_1, $m_2 = m_u$ or m_d, which adds $\beta_1 m_u + \beta_2 m_d$ in $H^{(o)}$ of eq. (13). This results in the replacement $E_{\alpha\beta} \rightarrow E_{\alpha\beta} + \eta_{\alpha\beta}(m_u+m_d)$, where $\eta_{\alpha\beta}$ is determined to be close to 0.6 for the model considered here, almost independent of the quark single particle orbits.

This induces a first order change ΔE of the Goldstone mode given by

$$(\Delta E)^2 = \frac{\sum_{\alpha\beta} E_{\alpha\beta} C_{\alpha\beta} \eta_{\alpha\beta}}{\sum_{\alpha\beta} C_{\alpha\beta}} (m_u + m_d) \tag{20}$$

The resulting relationship between the pion mass $m_\pi = \Delta E$ and the current quark masses is of the same functional form as obtained from current algebra and QCD sum rules [9)], namely $m_\pi^2 = \text{const.}\,(m_u + m_d)$. The numerical calculations reproduce the empirical pion mass for $m_u + m_d$ between 30 and 40 MeV, depending moderately on the size parameter c of the confining potential. The quadratic relationship between m_π and $m_u + m_d$ is reproduced with an extremely good accuracy. We note that for a larger model space, the $m_u + m_d$ required to reproduce m_π becomes smaller.

Typical results for the pionic wave function are presented in table 1 in terms of the amplitudes X and Y.

$(\alpha\beta)$	$E_{\alpha\beta}$ (MeV)	$X_{\alpha\beta}$	$Y_{\alpha\beta}$	$X^2_{\alpha\beta} - Y^2_{\alpha\beta}$
$(1s_{1/2})$	645	0.77	0.49	0.35
$(1p_{1/2})$	959	1.00	0.74	0.45
$(1p_{1/2})$	1108	0.69	0.53	0.20

Table 1: Amplitudes X and Y obtained with $M(r) = cr^3$, $c = 1.25$ GeV fm^{-3} and m_u+m_d=36 MeV. The resulting pion mass is $m_\pi = 140$ MeV. The normalization is $\Sigma(X^2 - Y^2) = 1$.

It should be noted that the appearence of the amplitudes Y implies that $(3q3\bar{q})$, $(5q5\bar{q})$... components are present in the wave function in addition to $q\bar{q}$-components. The situation here is analogous to strong ground state correlations in the RPA description of low-lying collective modes in many-body systems (see Fig. 2). This analogy is in fact quite far-reaching. The self-consistent calculation of the confining mean field M(r) corresponds to the Hartree-Fock approximation for each individual quark (inside the hadron) interacting with the Dirac sea. On top of this there exists an additional residual interaction in $q\bar{q}$-channels, analogous to that operating in particle-hole excitation channels in many-Fermion systems. The point is that chiral symmetry implies a very direct relationship between the scalar interaction generating the confining potential, and the pseudoscalar-isovector residual interaction acting selectively in the pion channel.

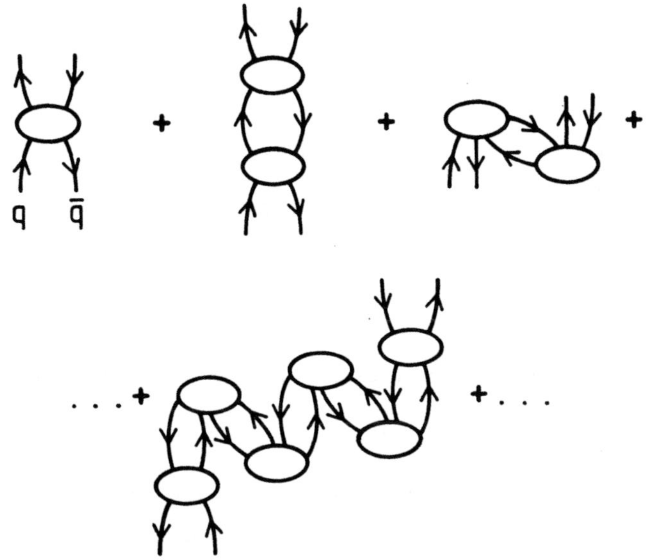

Figure 2: Illustration of the appearance of many-quark-many-antiquark configurations in the pion wave function by the RPA procedure described in the text.

To illustrate this further, we show in Fig. 3 the strength distribution of pseudoscalar-isovector states which measures the response with respect to an operator $\hat{O} = \int d^3r\ \bar{\Psi} i\gamma_5\tau_3\Psi$. The strength function is $S_N = |\langle O|\hat{O}|\pi_N\rangle|^2$, which satisfies an energy weighted sum rule. The collectivity of the pionic mode becomes obvious again by comparison of RPA and unperturbed strength distributions. Turning on the pseudoscalar-isovector residual $q\bar{q}$ interaction removes the $J^\pi = 0^-$, $I = 1$ strength almost completely from the unperturbed states and concentrates it in just one dominant low-lying state, the physical pion. An immediate consequence of this result is that, besides the pion, very little pseudoscalar-isovector strength should be observable in the meson spectrum at higher masses.

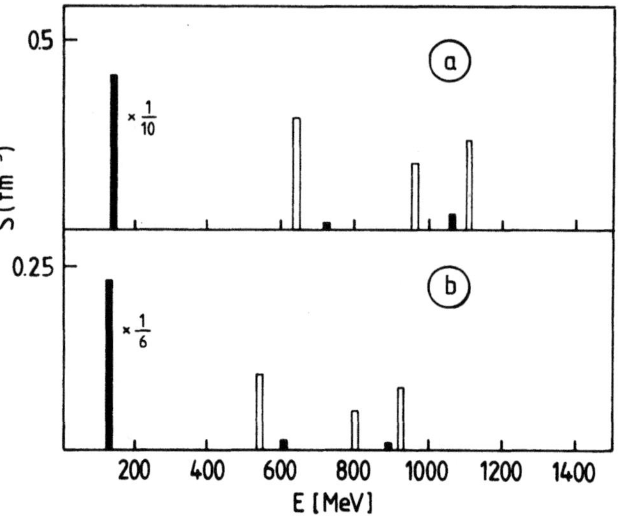

Figure 3: Strength distribution of pionic $q\bar{q}$-states. The model space is $\{1s_{1/2}\ 1p_{3/2}\ 1p_{1/2}\}$, and a confining potential $M(r) = cr^3$ is used. Upper part (a): $c = 1.25$ GeV fm^{-3}; lower part (b): $c = 625$ MeV fm^{-3}. Open bars: unperturbed pionic $q\bar{q}$ states (no residual interaction). Solid bars: full RPA calculation with $m_u + m_d = 40$ MeV (see text).

7. Pion Size and Pion Decay Constant

Given the pion wave function of eq. (19) one can calculate the pion rms radius $r_\pi \equiv \langle r^2\rangle^{1/2}$ and the pion decay constant f_π. The rms radius of the $q\bar{q}$-core of the pion can readily be expressed in terms of the amplitudes X and Y. The numerical value depends on the size parameter c; for example, for the value $c = 1.25$ GeV fm^{-3} used in fig. 1 one obtains $r_\pi = 0.46$ fm.

The relationship between the radius and the pion decay constant is of some interest. In terms of the amplitudes X and Y the pion decay constant becomes (with $\tilde{g} = g/r$, $\tilde{f} = f/r$):

$$f_\pi = \left[\sqrt{\frac{3N}{2m_\pi}} \cos\Theta_c/4\pi\right] \sum_{\alpha\beta} (X_{\alpha\beta} - Y_{\alpha\beta}) \sqrt{2j_\alpha + 1}\, [\tilde{g}_\alpha(o)\tilde{g}_\beta(o) + \tilde{f}_\alpha(o)\tilde{f}_\beta(o)] \qquad (21)$$

where Θ_c is the Cabbibo angle, N is a wave function normalization constant, and $\tilde{g}$ and $\tilde{f}$ are the upper and lower components of the spinors u or v (eq. (10)), respectively. It turns out that f_π is practically independent of the quark masses as long as $m_u + m_d \ll E_\alpha$, as expected; f_π depends, however, rather sensitively on the size parameter and therefore on the pion radius. In fig. 4 the connection between r_π and f_π is shown graphically. The experimental value $f_\pi = 93$ MeV is reproduced for $r_\pi = 0.43$ fm. This is not unreasonable, since r_π should be smaller than the measured charge radius of the pion, with additional contributions to the charge density presumably coming from meson cloud effects not included in the model. The fact that f_π can be made compatible with a rather small core size of the pion is due to the high degree of collectivity of the pion wave function. This is demonstrated in Fig. 4 by comparison with a "standard" model where the pion is a single $q\bar{q}$ pair in a pure $(1s\ 1/2)^2$ configuration.

The result obtained for f_π follows very closely an inverse proportionality, $f_\pi = \text{const}/r_\pi$. The constant is related to the "spectroscopic factor" Z for finding a $q\bar{q}$ pair in the pion, $|\pi\rangle = \sqrt{Z}|q\bar{q}\rangle + \ldots$ etc. Brodsky and Lepage [10], in their analysis of $\pi \rightarrow \mu\nu$ and $\pi^0 \rightarrow 2\gamma$ decays, obtain phenomenological values $r_\pi = 0.42$ fm and $Z \lesssim 1/4$. Our radius turns out to be very close to that of ref. 10). While our Z factor cannot be directly compared with theirs, we nevertheless find a value close to their upper limit. Again, in the present model the greatly reduced $q\bar{q}$-valence pair content of the pion is a necessary consequence of the collectivity of the pion mode. The spectroscopic factor will be further reduced by couplings e.g. to $2q2\bar{q}$ states not included in RPA. Consequently, we expect also that r_π given here is probably an upper limit.

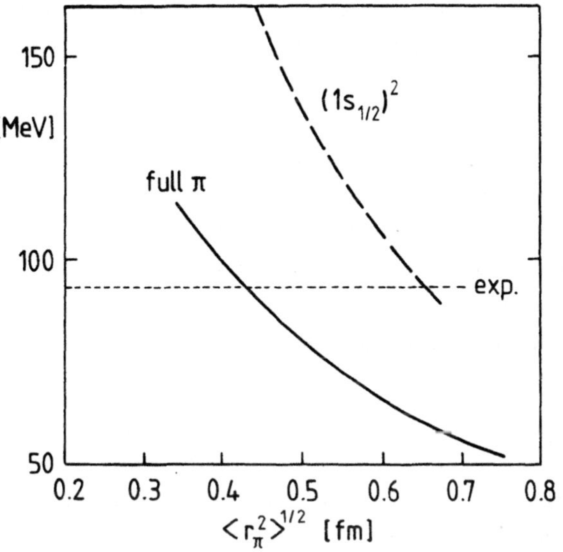

Figure 4: Dependence of the pion decay constant f_π on the rms radius of the pion core. Dashed curve: $q\bar{q}$-pion in pure $(1s\ 1/2)^2$ configuration. Solid curve: "collective" pion, represented by the amplitudes X and Y in a {1s,1p} model space (see also table 1). The full curve is very well approximated by $f_\pi = 0.2/r_\pi$.

8. Related topics

8.1 Pion cloud contributions to the proton charge form factor

The finite pion size provides a natural cutoff in calculations of the nucleon self-energy due to pion cloud effects, as shown in Fig. 5. While this procedure diverges for pointlike pion-quark coupling vertices in chiral bag models, it gives meaningful results if the pion has a finite radius.

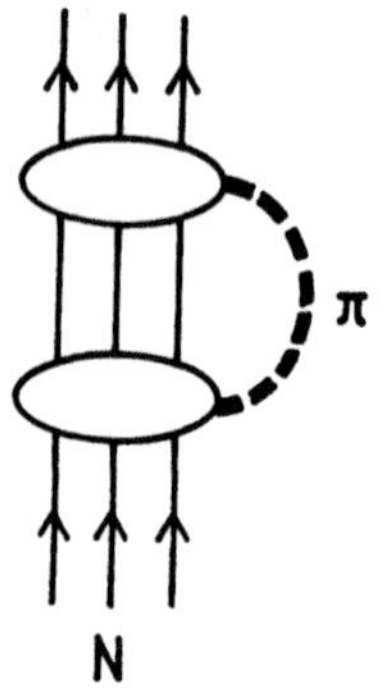

Figure 5: Nucleon self-energy due to the dressing of the three-quark core by a pion cloud.

The pion cloud contributes to the nucleon charge distribution. We show here a calculation [11] of the proton charge form factor which has been performed as follows: the three valence quarks are assumed to move in a confining potential $M(r) = cr^3$. The wave functions are projected onto fixed total momentum, as in refs. 7,8). A chiral pion quark coupling is introduced as in ref. 8), but with a Gaussian momentum space cutoff reflecting the finite size of the pion and adjusted to an rms radius of 0.4 fm. To obtain the charge form factor, a virtual photon is coupled, in a gauge invariant way, to any quark or charged pion line in Fig. 5. The procedure includes the ρ-meson pole structure of the $\gamma\pi\pi$ form factor. The summation over all intermediate three-quarks states turns out to be terminated by the finite pion size in such a way that it involves essentially N and Δ intermediate states, times a charge renormalization factor.

The result is shown in Fig. 6. For a pion radius of 0.4 fm, the pion cloud carries about 1/3 of the proton charge, while 2/3 remain in the three-quark core. The proton charge radius is in large part determined by the pion cloud, while at larger momentum transfers, the valence quark core starts to dominate, as expected. Also shown in Fig. 6 for comparison is the standard dipole fit, $G_E = (1 - q^2/0.71\ \mathrm{GeV}^2)^{-2}$.

7.2 Nucleon Axial form factor

It is useful to consider the axial form factor $G_A(q^2)$ (the Fourier transform of $A_3^3(x)$, see eq. (2) together with the charge form factor $G_E(q^2)$. The axial form factor measures essentially the spin distribution within the nucleon. The role of the pion field in $G_A(q^2)$ differs substantially from that in $G_E(q^2)$. In fact, the leading terms in the axial current to first order in the pion field, $f_\pi\ \partial^\mu\ \phi_\pi$, do not contribute to $G_A(q^2)$ at all as long as ϕ_π is a continuous function of position relative to the nucleon center-of-mass. To that order, $G_A(q^2)$ is entirely determined by the three-quark valence core [8].

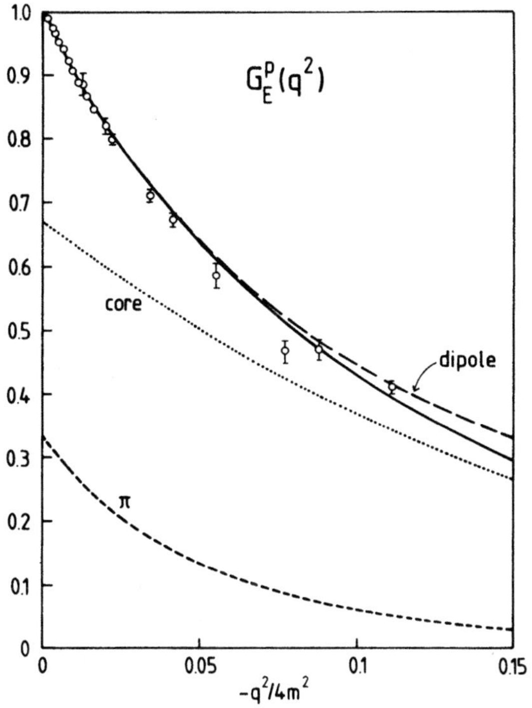

Figure 6:
Proton charge form factor showing separately the contributions from the three-quark valence core and from the coupling of the virtual photon to the pion cloud. (Input: confining potential $M(r)=cr^3$ with $\langle r^2\rangle^{1/2}=0.55$ fm for the quark core; Gaussian pion-quark vertex with radius $r_\pi=0.4$ fm.) The resulting proton charge radius is $\langle r_p^2\rangle^{1/2}=0.83$ fm (m is the nucleon mass).

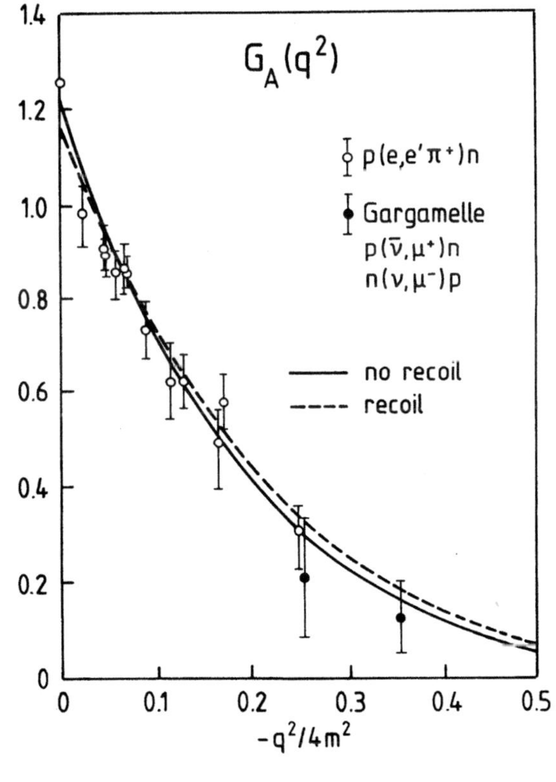

Figure 7:
The nucleon axial form factor calculated according to ref. 8) from the three-quark valence core alone. The input is the same as in $G_E(q^2)$. Here m is the nucleon mass. The solid curve: no recoil corrections; dashed curve: recoil corrections included according to the projection procedure described in ref. 8).

The result for $G_A(q^2)$ using the same quark-core input as in $G_E(q^2)$ is shown in Fig. 7. The degree of consistency between G_E and G_A is remarkable, but one should note that refinements may be necessary if terms non-linear in the pion field are added. The normalization of the axial form factor is $G_A(o) = g_A$, the axial charge, the empirical value being $g_A = 1.26$. The pion-nucleon coupling constant is

$$g_{\pi NN} = G_{\pi NN}(q^2 = m_\pi^2), \tag{22}$$

where $G_{\pi NN}(q^2)$ is the pion-nucleon form factor which is related, at $q^2 = 0$, to g_A by the Goldberger-Treiman relation

$$G_{\pi NN}(o) = \frac{m}{f_\pi} g_A . \tag{23}$$

The size associated with $G_A(q^2)$ implies a relatively soft $G_{\pi NN}(q^2)$ [8,12]. In fact, $G_{\pi NN}$ is well approximated by an exponential,

$$G_{\pi NN}(q^2) = G_{\pi NN}(o) \exp\,[q^2/\Lambda^2] \quad , \tag{24}$$

Using the same quark core wave functions as those determining the axial form factor, Fig. 7, one obtains $\Lambda \simeq 0.7$ GeV. Such an extended πN-vertex has to be confronted with the much shorter ranged cutoffs usually employed in phenomenological one-boson exchange models of the NN interaction.

References:

1) A. Chodos and C.B. Thorn, Phys. Rev. D 12 (1975) 2733

2) G.E. Brown and M. Rho, Phys. Lett. 82 B (1979) 177;
G.E. Brown, M. Rho and V. Vento, Phys. Lett. 84 B (1979) 383;
V. Vento, M. Rho, E. Nyman, J.H. Jun and G.E. Brown,
Nucl. Phys. A 345 (1980) 413

3) G.A. Miller, A.W. Thomas and S. Théberge, Phys. Lett. 91 B (1980) 192;
S. Théberge, A.W. Thomas and G.E. Miller,
Phys. Rev. D 22 (1980) 2838, Phys. Rev. D 24 (1981) 216;

4) Y. Nambu and G. Jona-Lasinio
Phys. Rev. 122 (1961) 345, Phys. Rev. 124 (1962) 246

5) R. Brockmann, W. Weise and E. Werner, Phys. Lett. 122 B (1983) 201;
W. Weise, in: Quarks, Mesons and Isobars in Nuclei,
R. Guardiola and A. Polls, eds., World Scientific (1983), p. 146;

6) V. Bernard, R. Brockmann, M. Schaden, W. Weise and E. Werner,
Nucl. Phys. A (1983), in print

7) R. Tegen, R. Brockmann and W. Weise, Z. Physik A 307 (1982) 339

8) R. Tegen, M. Schedl and W. Weise, Phys. Lett. 125 B (1983) 9;
R. Tegen and W. Weise, Z. Physik A (1983) in print

9) M.A. Shifman, A.I. Vainshtein and V.I. Zakharov,
Nucl. Phys. B 147 (1979) 385, Nucl. Phys. B 163 (1980) 43

10) S.J. Brodsky, in: Quarks and Nuclear Forces,
Springer Tracts in Mod. Phys. 100 (1982) 81;
S.J. Brodsky and G.P. Lepage, Phys. Scripta 23 (1981) 945

11) E. Oset, R. Tegen and W. Weise, preprint (1983)

12) P.A.M. Guichon, G.A. Miller and A.W. Thomas, Phys. Lett. 124 B (1983) 109

APPLICATION OF THE CLUSTER EXPANSION TO QCD

D. Schütte
Institut für Theoretische Kernphysik der Universität Bonn
Nußallee 14-16, D-5300 Bonn, West-Germany

1. Introduction

It is the purpose of this paper to present a new theoretical frame for treating QCD in an approximate way. The basic idea is to introduce a suitable cutoff version of the QCD-Hamiltonian and to solve the emerging many-body problems for the vacuum and for the hadrons by non--perturbative many-body techniques (a variational exp S-method). The hope is to obtain in this way more realistic models for hadrons (the only parameter for massless quarks being the QCD coupling-constant which has to be adjusted by some hadron mass) with the special structure that gluonic degrees of freedom are included. The theoretical significance of the results and the validity of confinement can be checked by studying the cutoff-dependencies.

We have organized our paper as follows: In sect. 2 the general structure of the cutoff-QCD Hamiltonian and its ground state solution is discussed and compared to standard many-body theory. The basic ansatz used in the variational method is given in sect. 3. The structure of the cluster expansion used for calculating expectation values is presented for the ground state in sect. 4 and for excited states in sect. 5.

Finally, we explain in sect. 6 the principles of performing renormalization by defining a suitable cutoff-dependent QCD coupling constant and discuss the structure which is to be expected in the case of the validity of confinement.

2. General structure of the cutoff QCD many-body theory

We introduce quark (antiquark) fermion operators $q_\alpha, q_\alpha^+ (\bar{q}_\alpha, \bar{q}_\alpha^+)$ and gluon boson operators b_k, b_k^+, where α, k are abbreviations for all quantum numbers needed to specify the corresponding s.p. states (momentum, spin, flavour, colour). The $t = 0$ field operators defining the QCD Hamiltonian may be expanded in terms of these fermion and boson operators. A cutoff version of QCD is then defined by restricting this expansion for the field operators to a finite number of terms. We use a phase space cutoff as introduced in ref.[1] which is given by the following prescription:

(a) Impose to the s.p. wave functions the boundary conditions of a finite box Ω. This makes the s.p. momentum spectrum discrete.

(b) Take only a finite number M of these s.p. states for the definition of the field operators.

The quantities Ω and M define the cutoff. As e.g. worked out by Pottinger et al.[2], the QCD Hamiltonian has then (in the Coulomb gauge) the structure (for simplicity, we specify only the Yang-Mills-case, i.e. we disregard quarks):

$$H = H_2(b^+b) + (H_3^0(b^+b^+b^+) + H_3^1(b^+b^+b) + \text{h.c.})$$
$$+ (H_4^0(b^+b^+b^+b^+) + H_4^1(b^+b^+b^+b) + H_4^2(b^+b^+bb) + \text{h.c.})$$
$$+ \text{higher order terms} \qquad (1)$$

($H_3^1(b^+b^+b)$ stands for an expression of the type $\sum_{0 \leq k_1, k_2, k_3 \leq M} H^1_{3,k_1k_2k_3} b^+_{k_1}, b^+_{k_2}, b_{k_3}$, etc.) Due to the cutoff, this Hamiltonian is a well-defined operator acting in the Fockspace generated from the Fockspace-vacuum $|0\rangle$ (defined by the condition $b_k|0\rangle$ for all k) by the operators b_k^+.

The many-boson-theory defined by this Hamiltonian is analogous to that of the standard non-relativistic (e.g. nuclear) case whose dynamics are described by an operator of the type

$$H_{st} = T(a^+a) + V(a^+a^+aa) \ . \qquad (2)$$

We mention the following structures:

(i) Both theories are defined in a Fockspace, only the statistics are different (we have fermions in the case of the nucleus).

(ii) A difference important for the interpretation of eigenstates is that the standard Hamiltonian H_{st} conserves the particle number which H does not. Thus the ground state of the fieldtheoretical Hamiltonian has to be interpreted as the "physical" vacuum, whereas that of H_{st} is the energetically lowest eigenstate of some particle number N (which has to be fixed before doing a calculation). But the mathematical analogy between the structures of the Hamiltonians (1) and (2) tells us that the vacuum "with interaction" is a state like nuclear matter (better bose-matter). This analogy with nuclear matter becomes very obvious in the case of fermions with selfinteraction, where the cutoff momentum plays the formal role of a fermi-momentum (see ref.[1]).

Eigenstates of the QCD-Hamiltonian yielding observable properties are given by "vacuum excitations": e.g. mesons correspond to p-h-excitations, glue-balls to two-particle, baryons to 3p excitations of the standard theory.

(iii) A fundamental (and for the volume-dependence of any property important) structure of the ground state ψ_o of H is that there exists a (unique) operator S

$$S = S_2(b^+b^+) + S_3(b^+b^+b^+) + \ldots \tag{3}$$

(again, $S_2(b^+b^+) := \sum S_{2,k_1k_2}\, b^+_{k_1} b^+_{k_2}$, etc.) such that

$$\psi_o = e^S |0\rangle \tag{4}$$

In the standard many-body case, the state $|0\rangle$ has to be replaced by a suitable slaterdeterminant of a given particle number. The only condition for the validity of this exponential form of ψ_o is that this state is not orthogonal to the Fockspace vacuum $|0\rangle$.

(iv) If perturbation theory is valid, there exists an expansion of the operator S in terms of linked (Goldstone) diagrams. A corresponding expansion of the ground state energy has important applications in standard many-body theory. In field-theory, these representations of vacuum properties play a minor role since the standard perturbation theory for Wightman functions in terms of Feynman diagrams yields at once all physical quantities relative to the vacuum.

3. Variational approach to the eigenstates of the cutoff QCD

In the case QCD, perturbation theory is assumed not to be valid and extended methods have to be applied in order to approximate the eigenvalue problem of H. We propose, herefore, to use the exponential form

$$\psi_o = e^S|0> \qquad S = S_1(b^+b^+) + S_3(b^+b^+b^+) + S_4(b^+b^+b^+b^+)+\ldots$$

with a suitably truncated form for the operator S (e.g. up to 4th order) as an ansatz for the ground state wave-function and to determine the coefficients defining S by the Ritz-variational principle.

In the same way, also excited states can be approximated by using an ansatz

$$\psi = De^{S'}|0>$$

$$D = D_2(b^+b^+) \qquad \text{for a glue-ball}$$

$$= D_2'(\bar{q}^+q^+) \qquad \text{for a meson}$$

$$= D_3(q^+q^+q^+) \qquad \text{for a baryon} \qquad (5)$$

Note that due to vacuum polarization effects, S' is expected to emerge from a variational calculation as different from S, the state exp S'|0> being then possibly interpretable as the vacuum modified through the formation of a bag.

We mention two cases where this variational method has been applied to QCD yielding nonperturbative results:

(a) The theory of Finger et al.[3] for chiral symmetry breaking. Here, only quark degrees of freedom are taken into account and the vacuum is defined as a BCS-state, i.e. $S = S_2(\bar{q}^+q^+)$. With the notations of ref.[1], this corresponds to a Hartree-Fock approximation of the vacuum state (which is then a filled Dirac--sea). The chiral-symmetry of the QCD-Hamiltonian for massless quarks yields then a self-consistent symmetry of the H.F. equations which is broken for strong enough interaction.

(b) The soliton-bag model of Friedberg and Lee[4], pursued in detail by Wilets et al.[5]: In this case the fact is used that the operator S has to be a spin-isospin-colour-scalar. It is, therefore, tempting to approximate S as being generated by an independent effective scalar quantum field. The QCD-Hamiltonian has then to be enlarged by introducing a phenomenological effective coupling of this new scalar field with itself and with the quarks and gluons. The ansatz (4) is then a coherent state which allows to treat S as a classical field. It has been shown by the above authors that the effective QCD-Hamiltonian can be chosen such that $S \neq 0$ for the vacuum and that S' for the hadron differs from S quite in the same way as the bag-model would predict.

In standard many-body theory[6,7] and in simple theories with mesonic degrees of freedom[8,9], some experience has been achieved with the ansatz (3) in a more general form (e.g. Brueckner-theory involves terms up to 4th order). In view of the structure of the QCD-Hamiltonian this leads to the suggestion that the following ansatz should be necessary (whether it is sufficient has to be checked by trying out refinements in a later state of the theory):

$$S = S_2(b^+b^+) + S_3(b^+b^+b^+) + S^4(b^+b^+b^+b^+)$$
$$+ S_o^2(\bar{q}^+q^+) + S_1^3(\bar{q}^+q^+b^+) \qquad (6)$$

(here, we have included quarks for definiteness).

4. Cluster expansion of the expectation values

For a complicated ansatz for S such as in eq. (6) it is not possible to obtain rigorous expressions for the expectation value

$$E(\psi_o) = \frac{\langle\psi_o H \psi_o\rangle}{\langle\psi_o \psi_o\rangle} \qquad (7)$$

However, there exists a powerful cluster expansion[6-9] of this quantity which states that one can define diagrams, involving elements related to S, S^+ and H (fig. 1), and Feynman rules such that we have a linked cluster theorem, i.e.

$$E(\psi_o) = \sum \text{all contributions from connected diagrams} \qquad (8)$$

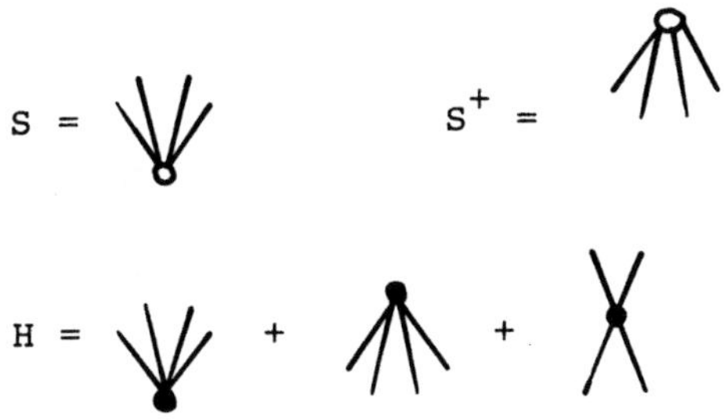

Fig. 1

Diagrammatic elements related to the operators S, S^+ and H for the case $S = S_4(b^+b^+b^+b^+)$, $H = H_4^o(a^+a^+a^+a^+) + H_4^2(a^+a^+aa) + H_4^4(aaaa)$

Fig. 2

Diagrams defining the linked cluster expansion of the expectation value $E(\psi_o)$ (eq. 7).

Some lowest order examples are given in fig. 2. Since (8) is a complicated infinite sum, the problem arises which technique should be used to approximate $E(\psi_o)$. Analogous to standard many-body theory[7-10], we propose as the simplest possibility the introduction of "occupation factors". This amounts to the following prescriptions:

Compute for each term of the Hamiltonian the lowest order contribution (in S).

Relate to each of these diagrams a well-defined diagram class by replacing each "free" line by a "dressed" line as depicted in an example in fig. 3 (more details are given in refs.[7,9]):

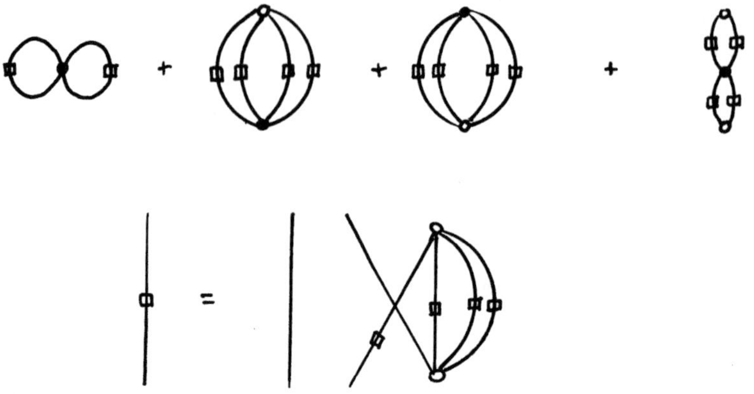

Fig. 3

Lowest order approximation of $E(\psi)$ including occupation factors.

We mention the following arguments which lead us to the hope that this "occupation factor procedure" might lead to a reliable estimate of $E(\psi_o)$. (Of course, the ultimate justification has to emerge from a numerical investigation of higher order contributions).

(a) The method is exact for $S = S_1(b^+b^+) + S_2^o(\bar{q}^+q)$

(b) It leads to the Brueckner-approximation for the standard many--body theory[7)]

(c) For the fieldtheoretical Lee-model it yields a "mesonic" Brueckner--theory consistent with renormalization[8)]

(d) The application of the method to be solvable version of the Lee model introduced in ref.[11)] leads to an approximation to $E(\psi_o)$ which is clearly valid beyond the perturbation theory limit (fig. 4).

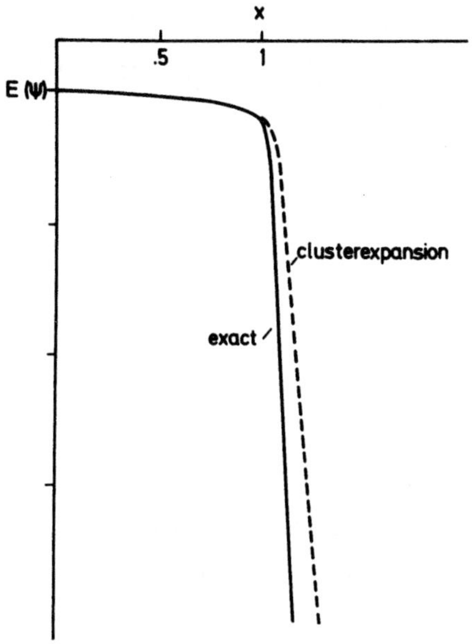

Fig. 4
Validity of the approximation given in fig. 3 for $E(\psi)$ for the solvable Lee-model. In the notation of ref.[11] we have put $N = 500$, $\alpha = 1$, $x = G\sqrt{N}$, $S = \lambda\tau_+ b^+$. Compared is the exact minimum of $E(\psi)$ ($\psi = e^S\phi$) to that of the approximate cluster expansion (fig. 3). The limit of perturbation theory is $x = 1$.

5. Cluster expansion for excitation energies

A notable general property of the cluster expansion is that it leads to an expression for excitation energies which is naturally split up into contributions from the kinetic and potential energy of the gluon in the glueball (or the quarks in the hadron) and into that of the "pressure of the vacuum on the bag": Herefore, we define

$$\psi_o' = e^{S'}|0\rangle$$

as the "vacuum modified by the bag". It is then easily proven that the energy difference $E(\psi)-E(\psi_o')$ is given by a completely linked expansion - by taking the difference all diagrams are cancelled where the operators D^+, D and H are not linked (fig. 5).

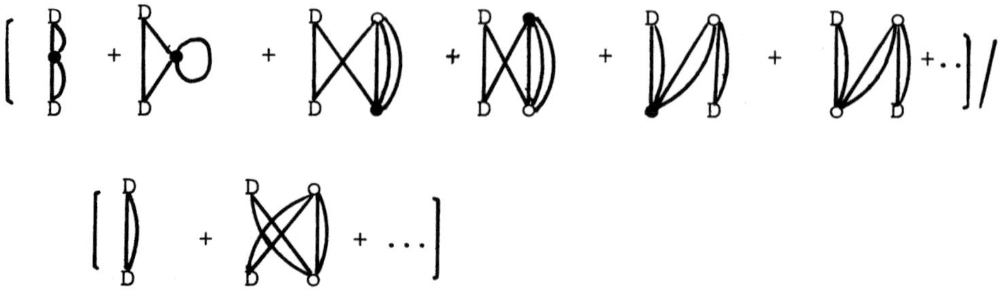

Fig. 5

Structure of the cluster expansion for the energy of a glueball ($D = D_2(b^+b^+) =$ D) relative to the energy of the "vacuum modified by the bag" (eq. 9) following from an analysis of the expression

$$E(\psi)-E(\psi_o') = [\langle\psi_o'|D^+HD|\psi_o'\rangle/\langle\psi_o'|\psi_o'\rangle]/[\langle\psi_o'|D^+D|\psi_o'\rangle/\langle\psi_o'|\psi_o'\rangle]$$
$$-\langle\psi_o'|H|\psi_o'\rangle/\langle\psi_o'|\psi_o'\rangle$$

The interpretation of this expansion clearly is that of "kinetic and potential energy" of the two gluons in the glueball (for $D = D_2(b^+b^+)$). Thus the physically observable glueball energy has to be written in the form $E(\psi)-E(\psi_o) = (E(\psi)-E(\psi_o'))+(E(\psi_o')-E(\psi_o))$ where the second term is interpretable as arising from the vacuum pressure as phenomenologically introduced in standard bag models.

6. Treatment of the cutoff-dependence and the confinement problem

Predictions for masses and other properties of hadrons should be made within our frame according to a "renormalization" scheme proposed in ref.[1] which is quite analogous to that of lattice QCD calculations:

For a given cutoff, the QCD coupling constant g should be adjusted to some hadronic mass (e.g. a meson or glueball mass). This fixes (for massless quarks) the QCD-Hamiltonian and allows then to predict in principle any other hadronic property. In this way, the coupling constant becomes a function of the cutoff (in fact $g = g(M)$, since g is adjusted to some energy-difference which is independent from the volume cutoff Ω). If our theoretical scheme is consistent, the prediction of hadronic properties should become independent from M if M is large enough. This is analogous to the continuum-limit in lattice QCD calculations. All this should hold for colour-singlett-quantities.

The confinement problem of QCD could be investigated within our frame by looking at the energies of colour non-zero states (the simplest case would be just gluons and quarks, defined by setting $D = b_k^+$ or $D = q_\alpha^+$). For given M and fixed coupling g(M), the mass of such a state is finite. Confinement should show up in the cutoff-dependence of these energies: When g(M) is fixed by some hadron-mass and we look at the limits for $M \to \infty$, we should obtain that the energies of colour zero-states are stable, whereas the energies of colour-nonzero states go to infinity.

We remark that all these statements about the cutoff-dependences represent expectations (for a simple case the consistency of our frame has been verified in ref.[1]) which have to be checked by working out the details of the expressions the energies $E(\psi_o)$, $E(\psi)$. Such investigations are being pursued at present in Bonn[12].

References

1) D. Schütte, On the structure of the relativistic Hartree-Fock theory, Nucl. Phys. A 411 (1983) 369
2) D.E.L. Pottinger and R.C. Warner, Nucl. Phys. B216 (1983) 336
3) M. Finger et al., Phys. Rev. D20 (1979) 3253
4) R. Friedberg and T.D. Lee, Phys. Rev. D16 (1977) 1096
5) R. Goldflam and L. Wilets, Phys. Rev. D25 (1982) 1951
6) J. da Providencia, Nucl. Phys. 54 (1963) 572
7) Y. Gerstenmaier and D. Schütte, Z. Nat. 35a (1980) 796
8) D. Schütte and J. da Providencia, Nucl. Phys. A338 (1980)
9) D. Schütte, Lecture Notes in Physics 138, Springer Verlag, 1980, 268
10) S. Rosati and S. Fantoni, Lecture Notes in Physics 138, Springer Verlag 1980, 1
11) D. Schütte and J. da Providencia, Nucl. Phys. A282 (1977)
12) W. Ferchländer, H. Nguyen and D. Schütte, to be published

THE CLOUDY BAG MODEL OF HADRONS

Gerald A. Miller
CERN, Geneva 23, Switzerland
and
Physics Department
University of Washington
Seattle, WA 98195 U.S.A.

I. INTRODUCTION

The Cloudy Bag Model[1] is a phenomenological representation of hadronic properties and meson-baryon scattering which includes both quark confinement and pionic cloud effects. The model has a partially conserved axial vector current (obeys PCAC) and in the limit of zero quark mass obeys the current algebra relations of Gell-mann and Levy.[2] It is easy to apply and no large computer calculations are required.

A contrasting approach to hadronic properties is the lattice approximation to quantum chromodynamics (QCD) discussed[3] at this meeting. Impressive results, based on the proposed fundamental theory of the strong interaction, have been obtained. However, the computer time required for these calculations is very great indeed. For example, a recent lattice gauge theory study of the hadronic spectrum carried out at CERN[4] required the equivalent of about 5000 hours of CDC 7600 time. There is nothing wrong with using the computer to learn about hadronic properties. However, we must remember that as nuclear physicists it is our goal to understand the properties of the nucleus. It is very difficult to imagine that studies of such large systems as nuclei could be carried out using present lattice theory techinques. Thus it is necessary that we develop and test simple models of hadrons that can be more easily applied to studies of the nucleus. Even though QCD is THE theory of the strong interaction, phenomenological models are needed. The Cloudy Bag Model is one such approach.

The outline of this talk is as follows:

1. A more detailed explanation of "What is a cloudy bag?".
2. Summary and update of early results.
3. Current algebra properties of the Cloudy Bag Model[5,6] with emphasis on recent studies of mesonic decays.[7]
4. Connection[8] between the axial vector form factor, pion-nucleon vertex function and the Goldberger-Treiman relation.

Let the story begin.

II. WHAT IS A CLOUDY BAG?

Our starting point is the much-discussed MIT bag model[9] which in its simplest version treats the baryon as three non-interacting, relativistic quarks confined in a static, spherical cavity. Although the so-called static cavity approximation violates the Lorentz invariance of the theory, its relativistic nature does provide some advantages. As a theory of massless quarks it might be a step closer to QCD than the constituent quark model. The relativistic Lagrangian formulation allows one to construct conserved currents, so that ultimately (with the inclusion of the pionic cloud) the constraints of current algebra can be satisfied. In addition, the asymptotic freedom and confinement aspects of QCD are implemented in an extremely simple fashion. The MIT bag model leads to generally good (but far from perfect) results for the energies and other properties of the hadronic ground states.

However there is a very significant (to a nuclear physicist, anyway) problem with the MIT bag model. It does not allow for the emission and absorption of a pion by a nucleon. This means that there is no one pion exchange potential (OPEP) between nucleons. The existence of OPEP is perhaps the most solid fact[10] that can be gleaned from all of the extensive analyses of nucleon-nucleon scattering, so that the MIT bag is in direct conflict with experiment. Furthermore, this problem is not restricted to the two nucleon system. All nucleons are identical so that if a pion can be emitted by one nucleon and absorbed by another, it can be emitted and absorbed on the SAME nucleon. This means that nucleons have a pion cloud. The above argument is older than I am, but it is just as correct as ever. Thus the MIT bag lacks a pion cloud that it should have.

There is a more fancy way to say the same thing. Even if the pion were massless in the MIT bag model, the Lagrangian fails to obey the very important property of chiral symmetry. An even more high-brow way to put it is that the model has no SU(2) x SU(2) (or any other) current algebra.

The breaking of chiral symmetry and its repair is intimately connected with the inclusion of pionic effects,[11] so I will discuss this subject in a bit more detail.

II.1 CHIRAL SYMMETRY BREAKING OF THE MIT BAG

The basic point is that quarks, q, carry an axial current $\underset{\sim}{A}^{MIT}$, such that

$$\underset{\sim}{A}_{\mu}^{MIT} = \frac{1}{2}\bar{q}\, \gamma_\mu \gamma_5 \underset{\sim}{\tau}\, q\, \Theta(R-r) \qquad (1)$$

where R is the bag radius. With the use of standard MIT bag model solutions one can easily show that

$$\hat{r}\cdot\vec{\underset{\sim}{A}}\,\Big|_{r=R} \neq 0 \qquad (2)$$

that $\hat{r}\cdot\vec{\underset{\sim}{A}}$ does not vanish at the surface of the bag even though the surface value of the normal component of the standard current, $\bar{q}\gamma_\mu q$, is zero is possible only through the relativistic nature of the quark wave functions. The upper and lower components allow for two different values of the currents at the bag surface. The consequence of Eq. (2) is that axial flux leaves the bag. When this is combined with the fact that in a static bag the axial charge is constant one is immediately led to the result:

$$\partial^\mu \underset{\sim}{A}_{\mu}^{MIT} \neq 0 : \qquad (3)$$

The axial current is not conserved. Eq. (3) has nothing to do with the pion mass; even when the pion mass is set to zero Eq. (3) holds.

II.2 CHIRAL SYMMETRY AND THE BAG

The solution of this problem was obtained by Brown & Rho and others.[12] One simply includes the effects of a pion field, $\underset{\sim}{\phi}$. Since the pion carries an axial current (that is responsible for the process $\pi \to \mu\nu$, one can obtain a conserved current by allowing a quark which hits the surface to transfer its axial current to a pion that can exist outside, Fig. 1. Thus if one regards the baryon as a system of three quarks in a bag that is surrounded by an extended pion cloud, no axial flux is lost. This is because the expectation value of $\underset{\sim}{\phi}$ vanishes for distances far from the bag center. Thus one can indeed construct

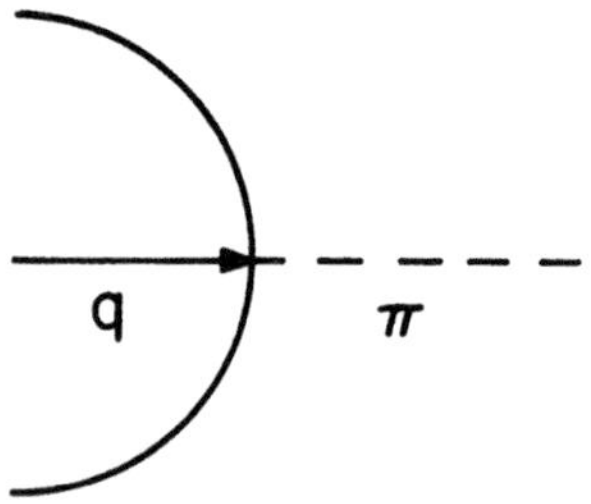

Figure 1. Pionic mechanism for enforcing conservation of axial vector current.

bag models for which

$$\partial_\mu A^\mu = 0 \tag{4}$$

with $m_\pi = 0$.

II.3 DO PIONS MATTER?

Pionic effects must be included to allow for the existence of OPEP and the resulting baryonic pion cloud, as well as to obtain a conserved axial vector current (and PCAC for non-zero values of the pion mass). Furthermore with pions one can show that the necessary current commutation relations are satisfied.[13] Thus pions must be included. The only operative question is: How much do pions change the properties of baryons?

In order to answer the above question consider a lowest-order perturbation theory calculation of the nucleonic expectation value of the pion number operator, N_π. As shown in Fig. 2, the process is of second order in the pion-nucleon coupling constant, which via the Goldberger-Treiman relation is proportional to 1/f, with f the pion decay constant ≅ 93 MeV. Since the expectation value of the pion number operator must be dimensionless, and the only other significant parameter is the bag radius, R, we must have

$$\langle N_\pi \rangle \propto \frac{1}{f_\pi^2 R^2} \ . \tag{5}$$

The coefficient of proportionality is about 1/14. The point of Eq. (5) is that for a small bag there are lots of pions, but for a large bag, there are not so many.

This brings us to the fundamental philosophical point behind the

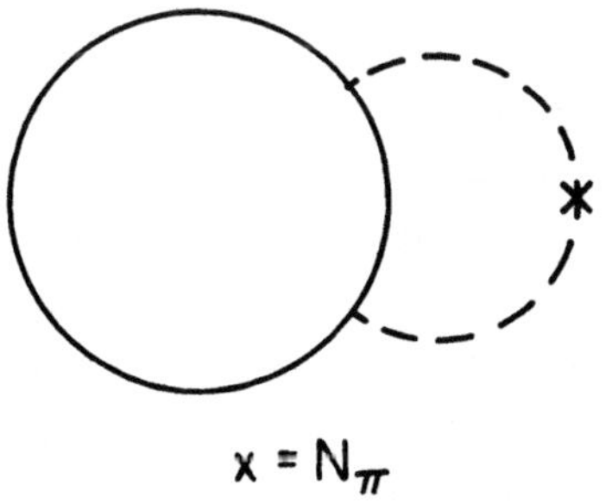

Figure 2. Expectation value of the pion number operator. The pion is represented as a dashed line. The pion number operator N_π is represented by the cross.

Cloudy Bag Model. It is our view that QCD forces are responsible for confinement phenomena. Although pions supply needed corrections, they do not cause confinement. This leads us to a picture of the nucleon in which the bag has a fairly large radius (close to that of the MIT bag, but perhaps a bit smaller), and the pionic effects are amenable to a perturbation-theoretic treatment. The implementation of this view is the subject of the next sub-section.

II.4 TECHNIQUE

The Lagrangian density of the Cloudy Bag Model is given by the expression

$$\mathcal{L}_{CBM}(x) = [i\bar{q}\partial q - B]\,\Theta_V - \frac{\Delta_s}{2}\,\bar{q}\,\exp(i\underset{\sim}{\tau}\cdot\underset{\sim}{\phi}\gamma_5/f)\,q$$
$$- \frac{1}{2}\,D_\mu\underset{\sim}{\phi}\cdot D^\mu\underset{\sim}{\phi} + \frac{m_\pi^2}{2}\,\phi^2 \qquad (6)$$

The quark field operator is q, B is the positive phenomenological energy density of the bag, Θ_V is a function that is unity for positions inside the bag and vanishes outside, and Δ_s is a surface delta function. The pion field operator is $\underset{\sim}{\phi}$ and, the original MIT bag model is regained if $\underset{\sim}{\phi}$ is set to zero. The pion-quark interaction occurs at the bag surface and the pion kinetic energy and mass are included in the last two terms of the right-hand-side of Eq. (6).

To treat the pion as an elementary field, as in Eq. (6) is to make a long wavelength approximation since effects of the pion's finite extent are neglected. One justification for this is that the pionic size (which is less than its measured r.m.s. charge radius[14]) is much smaller than its Compton wavelength. We do not deny that the pion has a finite size, but wish to use the above model only in low momentum transfer situations for which the size should be irrelevant.

In writing the pion kinetic energy as in Eq. (6) we allow the pion to exist inside as well as outside the bag. Pions inside the bag can be generated, for example, by residual non-confining forces between a quark and an anti-quark. Since these forces are particularly strong and attractive in the $q\bar{q}$ channel with the quantum numbers of the pion,[14,15] a correlated $q\bar{q}$ pair inside the bag that looks pretty much like a pion could exist. In any case, there is no fundamental or phenomenological reason to exclude pions from the interior of the bag as is done in Ref. 11. If m_π is set to zero $\mathcal{L}_{CBM}$ has a conserved axial current. Furthermore by allowing the pions to penetrate the bag, one avoids[16] a well-known problem[17] of obtaining too large a value of g.

As stated above we regard the bag radius, R, to be fairly large and the pion field to be fairly small. Thus we use a perturbative quantum field theoretic treatment of pionic effects. The zero'th order result is just the usual MIT bag model solution, so that the nice features of that model are incorporated from the beginning. Furthermore the terms of Eq. (6) non-linear in $\underset{\sim}{\phi}$ contain effects of pion loops. The finite pion size is expected to be significant in reducing the values of such diagrams, so that we believe that it is reasonable and sensible to neglect non-linear pionic effects. Since $\underset{\sim}{\phi}$ is small and non-linearities are ignored we obtain the approximations:

$$\exp\,(i\tau\cdot\phi\gamma_5/f) \Rightarrow 1 + i\underset{\sim}{\tau}\cdot\phi\gamma_5/f_\pi$$

$$D_\mu\phi \Rightarrow \partial_\mu\underset{\sim}{\phi} \qquad (7)$$

The resulting Lagrangian is simple and still obeys PCAC.

Another approximation is to allow quarks to exist only in the $s_{1/2}$ single particle MIT bag state of lowest energy. If one includes effects of the pion's finite extent in both the relative space and time variables,[18] one can show that effects of quarks in excited states are very small indeed.

Another procedural matter to be discussed is the treatment of the bag radius, R. It is determined from calculations of masses in the MIT bag model. However, computed values of masses are insensitive to the value of R. This is especially true if one considers uncertainties in some unmentioned parameters that go into the mass calculations. In our studies of pionic effects in calculations of various observables we regard R as a basically free parameter.

Values in the range between 0.8 fm and 1.1 fm result in good agreement with a variety of observables.

It is to be noted that pionic field effects do not dynamically influence the bag radius in our approach. This is undoubtedly an approximation. However, the inclusion of the finite size of the pion as well as the procedure of obtaining the self-energy as the solution of the appropriate transcendental equation (Eq. (3.19) of Ref. 19) leads to the result[13] that effects of the pionic pressure terms on the bag radius are fairly small.

There is one very serious problem with the MIT bag model that cannot be handled by including pionic effects. By requiring that the bag be centered at a point fixed in space one introduces a variety of spurious center-of-mass effects. This results in a theoretical uncertainty of at least ten percent in the calculated value of any observable.

III. SUMMARY OF EARLY CLOUDY BAG MODEL RESULTS

In this section some of the first calculations are reviewed. Subjects covered include pion-nucleon scattering in the (3,3) resonance region, static properties of nucleons, and the magnetic moment of the sigma hyperon, Σ^- .

III.1 PION-NUCLEON SCATTERING IN THE (3,3) RESONANCE REGION

It is well-known that there is a very large peak in pion-nucleon cross-section that occurs when the pion kinetic energy is about 200 MeV. This peak is referred to as the delta. One long standing question concerned the exact nature of the delta. In the Chew-Low-Wick[20] theory the delta is a composite system of a pion and a nucleon. The attraction is provided by the crossed Born term, which is iterated to produce a transition matrix in accord with the experimental data. (The iteration of the crossed Born term is the solution of the Low equation if one neglects the left-hand pion-nucleon cut.[21]) A more modern view[22] is that the delta is a three-quark state that is weakly coupled to the pion-nucleon continuum. The Cloudy Bag Model has both sets of terms. The procedure is to vary R, the only free parameter, to obtain the best fit to pion-nucleon scattering data. This gives a first handle on the value of R. Furthermore one also can examine the solution to see which of the Chew-Low series or the three-quark delta is the more important. This enables us to answer the question: What is a delta?

The answer is shown in Fig. 3. (More details are available in Ref. 19.) We see that the best fit is obtained with R = 0.82 fm. This was one of the earliest Cloudy Bag Model determinations of the bag radius. Later work by Théberge[23,24] showed that for 0.8 fm < R < 1.1 fm one could obtain an essentially equivalent fit. The next step is to study the relative importance of the crossed Born series and the elementary delta terms by setting either the πNN($f_{\pi NN}$) or the $\pi N\Delta$, vertex function ($f_{\pi N\Delta}$) equal to zero. If the $f_{\pi N\Delta}$ is set to zero, the resonance disappears as shown by the dotted curve. On the other hand, if $f_{\pi NN}$ = 0 the resonance peak still exists, it moves up by a relatively small amount of about 50 MeV, dash-dotted line. Thus the three-quark delta terms are dominant.

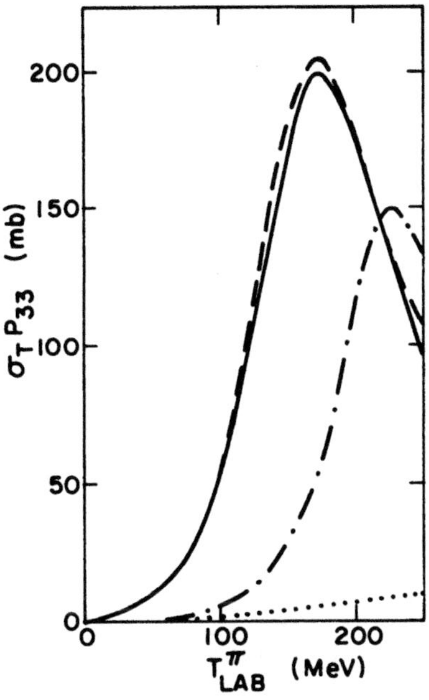

Figure 3. Best fit in the cloudy bag model (dashed curve) to the experimental P33 total cross section (solid). The dash-dotted (dotted) line shows the effect of arbitrarily setting the πNN (πNΔ) coupling constant to zero.

III.2 STATIC PROPERTIES OF THE NUCLEON

Values of the bag radius such that are consistent with pion-nucleon scattering in the (3,3) resonance region. The spectrum of hadronic ground states (except the pion) can also be reproduced with such values.[25,26] Thus having constrained the bag radius, we obtain a parameter-free model which can be tested by computing various nucleonic properties.

Let's discuss two of these in a qualitative fashion. The first is the neutron's charge distribution, which is known to be non-zero. For example, see Galster et al.[27] In the Cloudy Bag Model the neutron sometimes consists of a proton and a π^- cloud. Since the positive charge is confined by the bag radius, while the π^- can extend outside by a distance of about its Compton wavelength, one expects the mean-squared-charge radius to be negative. This is indeed the experimental result. Within the framework of the MIT bag model the non-zero charge distribution of the neutron can be caused by gluonic exchange effects. However, a variety of calculations show that this effect is far too small.[28]

Another quantity of interest is the nucleon magnetic moment. The current of the moving charged pions gives rise to a contribution to the nucleonic magnetic moment. The electromagnetic properties of the nucleon are obtained from the Feynman diagrams of Fig. 4.

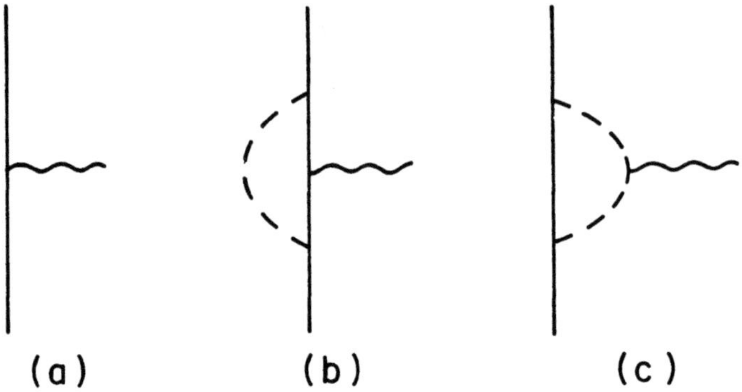

Figure 4. Electromagnetic properties of the nucleon. The baryon is a solid line, the pion a dashed line and the photon a wiggly line. The intermediate baryonic states are either the nucleon or the delta.

We note that the perturbation theory approach to calculating these quantities is valid[24,25] for bag radii such that $R \gtrsim 0.7$ fm.

A typical set of results is shown below in the table. The Cloudy Bag Model, the MIT Bag Model and the experimental values of various observables are compared. The bag radius is taken to be 1.0 fm to facilitate the comparison between our work and that of the MIT group.

OBSERVABLE	CLOUDY BAG	EXPERIMENT	MIT BAG
CHARGE RADII			
$\langle r^2_{ch}\rangle_p^{1/2}$	0.81 fm	0.83 fm	0.73 fm
$\langle r^2_{ch}\rangle_n$	-0.13 fm^2	-0.12 fm^2	0.00 fm
MAGNETIC MOMENTS			
μ_p	2.65 (2.4)	2.79	2.3 (1.9)
μ_n	-2.05 (-1.9)	-1.91	-1.6 (-1.4)

Table 1. NUCLEONIC PROPERTIES. The bag radius is 1 fm. The magnetic moments are given in units of nucleon magnetons. The terms in parentheses are computed without the Donoghue-Johnson effect.

In every case inclusion of pionic effects leads to improvement in the comparison with experiment. The result for the charge radius of the neutron is especially nice. The calculated magnetic moments include the Donoghue-Johnson[29] correction for the center-of-mass motion. Values shown in parentheses omit this questionable effect.

Using again R = 1.0 fm we can determine the composition of the physical nucleonic state. It is a nucleon bag 71% of the time, nucleon bag plus a pion 17% of the time, and a delta bag plus a pion 12% of the time. Thus in our picture the average number of pions in a nucleon is 0.29. There are not a great number of pions.

III.3 MAGNETIC MOMENTS OF THE BARYON OCTET-Σ^- MOMENT

Théberge and Thomas[30] studied the ground state baryon octet which consists of P,n Λ, $\Sigma^{\pm,0}$, $\Xi^{0,-}$ and examined the pion cloud effects. Pions do not interact with the strange quarks so that, generally speaking, pionic effects are small for strange hadrons.

However the Σ^- is an interesting exception. As noted by Pilkuhn and Eeg[31] there are two possible terms involving the emission of a π^-, as shown in Fig. 5. Thus pionic effects are about twice as large as usual for the Σ^-, so that it provides an interesting testing ground for various bag models.

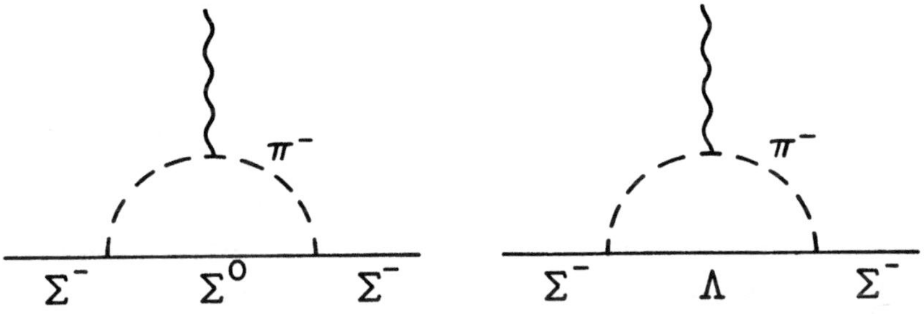

Figure 5. Magnetic moment contributions to the Σ^-.

The theoretical values of the Σ^- magnetic moment in nuclear magnetons for the MIT bag, cloudy bag and little bag are listed below:

THEORY	μ_{Σ^-}
Cloudy Bag	-1.08
MIT Bag	-0.81
Little Bag[32]	-0.58

For a long time the experimental situation[33] was cloudy, with an old atom experiment giving the result -1.41 ± 0.27 and a Σ^- beam experiment finding -0.89 ± 0.14. A new generation of Σ^- atom experiments performed by the William and Mary group at Brookhaven give -1.09 ± 0.03.

The agreement with the Cloudy Bag Model calculation is rather good. The early Little Bag prediction seems to be completely off. (Both of these calculations were made before the William and Mary experiment.)

Having briefly reviewed the earlier work, we turn to newer results.

IV. CURRENT ALGEBRA AND THE CLOUDY BAG MODEL

In this section some of the consequences that current algebra presents are discussed. It has long been known that the use of current algebra enables one to predict many relationships between observables (see e.g. Ref. 35) but in the Cloudy Bag Model one has a new situation. The current algebra predictions are maintained, but are embedded in a Lagrangian which enables one to predict absolute values of observables.

The beginning of the section deals with a version of the Cloudy Bag Lagrangian in which its current algebra aspects are more explicit than in Eq. (6). As a result there are some new applications to mesonic (as opposed to baryonic) systems obtained and discussed in section III.2.

IV.1 THOMAS TRANSFORMATION OF THE CLOUDY BAG MODEL

Although the Cloudy Bag Lagrangian of Eq. (6) has conserved axial and vector currents and obeys[13] the SU(2) x SU(2) current algebra it was difficult to identify the explicit terms and dynamics which led to the well-known current algebra results. For example, it was not clear how Eq. (6) could be used to obtain s-wave pion-nucleon scattering.

This problem was solved by Thomas[5] (see also Ref. 6) who obtained a transformation on the quark field operators which not only made the current algebra results more explicit, but also simplified the form of the Lagrangian and improved its convergence properties.

The basic idea can be understood by considering the surface delta term of Eq. (6)

$$\bar{q} \exp (i\underset{\sim}{\tau}\cdot\underset{\sim}{\phi}\gamma_5/f)\, q\, \Delta_s \qquad (8)$$

Thomas (using a technique similar to that of Weinberg[35]) redefined the quark fields according to

$$q_W = \exp\,(i\underset{\sim}{\tau}\cdot\underset{\sim}{\phi}\gamma_5/2f)\; q \qquad (9)$$

so that the surface term of Eq. (8) becomes

$$\bar{q}_W\, q_W\, \Delta_s \qquad (10)$$

From Eq. (10) it is clear that the surface boundary conditions of the Cloudy Bag Model are much simpler when expressed in terms q_W. Indeed, the linear boundary condition of the MIT bag model holds for q_W, which is a nice simplification.

Of course, the transformation of Eq. (9) must be applied to the entire Cloudy Bag Model Lagrangian. For the quark kinetic energy term we have that

$$i\bar{q}\partial\!\!\!/ q\Theta_v \Rightarrow i\bar{q}_W \exp(i\underset{\sim}{\tau}\cdot\underset{\sim}{\phi}\gamma_5/2f)\ \partial\!\!\!/\ \exp(i\underset{\sim}{\tau}\cdot\underset{\sim}{\phi}\gamma_5/2f) q_W\Theta_v \qquad (11)$$

An exact expression for the right-hand-side of Eq. (11) has been given by Thomas. It is my intention here to exhibit the essential features of Eq. (11) and, for this it is sufficient to make an expansion in powers of 1/f, keeping terms to order $1/f^2$. The result is

$$i\bar{q}\partial\!\!\!/ q\Theta_v \Rightarrow i\bar{q}_W\partial\!\!\!/ q_W\Theta_v + \frac{1}{2f}\,\bar{q}_W\gamma_\mu\gamma_5\underset{\sim}{\tau}q_W\partial^\mu\underset{\sim}{\phi}\Theta_v$$

$$- \frac{1}{2f_\pi^2}\,\bar{q}_W\,\frac{\gamma_\mu\underset{\sim}{\tau}}{2}\,q_W\cdot(\underset{\sim}{\phi}\times\partial^\mu\underset{\sim}{\phi})\ \Theta_v \qquad (12)$$

In the term that is independent of $\underset{\sim}{\phi}$, one simply recovers the quark kinetic energy. The next term is linear in $\underset{\sim}{\phi}$ and is responsible for the familiar terms that cause emission and absorption of a single pion. It is of a volume pseudovector form, so that at first glance one might believe that the pion-nucleon and pion-nucleon-delta vertex functions are different than the ones obtained from Eq. (6). However, this is not the case. For the systems of interest here, we have that

$$\int \frac{d^3x}{2f}\,\bar{q}_W\gamma_\mu\gamma_5\underset{\sim}{\tau}\,q_W\,\partial^\mu\,\phi\,\Theta_v = \frac{i}{2f}\int d^3x\,\bar{q}_W\gamma_5\underset{\sim}{\tau}q_W\cdot\underset{\sim}{\phi}\,\Delta_s \qquad (13)$$

the relation (13) holds for systems in which all quarks are in the same orbital.

The proof is very simple so that I reproduce it. Observe first that the term with $\mu = 0$, vanishes by virtue of parity conservation. The remaining term involving the gradient of the pion wave function is simplified by an integration by parts. The use of field equation satisfied by the quarks shows that terms involving gradients of the quark (or adjoint) wave functions cancels as long as the q_W and the $\bar{q}_W$ refer to quarks of the same energy. The remaining term is of a surface delta form. The last step is to use the linear boundary condition to obtain the form (13).

The term involving $\underset{\sim}{\phi}\times\partial^\mu\underset{\sim}{\phi}$ is the one of particular interest to Thomas.[6] It is responsible for s-wave pion baryon scattering. In particular, it yields the well-known result of Tomazowa[36] and Weinberg. As shown in the next section it also has some nice consequences regarding the various decay channels of the ρ meson.

IV.2 CLOUDY BAG MODEL AND MESONIC PROPERTIES

I'd like to discuss some recent work done by Paul Singer and myself[7] in which the Cloudy Bag Model is applied in the understanding of various decay widths of the vector mesons.

Start by considering the rho meson which decays via the strong interaction to two pions and via the electromagnetic into a e^+e^- pair. The very same term that is responsible for the s-wave pion nucleon scattering also explains both of these rho meson decay widths and the relationship between them. The relevant matrix element is repeated below (we leave out the subscript W in this subsection.)

$$\frac{1}{2f^2} \langle\rho|\bar{q}\, \frac{\gamma_\mu}{2}\, \underset{\sim}{\tau}\, q\, \Theta_V|0\rangle (\underset{\sim}{\phi}\times\partial_\mu\underset{\sim}{\phi}) \tag{14}$$

The quark isovector current connects the rho meson with the vacuum and the $\underset{\sim}{\phi}\times\partial_\mu\underset{\sim}{\phi}$ term creates two pions. The result is the transition matrix element for the decay $\rho \to \pi\pi$ which is defined as $f_{\rho\pi\pi}$. The matrix element connecting the rho to the vacuum also appears in the decay $\rho \to e^+e^-$. It is defined as m_ρ^2/f_ρ. By taking the matrix element of the term of (14) between a physical rho state and a two-pion state one obtains the result

$$\frac{1}{2f^2}\ \frac{m_\rho^{\ 2}}{f_\rho} = f_{\rho\pi\pi} \tag{15}$$

This is the well-known KFSR relation which is well-satisfied experimentally.

This analysis of the Cloudy Bag Model Lagrangian shows that it does satisfy the KFSR relation. But one can go even further, since the bag model allows one to compute the constant, f_ρ. It is a straightforward procedure to evaluate the matrix element $<\rho|\bar{q}\gamma_\mu \tau q|0>_{Bag}$. However f_ρ is defined in terms of the physical rho meson which is an eigenstate of momentum. To bridge the gap it is necessary to relate the rho bag state to momentum eigenstates via a wave packet expansion. We use Wong's[38] version of the Peierls-Yoccoz[39] projection procedure. The calculation is described in Ref. 7, so we only present the results here. It is true that the procedure used is very crude (as is any of the current attempts to fix the static cavity approximation's lack of Lorentz invariance). However the severity of the problem is mitigated somewhat by the relatively large mass of the rho meson. Indeed were the pion mass equal to the rho mass the discrepancy between the Donoghue-Johnson[29] and Wong computations of the pion decay constant, f, would vanish.

The calculation yields a value $f_\rho = 5.4$ which is in good agreement with the value 5.09 ± 0.31 extracted from the partial decay width for the rho to decay into an electron-positron pair.[40] The next step is to compute $f_{\rho\pi\pi}$ and the resulting width for $\rho \to \pi\pi$. We use the KFSR relation Eq. (15) along with the calculated value of f_ρ to get $f_{\rho\pi\pi} = 6.3$. By squaring this number and multiplying it by the appropriate phase space factors one obtains $\Gamma(\rho \to \pi\pi) = 164$ MeV which is in excellent agreement with the experimental value of 154 ± 5 MeV.

Thus the decay widths for the processes $\rho \to \pi\pi$ and $\rho \to e^+e^-$ as well as the algebraic relationship between their amplitudes are well-understood in the Cloudy Bag Model.

Another application to mesonic systems involves calculating the vertex functions for absorption or emission of a single pion. In particular we may compute the coupling constants for the processes $\omega \to \rho\pi$, $K^* \to K\pi$ and $K^* \to K^*\pi$ using the term linear in ϕ of the Lagrangian of Eq. (6) or the one of Ref. (5). One uses standard bag wave functions, with bag radii of Ref. 26, to compute the necessary matrix elements. For example we compute

$$\frac{1}{2f} <K\pi| \int d^3x\, \bar{q}\, \gamma_5 \tau \cdot \phi q \Delta_s |K^*> \tag{16}$$

The resulting coupling constants when expressed in terms of dimensionless units[7] are:

$$f^{\omega\rho\pi} = 10.3,\ f^{K^*K\pi} = 5.8,\ f^{K^*K^*\pi} = 10.2 \tag{17}$$

Experimental information is available from the decays $\omega \to \pi\pi\pi$ and $K^* \to K\pi$ so that one can extract the values

$$f^{\omega\rho\pi}_{EXPT} = 12.0 \pm 0.4 \qquad f^{K^*K\pi}_{EXPT} + 6.5 \pm 0.2 \tag{18}$$

A comparison of the result for $f^{K^*K^*\pi}$ with experiment is difficult. There is only an experimental upper limit available,[40] $\Gamma_{K^*}(K\pi\pi) < 36$ keV. Our value is consistent with the SU(3) prediction $f^{K^*K^*\pi^0} = \frac{\sqrt{3}}{2} f^{\omega\rho\pi}$ (corrected in our model by an SU(3) breaking factor M_{K^*}/M_ω), which gives[41] for this decay a rate approximately equal to the present upper limit.

A comparison between Eqs. (17) and (18) shows that the Cloudy Bag Model provides an excellent description of the single pion emission vertices. When combined with the results for the decays of the rho meson, we see that both one and two pion emission vertices are well explained.

Another application, still in progress, involves computing the electromagnetic decays of a vector meson into a pseudoscalar meson and a single photon. These transitions occur via M1 transitions. For example, consider the processes $K^{*+} \rightarrow K^+\gamma$ and $K^* \rightarrow K\gamma$. SU(3) symmetry predicts that the width for the former process is only one-fourth of the latter, whereas the experimental values for the two widths are almost equal.[42] The effects of the strange quark mass are not sufficient to account for the rates.[42] However, including the pion cloud contribution shown in Fig. 6 does lead to an explanation of the data.[43] The present case is an exception to the general rule that pionic effects are small for strange systems because of the relatively unimportant contribution of the ordinary term (Fig. 6a) in the process $K^{*+} \rightarrow K^+\gamma$. Thus it appears that the Cloudy Bag Model will provide a reasonable explanation of these M1 transitions.

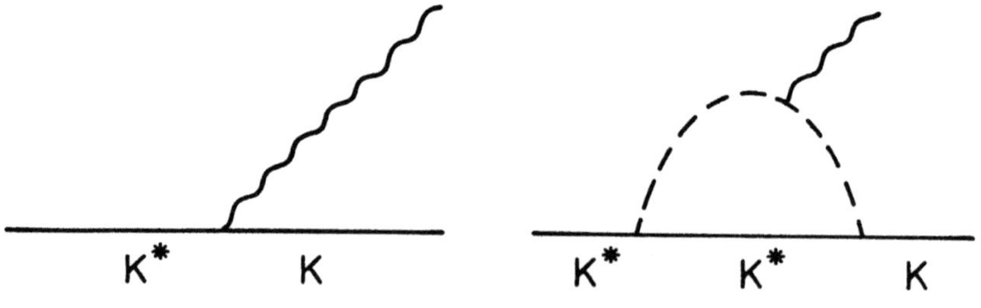

Figure 6. M1 Decays of the K Mesons. a) photon-quark interaction term. b) pion cloud contribution. The wavy line represents the photon, and the dashed line the pion.

V. THE AXIAL FORM FACTOR OF THE NUCLEON, PION-NUCLEON VERTEX FUNCTION AND THE GOLDBERGER-TREIMAN (GT) RELATION

In this section the axial form factors $g_A(q^2)$, vertex function $g_{\pi NN}(q^2)$ and the relation between them are discussed. An application is made to the understanding of the error in the GT relation.

As mentioned above there has been considerable discussion of the calculation of the axial charge of the nucleon in various chiral bag models. However little attention has been given to the axial-vector form-factor, $g_A(q^2)$, which is known from measurements of muon production cross-sections ($\nu_\mu + n \to \mu^- + p$) and pion electroproduction. Phenomenologically the best fit is a dipole with mass $m_A = 0.95 \pm 0.14$ GeV. In fact Amaldi et al[44] quote an error of ± 0.07 GeV from the neutrino reaction, and ± 0.14 GeV from electroproduction. We take the more conservative error in order to improve the overlap with the pion production data from the ZGS, which yielded values of m_A between 1.0 and 1.4 GeV.[45] Of course these data are most sensitive to the low $-q^2$ behavior, which can be parametrized as

$$g_A(q^2)/g_A(0) = 1 - q^2 r_A^2/6 \ , \tag{19}$$

with $r_A = 0.72 \pm 0.12$ fm.

In the Cloudy Bag Model the pion field is allowed to penetrate the interior of the MIT bag. The appropriate expression for the axial-vector current density is

$$\vec{A}_i(\underline{x}) = 1/2\ \bar{q}(\underline{x})\ \vec{\gamma}\gamma_5\tau_i\ q(\underline{x})\ \ \theta(R-r) - f_\pi\ \vec{\nabla}\ \phi_i(\underline{x}) \tag{20}$$

For this model $g_A(q^2)$ is entirely determined as the (coefficient of the term proportional to $\sigma\tau_i$ of the) expectation value of the Fourier transform of the first term in Eq. (20) in the physical nucleon.

There is also a term proportional to $\vec{\sigma}\cdot\hat{q}\hat{q}$ in the above Fourier transform, but it is not part of $g_A(q^2)$. One interesting thing about the Cloudy Bag Model is that the $\vec{\nabla}\underset{\sim}{\phi}$ term of Eq. (20) does not contribute to $g_A(q^2)$. This is because the pion is allowed to penetrate the bag so that a plane wave expansion of $\underset{\sim}{\phi}$ is justified. Upon taking the Fourier transform of (20) the $\vec{\nabla}\underset{\sim}{\phi}$ term becomes one proportional to $\hat{q}$ which does not contribute to g_A. Thus in the Cloudy Bag as in the MIT Bag the value of $g_A(q^2{-}0) \equiv g_A$ is 1.09. This is in contrast with other bag models[11] which by excluding pions from the bag interior obtain an enhanced value of g_A of $\frac{3}{2}$ (1.09) which is much too big. (The difference between 1.09 and the experimental value of 1.26 can be obtained in a variety of ways. See, for example, Ref. 46.

Taking the expectation value of the Fourier transform of Eq. (20) leads to the result

$$g_A^{CBM}(q^2)/g_A^{CBM}(0) = 1.5\, j_0^{-2}(\omega) \int_0^R dr\, r^2 \Big\{[j_0^2(\tfrac{\omega r}{R}) - j_1^2(\tfrac{\omega r}{R})]\, j_0(qr) + 2\, j_1^2(\tfrac{\omega r}{R})\, j_1(qR)/(qR)\Big\}. \tag{21}$$

As usual, $\omega = 2.04$ is the eigenfrequency for massless quarks.

Expanding Eq. (21) to order q^2 for comparison with Eq. (19) we find $r_A^{CBM} = 0.62$ R, and hence $R = 1.16 \pm 0.20$ fm. (Of course we must add a caution that we have also omitted corrections arising from possible spurious centre of mass motion and recoil.) The result of these considerations is that the axial vector form factor is very soft. It is clear that better data would be much appreciated.

Another property of the nucleon which has been intensely discussed in the context of chiral bag models is the form-factor for coupling to a pion, $g_{\pi NN}(q^2)$. As shown in Eq. (21), in such models $g_A(q^2)$ arises from an integral over the bag volume. On the other hand, $g_{\pi NN}(q^2)$ is associated with the surface of the bag, and one might intuitively expect it to fall more rapidly with q^2. In the CBM this is certainly the case, because the form-factor there is

$$\begin{aligned} g_{\pi NN}(q^2) &= 3j_1(qR)/(qR) \\ &= 1 - q^2(0.6R^2)/6\ . \end{aligned} \tag{22}$$

In analogy with the treatment of r_A we can define r_π through the relation

$$g_{\pi NN}(q^2) \equiv 1 - q^2 r_\pi^2/6 + 0(q^4). \tag{23}$$

This gives $r_\pi = 0.77$ R. This should be compared with the value of r_A(=0.62R) calculated from Eq. (21). That is, in the CBM (r_A/r_π) equals 0.80.

Thus the πNN form factor is even softer than the axial form factor. This is because the π-quark interaction is at the surface with $r = R$, whereas the axial vector interaction occurs throughout the bag volume where $r \leq R$.

It is useful to observe that the form factor $3j_1(qR)/(qR)$ has a

very simple interpretation in the Cloudy Bag Model. The pion is in a plane wave with angular momentum one. Hence one has a j_1 function. The pion interacts with quarks only at the bag surface so that the argument of the spherical Bessel function must be qR. The factor 3/qR is included to obtain the normalization $g_{\pi NN}(q^2 = 0) = 1$.

At this stage we recall the infamous discrepancy in the Goldberger-Treiman relation

$$5.97 \pm 0.03 = g_A(0)\, m_N \neq g_{\pi NN}\, f_\pi = 6.36 \pm 0.10 \,, \tag{24}$$

where we have used the measured value of the NNπ coupling constant[47] ($g^2/4\pi = 14.3 \pm 0.4$) at the nucleon pole ($q^2 = -m_\pi^2$). On the other hand the derivation of Eq. (24) requires that one should use $g_{\pi NN}$ at $q^2 = 0$. It is amusing to ask what radius (r_π^{GT}) would give enough variation between $q^2 = -m_\pi^2$ and $q^2 = 0$ to resolve all of this discrepancy. Clearly that would be $(r_\pi^{G-T})^2\, m_\pi^2/6$ equal to 6.1 ± 1.7%, or

$$r_\pi^{G-T} = 0.86 \pm 0.12 \text{ fm.} \tag{25}$$

Using the experimental value for r_A given in Eq. (1), and the CBM ratio $r_A/r_\pi = 0.80$, we find $r_\pi = 0.90 \pm 0.15$ fm. (We hope that the use of the ratio of these quantities will reduce the unknown error associated with centre of mass motion.)

Thus, within the rather large experimental errors, the variation of the NN_π form-factor with q^2 in the Cloudy Bag Model is enough to remove all of the Goldberger-Treiman discrepancy.

Let us summarize briefly. We have calculated the q^2-dependence of the axial-vector form-factor of the nucleon, and the NNπ vertex function, using the Cloudy Bag Model. These calculations use a static bag, without c.m. or recoil corrections. All of these approximations should be improved in future work. Nevertheless our central result is very simple, and probably correct independent of these approximations. That is, the NNπ form-factor is considerably softer than the axial form-factor. (The corresponding r.m.s. radius is at least 25% larger.) If one then uses the experimental axial form-factor, the q^2 dependence of the resulting NNπ form-factor is enough to entirely remove the Goldberger-Treiman discrepancy -- within the rather large experimental uncertainty. At this stage we might remark that the NNπ form-factor obtained in this fashion is rather soft. Clearly it would be very valuable to have more precise data on $g_A(q^2)$, which is of fundamental importance in chiral bag models. Finally we must point out that there may well be additional contributions to the Goldberger-Treiman discrepancy -- see, for example, Refs. 48 - 50. Our point is merely that given the rather large experimental errors at present, these additional corrections are not essential.

VI. SUMMARY

The Cloudy Bag Model is a phenomenological treatment of hadronic structure. Quarks confined in an MIT bag are surrounded by a cloud of pions. The pionic effects are incorporated in such a way that the properties of chiral symmetry and the corresponding consistency with current algebra relations are maintained. As a result one is able to describe a great many phenomena. The root mean square charge radii of nucleons; magnetic moments of baryons; pion-nucleon scattering; strong and electromagnetic decays of vector mesons; and g_A are all computed in agreement with experiment. The discrepancy in the Goldberger-Treiman relation is also explained. To achieve these results a bag radius in the range 0.8 fm to 1.1 fm is required. Thus fairly large bags are needed. This is in agreement with the considerations of the MIT and many other bag models. The interesting implication, in my opinion, is that with such bag sizes nucleons overlap in nuclei, and that one should expect to encounter nuclear phenomena that depends on quark degrees of freedom.

I would like to thank my colleagues, A. W. Thomas and S. Théberge. Without them there would be no Cloudy Bag Model. I'd also like to thank my collaborators on the more recent projects discussed in this talk: G. A. Crawford, P. Guichon, M. Morgan and P. Singer. This work was partially supported by the U. S. Department of Energy.

REFERENCES

1. G. A. Miller, A. W. Thomas and S. Théberge, Phys. Lett. 91B, 192 (1980);
   S. Théberge, A. W. Thomas and G. A. Miller, Phys. Rev. D 22, 2838 (1980); D 23, 2106 (E) 1981;
   A. W. Thomas, S. Théberge and G. A. Miller, ibid 24, 216 (1981);
   A. W. Thomas, Chiral Symmetry and the Bag Model, CERN preprint TH.3368 (1982) and TRIUMF TRI-PP-82-29 (1982), to appear in Advances in Nuclear Physics, Vol. 13, eds. J. Negele and E. Vogt (Plenum, New York, 1983).
   G. A. Miller, S. Théberge and A. W. Thomas, Comm. Nucl. Part. Phys 10 (1981) 101.

2. M. Gell-Mann and M. Lévy, Nuov. Cim. 16 (1960) 53.

3. R. Scheerholtz, this meeting.

4. F. Fucito, G. Martinelli, C. Omero, G. Parisi, R. Petronzio and F. Rapuano, Nucl. Phys. B210, 407 (1982).

5. A. W. Thomas, J. Phys. G7, L283 (1981).

6. A. Szymacha and S. Tatur, Z. Phys. C7, 311 (1981).

7. G. A. Miller and P. Singer, CERN-TH3609, preprint 1983 and to appear in Phys. Lett. B.

8. P. A. M. Guichon, G. A. Miller and A. W. Thomas, Phys. Lett. 124B, 109 (1983).

9. A. Chodos et al., Phys. Rev. D 9, 3471 (1974);
   A. Chodos et al., Phys. Rev. D 10, 2599 (1974);
   T. DeGrand et al., Phys. Rev. D 12, 2060 (1975).

10. G. Breit and R. D. Haracz in "High- Energy Physics" Vol. I., E. H. S. Burhoe ed.) Academic Press Inc., NY 1967.

11. G. E. Brown and M. Rho, Phys. Lett. 82B, 177 (1979);
    G. E. Brown, M. Rho and V. Vento, Phys. Lett. 94B, 383 (1979).

12. T. Inoue and T. Maskawa, Prog. Th. Phys. 54, 1833 (1975);
    A. Chodos and C. B. Thorn, Phys. Rev. D 12, 2733 (1975).

13. M. Morgan, Univ. of Washington Ph.D. Thesis, 1984; and M. Morgan and G. A. Miller, to be published.

14. W. Weise, this meeting.

15. T. J. Goldman and R. W. Haymaker, Phys. Rev. D 24, 724 (1981).

16. A. W. Thomas, S. Théberge and G. A. Miller, Phys. Rev. D 24, 216 (1981);
    C. DeTar, Phys. Rev. D 24, 752 (1981); D 24, 762 (1981).

17. R. L. Jaffe in "Pointlike Structures inside and Outside Hadrons", ed. A. Zrchichi, Plenum Press NY 1982.

18. G. A. Crawford and G. A. Miller, CERN TH3645, preprint 1983 and to appear Phys. Lett. B.

19. S. Théberge, A. W. Thomas and G. A. Miller, Phys. Rev. D 22, 2838 (1980); D 23, 2106 (E) (1981).

20. G. F. Chew, Phys. Rev. 94, 1748 (1954);
G. F. Chew, Phys. Rev. 94, 1755 (1954);
G. F. Chew and F. E. Low, Phys. Rev. 101, 1570 (1956);
G. C. Wick, Rev. Mod. Phys. 27, 339 (1955).

21. G. A. Miller, in Meson-Nuclear Physics 1979, ed. E. V. Hungerford III (AIP, New York, 1979), p. 561, AICP 54;
G. A. Miller and E. M. Henley, Ann. Phys. (NY) 129, 131, (1980).

22. M. Gell-Mann and Y. Ne'eman, The Eightfold Way (Benjamin, New York, 1964).

23. S. Théberge, G. A. Miller and A. W. Thomas, Can. J. Phys. 60, 59 (1982).

24. S. Théberge, Univ. of B.C. Ph.D. Thesis, 1982.

25. F. Myhrer, G. E. Brown and Z. Xu, Nucl. Phys. A362, 377 (1981).

26. P. Mulders and A. W. Thomas, CERN preprint TH.3443, to be published, J. Phys. G.

27. S. Galster, H. Klein, J. Mortiz, K. H. Schmidt, D. Wegener and J. Bleckwenn, Nucl. Phys. B32, 221 (1971).

28. R. Maxwell, V. Vento, Nucl. Phys.;
F. Close, Nucl. Phys.

29. J. F. Donoghue and K. Johnson, Phys. Rev. D 21, 1975 (1980).

30. S. Théberge and A. W. Thomas, Nucl. Phys. A383, 252 (1983).

31. J. O. Eeg and H. Pilkuhn, Z. Physik A287, 407 (1978).

32. G. E. Brown, M. Rho and V. Vento, Phys. Lett. 97B, 423 (1980).

33. B. L. Roberts et al., Phys. Rev. D 20, 2154 (1979);
G. Dugan et al., A 254, 396 (1975);
T. Hansl et al., ibid. B 132.

34. R. Welsh et al., private communication and to be published.

35. S. Adler, R. Dashen, Current Algebra, Benjamin, NY (1968).

36. S. Weinberg, Phys. Rev. Lett. 17, 616 (1966); Phys. Rev. Lett. 18, 188 (1967);
Y. Tomazowa, Nuovo Cimento 46A, 707 (1966).

37. K. Kawarabayashi and M. Suzuki, Phys. Rev. Lett. 16, 255 (1966); 384 (E);
Riazuddin and Fayyazuddin, Phys. Rev. 147, 1071 (1966).

38. C. W. Wong, Phys. Rev. D 24, 1416 (1981).

39. R. E. Peierls and J. Yoccoz, Proc. Phys. Soc. 70, 381 (1957).

40. Particle Data Group, Phys. Lett. 111B, 1 (1982).

41. F. A. Constanzi, Phys. Rev. 182, 1571 (1969).

42. P. J. O'Donnell, Rev. Mod. Phys. 53, 673 (1981).

43. G. A. Miller and P. Singer, to be published.

44. E. Amaldi, S. Fubini and G. Furlan, Pion-electroproduction at low energy and hadron form factors, Springer Tracts in Modern Physics, Vol. 83 (Springer-Berlin, 1979).

45. J. Bell et al., Phys. Rev. Lett. 41, 1008 (1978); ibid 1012.

46. S. A. Chin and G. A. Miller, Phys. Lett. 121B, 232 (1983).

47. M. M. Nagels et al., Nucl. Phys. B147, 189 (1979); G. Hohler et al., Phys. Data 12-1 (1979).

48. H. Pagels, Phys. Rep. 16, 219 (1975).

49. C. A. Dominguez, Phys. Rev. D 25, 1937 (1982).

50. S. A. Coon and M. D. Scadron, Phys. Rev. C 23, 1150 (1981).

Recent Developments in Chiral Bag Models

J. Wambach

Institut für Kernphysik
Kernforschungsanlage Jülich
D-5170 Jülich, West Germany

1. Introduction

Although not yet derivable from QCD, Chiral Bag Models (CBM) are believed to represent the main low energy features of the theory. As now confirmed by exact solutions on a finite space-time lattice[1)] QCD undergoes at least two phase transitions, such that the physical world has little to do with the perturbative solutions of QCD. The two transitions are

(i) the "confinement phase transition" in which the system of quarks and gluons switches from the stronly interacting situation in which physical states are color singlets to a 'perturbative situation' in which the mutual interactions are weak

(ii) the 'PCAC-phase transition' associated with a global SU(N) x SU(N) chiral symmetry in N flavours. In one phase this symmetry is unbroken in the ground state (Wigner-Weyl phase) while in the other spontaneous symmetry breaking occurs (Goldstone phase). This phase is accompanied by an N^2-1 plet of massless pseudoscalar bosons.

In CBM this two phase structure of the vacuum is implemented ad hoc by deviding the physical space into two regions each of which represents a different phase (usually it is assumed that the confinement - and PCAC transition occur simultaneously). Chiral invariance and colour confinement are imposed via boundary conditions on the elementary fields in the two space-time regions.

In the first part of my talk I am going to give an overview of the CBM-version that has been developed by the Stony Brook-Saclay group[2,3,4,5)]. Since I was not actively involved in most parts of this work my report may not be free from misinterpretations and I apologize in advance for possible misquotes and errors. First the basic CBM-ideas are repeated by introducing an effective Lagrange density which defines the dynamics. Next a set of exact solutions is discussed in which spin and isospin are combined into a hybridized operator ('hedgehog solutions')[5)]. These solutions show instability towards collapse unless an extra stabilizing term is added. Such a term has been introduced by Skyrme[6)] in studiing a nonlinear chiral invariant field theory. The equations of motion of this theory allow for soliton solutions which obey Fermi statistics. The features of these solutions will be discussed in the 'hedgehog'-ansatz in sect. 4. Recently attempts have been made[4)] to incorporate the soliton picture in the CBM. The consequences, especially baryon number fractionalization are briefly discussed in sect. 5.

The second part (sect. 6) of the talk reports on work together with Z.Y. Ma in which we tried to go beyond the spherical solutions of the CBM equations. With the full nonlinearities this is a formidable task. So we first tried to see what a pertubation expansion around the free fields would give. Retaining axial-symmetry in the bag shape we have determined the equilibrium deformation for nucleons and isobars to second order in the quark-pion coupling.

2. The effective CBM-Lagrangian

Bag models attempt to describe QCD phenomena in the low energy and low momentum regime. With a few parameters they have been quite successful in hadron spectroscopy[7] and low energy dynamics such as pion-nucleon scattering[8]. The formation of a 'bag' is not followed dynamically from the QCD-equations but postulated by adopting two space-time regions V^+ and V^- (fig. 1)

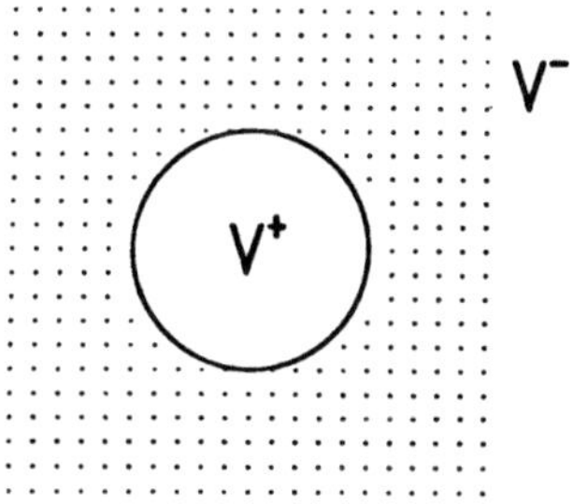

Fig. 1: Two phases of the QCD vacuum, spacially separeted to form a bag. V^+ denotes the 'perturbative' and V^- the 'nonperturbative' vacuum.

In the perturbative region V^+ quarks (usually massless) and gluons are quasifree, i.e.

(i) quarks are free inside V^+ but not allowed to exit V^+ via a boundary condition on the bagsurface δV. This prevents color leakage from the bag.

(ii) gluons behave like free electromagnetic fields inside a cavity V^+ where the boundary conditions are those of a color superconductor in V^- with E and B fields reversed.

(iii) the quark-gluon coupling is perturbative, i.e. of order α_s where α_s is the color fine structure constant.

In the 'non-perturbative' region V^- chiral symmetry (usually SU(2) x SU(2) is considered only) is spontaneously broken giving rise to non linear selfinteracting pion fields. In the limit of an exact symmetry the pions are massless. At the bag boundary the quark- and pion fields are coupled to ensure confinement, continuity of the axial current and pressure balance.

The total Lagrange density is therefore given by three terms

$$\mathcal{L}(x) = \mathcal{L}_q(x) + \mathcal{L}_\pi(x) + \mathcal{L}_{int}(x) \tag{2.1}$$

$\mathcal{L}_q$ describes the quark-gluon dynamics in V^+. Denoting quark fields by ψ and the eight color charged gluon vector fields by A_a^μ (a: 1···8) $\mathcal{L}_q$ is given by[9]

$$\mathcal{L}_q = 1/2(\bar{\psi}\not{\partial}\psi - \mathcal{B} - 1/4 F_a^{\mu\nu} F_{a\mu\nu} - \sqrt{4\pi}\alpha_c \bar{\psi}\not{A}_a \lambda^a \psi)\Theta_{V+} - 1/2\bar{\psi}\psi\, n^\mu n_\mu \delta_V \tag{2.2}$$

($\not{\partial} \equiv \partial^\mu \gamma_\mu$ etc, n^μ denotes the outward surface normal and δ_V a δ-function at the bag surface. λ^a are the 8 SU(3)-colour matrices). $\mathcal{B}$ denotes the bag constant. To order α_c the field strength tensor $F_a^{\mu\nu}$ is linear i.e. one has as in ordinary QED

$$F_a^{\mu\nu} = \partial^\mu A_a^\nu - \partial^\nu A_a^\mu \tag{2.3}$$

In V^- acts the classical nonlinear pion-Lagrangian (non-linear σ-model[10])

$$\mathcal{L}_\pi = -1/2\{(1 + 1/f_\pi^2 \vec{\phi}^2) D_\mu \vec{\phi} D^\mu \vec{\phi} - 1/(4f_\pi)^2 (D_\mu \vec{\phi})^2\}\Theta_{V^-} \tag{2.4}$$

(the arrow on $\vec{\phi}$ denotes a vector in isospin space). Here D_μ is a nonlinear derivative

$$D_\mu (1 + 1/f_\pi^2 \vec{\phi}^2)^{-1} \partial_\pi \tag{2.5}$$

Quark and pion-fields are coupled at the surface to give

$$\mathcal{L}_{int} = \{1/(2f_\pi) \overline{\psi \vec{\phi} \vec{\tau}} \gamma^5 \psi (1 + 1/f_\pi{}^2 \vec{\phi}^2)^{-1/2}\} n^\mu n_\mu\, \delta_V \tag{2.6}$$

The action

$$S = \int \alpha^4 x\ (x) \tag{2.7}$$

has to be stationary ($\delta S = o$). Varying with respect to ψ, A_a^μ and $\vec{\phi}$ leads to three equations of motion.

$$i\not{\partial}\psi = \sqrt{4\pi}/2\ \alpha_c \not{A}_a \lambda^a \psi \qquad \text{in } V^+ \tag{2.8a}$$

$$\partial^\mu Fa_{\nu\mu} = -\sqrt{4\pi}/2\alpha_c \bar{\psi}\gamma_\nu \lambda_a \psi \qquad \text{in } V^+ \tag{2.8b}$$

$$D_\mu D^u \vec{\phi} = 0 \qquad \text{in } V^- \tag{2.8c}$$

and three boundary conditions

$$i\not{n}\psi = (1 + 1/f_\pi^2\ \vec{\phi}^2)^{-1/2}\ [1 \hat{+} i/f_\pi \vec{\phi}\vec{\tau}\gamma^5] \tag{2.9a}$$

$$n\mu F_a^{\mu\nu} = 0 \tag{2.9b}$$

$$n^\mu \partial_\mu\ \vec{\phi} = -1/(2f_\pi)(1 + 1/f_\pi^2\ \vec{\phi}^2)\ \bar{\psi}\gamma^5 \vec{\tau}\vec{\phi}\psi \tag{2.9c}$$

The last boundary condition ensures continuity of the axial current. Finally, varying S with respect to the bag boundary leads to an extra nonlinear boundary condition which has to be satisfied locally at each point on the surface. I will not give this 'pressure balance equation' here but refer for the explicit form to ref. 9.

3. Non perturbative solutions to the CBM-equations

Exact solutions to the CBM-equations (2.8a)-(2.9c) are very complicated and difficult to obtain. As a first approximation gluon fields are usually neglected and later added as a perturbation correction to the quark-pion solutions. Even then non-perturbative solutions for the ground state are only available sofar in the static 'hedgehog' ansatz for the quark- and pion fields[9]. In this ansatz $\vec{\phi}$ is proportional to the unit vector in radial direction and characterized by a single dynamical variable, the chiral angle Θ which depends on the radial coordinate only

$$\vec{\phi}(\underset{\sim}{r}) = f_\pi \tan \Theta(r)\, \hat{r} \tag{3.1}$$

Note that the isospin always points in the radial direction (hence a 'hedgehog'). A similar artificial coupling between ordinary-, spin- and isospin space has to be introduced in the quark sector. One defines a combined operator $\vec{K}$ as the sum of the total angular momentum operator $\underset{\sim}{J} = \underset{\sim}{L} + \underset{\sim}{S}$ and the isospin operator $\vec{T}$

$$\vec{K} = \underset{\sim}{J} + \vec{T} \tag{3.2}$$

The quark 'hedgehog' solutions can be classified according to the 'good' quantum numbers K^2, K_z. For the ground state $K^2=K_z=0$ the quark solution of the SU(2)xSU(2)-CBM equations is spherically symmetric. The single valence quark wave function is given by

$$\psi_0 = N/\sqrt{4\pi} \begin{pmatrix} i j_0(\omega_0 r) \\ -j_1(\omega_0 r)\underset{\sim}{\sigma}\cdot\hat{r} \end{pmatrix} v e^{-i\omega_0 t} \tag{3.3}$$

were N is a normalization constant[2] and ω_0 the quark energy (quarks are massless). Since the orbital angular momentum is zero the 'combined Pauli spinor' v is determined by the condition

$$(\underset{\sim}{\sigma} + \vec{\tau})\, v = 0 \tag{3.4}$$

Therefore v is a linear combination of u and d quarks with spin up ↑ and spin down ↓

$$v = 1/\sqrt{2}(d\uparrow - u\downarrow) \tag{3.5}$$

The equation of motion for the 'hedgehog'-pion field in V^- leads to a pendulum-like nonlinear differential equation in the chiral angle

$$\ddot{\Theta} + 2/r\dot{\Theta} - 1/r^2 \sin 2\Theta = 0 \tag{3.6}$$

Here the dot stands for differentiation with respect to r. Asymptotically

$$\lim_{\to\infty} \Theta \propto 1/r^2 \qquad (3.7)$$

i.e. the Yukawa limit for massless pions is obtained. The two linear boundary conditions (2.9a) and (2.9c) also acquire simple forms. Eq. (2.9a) leads to

$$\tan \Theta(r)\big|_{r=R} = (1 - y^2)/2y \qquad (3.8a)$$

were y denotes the ratio of the small to large component of the quark wave function (3.3) at the boundary r=R

$$y = j_1(\Omega_o)/j_o(\Omega_o) = \cos\Theta/(1+\sin\Theta)\big|_R \qquad (3.8b)$$

($\Omega_o = \omega_o R$). Continuity of the axial current gives[11)]

$$8\pi\ f_\pi^{\ 2}\ R^3\ \ \dot{\Theta}\sin\Theta = -3g_o \qquad (3.9a)$$

were

$$g_o = (1 - y^2)/(1 + y^2 - 2y/\Omega_o) \qquad (3.9b)$$

In the MIT-bag without pions[12)] y=1 which leads to Ω_o = 2.04 independent of the bag radius R. As the pion quark coupling is turned on however y can vary with R (eq. (3.8a)) and therefore the quark energies also depend on the bag size (upper part of fig. 2).

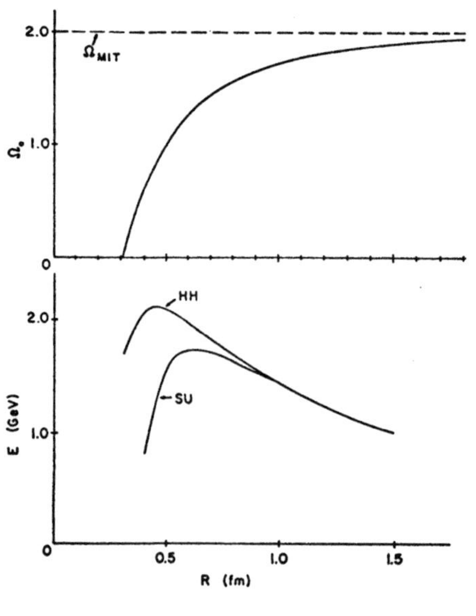

Fig. 2: The parameter Ω_o (upper part) for the lowest energy quark wave function (hedgehog). For large radius R Ω approaches the MIT value. The lower part gives the CBM-groundstate energy as a function of R. HH denotes the hedgehog and SU the lowest order result from a linearized CBM.

For later discussions the point $\Omega_0 = 0$ is of particular interest. At this (finite) bag radius R_c the small component vanishes and hence, the quark motion becomes completely non relativistic. The chiral angle is at this point 90^0.

The ground state energy of the hedgehog bag can be obtained in a closed form[5])

$$E_{hh}(R) = n\Omega_0/R+4\pi/3R^3\mathcal{B}+2f_\pi^2(R^3\dot{\Theta}^2-2R\sin^2\Theta). \tag{3.10}$$

The first term denotes the quark kinetic energy (n is the number of quarks in the bag), the second term is the volume energy associated with the formation of a bag and the last term contains the quark-pion interaction energy and the kinetic energy of the pion field in V^-. At R_c the quark kinetic energy vanishes. Expression (3.10) clearly shows the non perturbative character of the hedgehog solution. At fixed R $E_{hh}(R)$ grows with the inverse of the square of quark-pion coupling constant (fig. 3) $g = 1/f_\pi$ while perturbation theory gives a quadratic dependence of E on g.

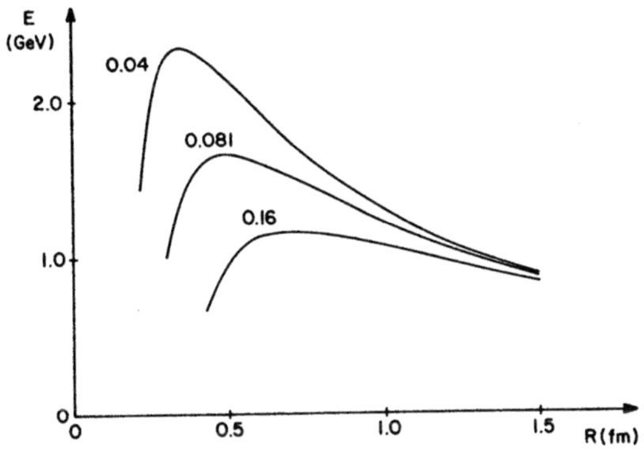

Fig. 3: CBM-groundstate energy as a function of R for various values of the inverse coupling constant $g^{-1} = f_\pi/4_\pi$

The behaviour of E_{hn} at small R (lower part of fig. 2) shows instability towards collapse, which it has in common with perturbative expansions of $\mathcal{L}$. (the pertubative E(R) is denoted by SU in fig. 2). The small R behaviour becomes more transparent if the pressure balance equation, which gives a relation between B and Θ, is utilized. One finds the simplified expression[11])

$$E_{hh}(R) = 2n\Omega_0/R - 8\pi/3\ R^3\mathcal{B}. \tag{3.11}$$

Since below R_c Ω_0 becomes more and more negative E_{hh} decreases without bound as R approaches zero.

4. Stability of the pion Lagrangian: the 'Skyrmion'

The instability of the CBM-groundstate energy is an intrinsic problem of $\mathcal{L}_\pi$ as it is specified in (2.4) and (2.5). This has been realized by Skyrme about twenty years ago[6]). He proposed stabilization by adding an extra terms to $\mathcal{L}_\pi$ such that it supports stable solitonic solutions in the physical 3+1-dimensional space. To discuss Skyrme's ideas let me switch from the nonlinear - to the linear σ-model by the transformation

$$\sigma(x) = f_\pi \quad (1+\vec{\phi}(x)/f_\pi{}^2)^{-1/2} = f_\pi \cos\Theta \tag{4.1a}$$

$$\vec{\pi}(x) = \vec{\phi}(x) \; (1+\vec{\phi}(x)/f_\pi{}^2)^{-1/2} = f_\pi \sin\Theta\hat{\pi} \tag{4.1b}$$

($\hat{\pi}$ denotes the unit vector in isospin space)
which holds in the case that the σ-mass goes to ∞. In this case

$$\vec{\pi}^2(x) + \sigma^2(x) = f_\pi{}^2 \tag{4.2}$$

and Θ denotes the position of the system on the chiral circle (fig. 4).

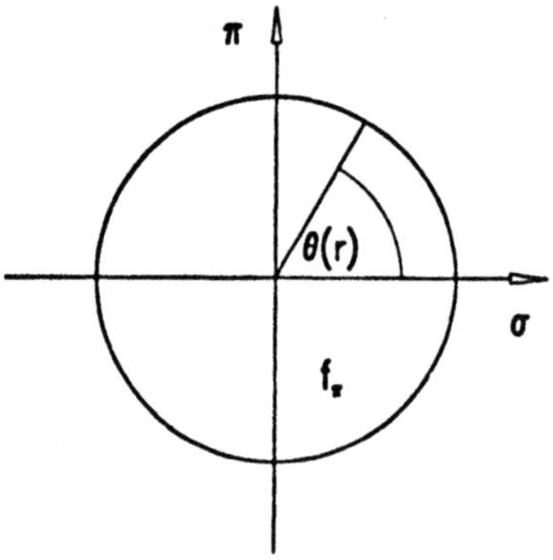

Fig. 4: The 'magic circle' for infinitely massive σ-fields. Θ denotes the chiral angle.

For small Θ, i.e. at large distances $\sigma \approx f_\pi$ and $\pi \approx f_\pi\, \Theta$ i.e. $\pi \approx f_\pi\; 1/r^2\; \hat{\pi}$ which is the Yukawa limit. In terms of a unitary field

$$U(x) \equiv 1/f_\pi \;(\sigma(x) + i\; \vec{\tau}\,\vec{\pi}\,(x)\,); \quad U^+U = 1 \tag{4.3a}$$

and the derivative

$$L_\mu(x) \equiv U^+(x)\; \partial_\mu\; U(x) \tag{4.3b}$$

$\mathcal{L}_\pi$ has a very simple form

$$\mathcal{L}_\pi = - f_\pi^2/4 \;\; \mathrm{Tr}\; [L_\mu, L_\mu] \tag{4.3c}$$

A possible way to obtain solitons in 3-space dimensions is to add to $\mathcal{L}_\pi$ higher powers in L_μ. Skyrme proposed a quartic term

$$\mathcal{L}_4 = - \varepsilon^2/4 \;\mathrm{Tr}\; [L_\mu,\, L_\nu]^2 \tag{4.4}$$

where ε is a phenomenological constant. In the 'hedgehog' ansatz for the static case

$$U(\underline{r}) = \exp\;(i\; \vec{\tau}\; \hat{r}\; \Theta(r)) \tag{4.5}$$

the Euler equations reduce again to a nonlinear second order differential equation in the chiral angle[3)]

$$\ddot{\Theta}(1+Ae^{-2\tau}\sin^2\Theta) + \dot{\Theta}(1-Ae^{-2\tau}\sin^2\Theta)$$
$$- \sin 2\Theta\{1-1/2Ae^{-2\tau}(\dot{\Theta}-\sin^2\Theta)\} = 0 \quad (4.6)$$

were A is defined as $(16\varepsilon^2/(f_\pi^2 r_o^2)$ with r_o being an arbitrary scale length. The variable τ is defined as $\tau \equiv \ln(r/r_o)$. For ε=o (4.6) is identical to (3.6) as it should be. The full solution Θ_{SK} (fig. 5) which has to be obtained numerically approaches at large distances the solution without quartic coupling (Θ_{hh} in fig. 5).

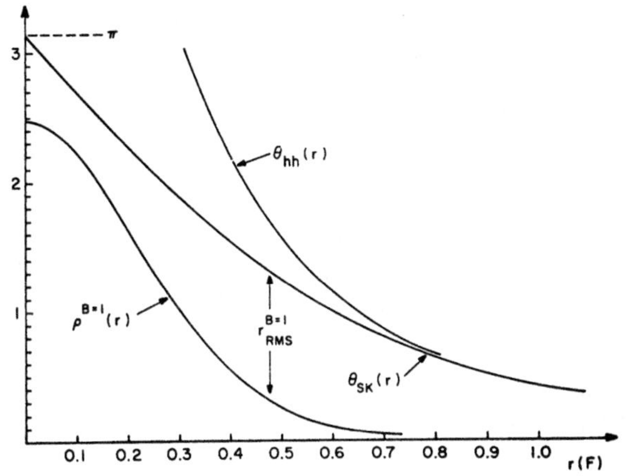

Fig. 5: Chiral angle for the soliton solution of $\mathcal{L}_{SK}$ (Θ_{SK}) and the ordinary hedgehog of the nonlinear σ-model (Θ_{hh}) as a function of r. The baryon number density $\rho^{\mathcal{B}}$ in arbitrary units is also given.

As r goes to 0 $\Theta_{hh}(r)$ diverges as 1/r as can be concluded also from the r-dependence of $E_{hh}(r)$ in (3.11) and the relation to the chiral angle Θ (eq. (3.10)). Θ_{SK} however stays finite and at r=0 is exactly a multiple of π

$$\Theta_{SK}(0) = n\pi \qquad n = 1,2,\cdots\cdots \quad (4.7)$$

Skyrme furthermore showed, that n has a physical meaning. He postulated the baryon current corresponding to the full Lagrangian $\mathcal{L}_{SK} = \mathcal{L}_\pi + \mathcal{L}_4$ to be given by

$$B_\mu \equiv 1/48\pi^2 \, \varepsilon_{\mu\nu\lambda\rho} \, \mathrm{Tr}\,[[L_\mu, L_\lambda] L_\rho] \quad (4.8)$$

($\varepsilon_{\mu\nu\lambda\rho}$ is the antisymmetric ε-tensor)
with the conserved baryon number B

$$B = \int B_o(x)d^3x = \int \rho^{\mathcal{B}}(r)r^2\,dr \quad (4.9a)$$

Here $\rho^{\mathcal{B}}$ is the baryon number density[3)]

$$\rho^{\mathcal{B}}(r) = -1/(2\pi^2)\ \sin^2 \Theta/r^2\ \dot{\Theta} \tag{4.9b}$$

displayed in fig. 5. From relations (4.7) and (4.9) one can now prove that

$$n = B \tag{4.10}$$

i.e. the soliton solutions ('Skyrmions') can be classified according to a conserved quantity, the baryon number. The miracle has happened, that out of interacting nonlinear boson fields stable fermions have been created. (That in fact the Skyrmions are fermions has been proven by several authors[13,14)]).

Skyrme's model has interesting consequences for low energy physics especially many body forces as discussed by Jackson and Rho[3)]. To compute physical properties of the Skyrmion, like size, mass etc. the coupling constant ε has to be specified. Since $\mathcal{L}_{SK}$ is chiral invariant the Goldberger-Treimann relation can be used. This gives $\varepsilon = 5.52 \times 10^{-3}$ a value consistent with limits from $\pi\pi$ scattering[14)]. The baryon energies E_B obtained with this value which are listed in table 1

B	1	2	3
E_B	1.425	4.250	8.445

table 1: Skyrmion energies in [GeV] with baryon number 1,2 and 3 as calculated in ref. 3.

allow to estimate the soliton-soliton interaction at zero seperation. (Note that the Skyrmion is still a hedgehog, i.e. a linear combination of nucleons and isobars). For two solitons (B=2) the interaction energy

$$V_2(o) = E_{B=2} - 2E_{B=1} \tag{4.11}$$

is equal to 1.43 GeV, i.e. one obtains strong short range repulsion. For three solitons at zero separation the net interaction energy

$$V_3(o) = E_{B=3} - 3E_{B=1} - 3V_2(o) \tag{4.12}$$

is very small (-.12 GeV) indicating that 3-body forces are weak. Another encouraging result of Skyrme's model concerning magnetic moments has been reported recently by Adkins et al[15)] who have quantized $\mathcal{L}_{SK}$ and obtain the physical states, i.e. proton, neutron Δ^+ etc. rather than a mixture. One of the problems of pertubative CBM has always been, that for small bagsizes most of the magnetic moment comes from the pion cloud[16)]. This is purely isovector such that μ_p/μ_n goes to -1 in contradiction to experiment. Isoscalar contributions had to be put in by hand to fit nonstrange- and strange baryon moments[17)]. Adkins et al. obtain the markable result that in Skyrme's model $\mu_n/\mu_p = -.7!$ $(\mu_n/\mu_p{}^{expt} = -.69)$. Apparently nonlinear effects provide in a consistent way the isoscalar part missing in the linearized pion cloud.

5. A topological soliton bag model

Skyrme's model creates baryons out of interacting nonlinear fields, i.e. the QCD-fields have been 'integrated out', to describe the infrared behaviour of QCD by effective fields. Probing hadrons deeply inside deep inelastic lepton scattering however shows scaling behaviour and thus reveals bound quasi free quarks and gluons An attempt to marry these ultraviolet features with the infrared behaviour summarized by $\mathcal{L}_{SK}$ has been made recently by Rho et al.[4]. The idea is to introdue a (spherically symmetric) defect V^+ into the soliton field U. In V^+ quarks and gluons exist quasi free and interact only weakly. This 'soliton bag model' (SCBM) has very interesting consequences on the baryon number. It fractionalizes in V^+ and V^- leaving the total B conserved. Let me try to devellop the argument. The SCBM-Lagrangian is the usual CBM-Lagrangian (2.2) - (2.6) suplemented by the quartic term $\mathcal{L}_4$ (eq. (4.4)). Expecially the boundary conditions stay formally the same, and in the hedgehog Θ_{hh} is simply replaced by Θ_{SK} obtained from the equations of motion of $\mathcal{L}_{SK}$ in V^-. For the k=0 quark hedgehog we have in analogy to (3.8a)

$$\tan \Theta_{SK}(r)/_R = 1- y^2/2y \qquad (5.1)$$

($y = j_1(\Omega_o)/j_o(\Omega_o)$). For simplicity let the defect radius be such that $\Theta_{SK}(R) = \pi/2$ (the conclusions will hold for any R). This choice is certainly valid since $0<R<\infty$ implies that $n\pi > \Theta_{SK}(R) > o$ (see fig. 5). As discussed in sect. 3 at $\Theta_{SK} = \pi/2$ the quark kinetic energy is zero and the small component of the quark wave function identically vanishes inside the defect. The large component on the other hand spreads uniformly over the volume V^+. Thus the wave function is particularly simple

$$\psi_o(r) = N/\sqrt{2} \left[\begin{pmatrix}1\\o\\o\\o\end{pmatrix}_\sigma \begin{pmatrix}o\\1\end{pmatrix}_\tau - \begin{pmatrix}o\\1\\o\\o\end{pmatrix} \begin{pmatrix}1\\o\end{pmatrix}_\tau \right] \qquad r \leq R \qquad (5.2a)$$

$$\psi_o(r) = 0 \qquad r > R \qquad (5.2b)$$

with $N^2 = 3/4\pi\ R^{-3}$. The four component spinor corresponds to the Dirac spinor and the two component spinor to isospin (see also (3.5)). Such a wave function is selfconjugate, i.e. under charge conjugation $C = \begin{pmatrix}\sigma_2 & o\\ o & \sigma_2\end{pmatrix}\tau_2$

$$C\psi_o = \psi_o \qquad (5.3)$$

(note that under charge conjugation a state of positive energy ω goes over to a state of negative energy-ω) Jackiw and Rebbi[8] showed that each quark in a 'zero mode' carries fractional baryon number

$$B_q = 1/2\ N_c^{-1} \qquad (5.4a)$$

where N_c is the number of colours. For $N_c = 3$ $B_q = 1/2$, i.e. the quarks in the defect carry only half the baryon number of the full object. The other half (B_{SK}) has to be in the soliton cloud in order B to be conserved. To see this one has to realize that B_{SK} has a very simple relationship to Θ_{SK}[11])

$$B_{SK} = 1/\pi \int_{r=\infty}^{r=R} (1-\cos 2\Theta)d\Theta$$
$$= 1/\pi \{\Theta(R) - \Theta(\infty) - 1/2\ (\sin 2\Theta(R) - \sin 2\Theta(\infty))\} \tag{5.5}$$

Since $\Theta(R) = \pi/2$ and $\Theta(\infty) = 0$ $B_{SK} = 1/2$ such that

$$B = B_q + B_{SK} = 1 \tag{5.6}$$

as it has to be.

6. A deformable chiral bag

In order to be able to solve the full CBM- or SCBM Euler equations the geometry of the bag has to be spherical. Therefore these theories so far do not tell us anything about possible deformations of hadrons. Such deformations especially for nucleons and isobars have however been anticipated[19]) in connection with the axial vector coupling constand g_A which is usually to large in a spherical CBM especially with pions expelled from the bag interiour. The γ_5-coupling between quarks and pions at the boundary produces an inhomgeneous pressure on the bag which is largest along the direction of the spin and zero perpendicular to it. From both g_A and the (D+F)/(D-F) ratio one predicts large oblate deformations[20]). Whether or not such deformations are present in the full CBM- or SCBM solutions can not be ansered up to now because of the numerical difficulties. A question, that can be answered however, is whether the linearized CBM deforms. In the following I wish to convince you, that the answer is yes. For arbitary bag shapes R (ϑ,φ) the linearized static CBM-equations to be solved are

$$\gamma_\mu \partial_\mu \psi = 0 \qquad r < R(\vartheta,\varphi) \tag{6.1a}$$

$$\partial^2 \vec{\pi} = 0 \qquad r > R(\vartheta,\varphi) \tag{6.1b}$$

$$\underset{\sim}{n}\underset{\sim}{\gamma}\psi = (1+i\gamma_5 \vec{\tau}\vec{\pi}/f_\pi)\psi \qquad r = R(\vartheta,\varphi) \tag{6.1c}$$

$$\underset{\sim}{n}\underset{\sim}{\nabla}\vec{\pi} = i\bar{\psi}\gamma_5\vec{\tau}/(2f_\pi)\psi \qquad r = R(\vartheta,\varphi) \tag{6.1d}$$

The outward surface normal $\underset{\sim}{n}$ is angular dependent. To obtain solutions I will proceed similarly as Chin[21]).

He suggests to use the free field solutions in the absence of coupling at the boundary to compute matrix elements of the linearized interaction term (2.6)

$$\mathcal{H}_{int} = -\mathcal{L}_{int} = i/(2f_\pi)\bar{\psi}\gamma_5\vec{\tau}\vec{\pi}\psi\ \delta(r-R(\vartheta,\varphi)) \tag{6.2}$$

in perturbation theory. In the decoupled case ($f_\pi \to \infty$) the quark field ψ obeys the familiar MIT boundary condition

$$\underset{\sim}{n}\underset{\sim}{\gamma}\psi = \psi \qquad\qquad r < R(\vartheta,\varphi) \tag{6.3a}$$

instead of (6.1c) and the pion field

$$\underset{\sim}{n}\underset{\sim}{\nabla}\pi = 0 \qquad\qquad r > R(\vartheta,\varphi) \tag{6.3b}$$

instead of (6.1d). With those solutions the second order groundstate energy shift $\Delta E_0^{(2)}$ is given by

$$\Delta E_0^{(2)} = 1/(2f_\pi^2)\int_{-\infty}^{\infty} dt\ d\underset{\sim}{r}_1\ d\underset{\sim}{r}_2\ \Delta(\underset{\sim}{r}_1,\underset{\sim}{r}_2,t)$$

$$\langle\phi_0|\bar{\psi}(x_1)i\gamma_5\vec{\tau}/2\psi(x_1)\bar{\psi}(x_2)i\gamma_5\vec{\tau}/2\psi(x_2)\delta(r_1-R)\delta(r_2-R)|\phi_0\rangle \tag{6.4}$$

$$t = t_1-t_2$$

The pion propagator is given in terms of the free pion wavefunctions as

$$\Delta(\underset{\sim}{r},\underset{\sim}{r}',\omega) = \int_{-\infty}^{\infty} dt\ e^{i\omega t}\ \Delta(\underset{\sim}{r},\underset{\sim}{r}',t) = \int d\vec{k}\ \frac{\phi^*_{\vec{k}}(\vec{\underset{\sim}{r}})\phi_{\vec{k}}(\vec{\underset{\sim}{r}}')}{\omega^2-k^2+i\eta} \tag{6.5}$$

where $\phi_{\vec{k}}$ are solutions to the Helmholtz equation corresponding to (6.1b)

$$(\underset{\sim}{\nabla}^2 + k^2)\phi_{\vec{k}}(\vec{r}) = 0 \qquad\qquad r > R(\vartheta,\varphi) \tag{6.6a}$$

with Neumann boundary condition

$$\underset{\sim}{n}\cdot\underset{\sim}{\nabla}\ \phi_{\vec{k}}(\vec{r}) = 0 \qquad\qquad R = R(\vartheta,\varphi). \tag{6.6b}$$

For arbitary bag shapes $R(\vartheta,\varphi)$ it is very complicated to solve the free field equations. One only gets numerical solutions[22].

Restriction to ellipsoidal shapes however simplifies the problem and the solutions can be obtained mostly analytically. An ellipsoid is characterized by two parameters which are chosen as R and D such that the two axes are RD and R/√D respectively (fig. 6).

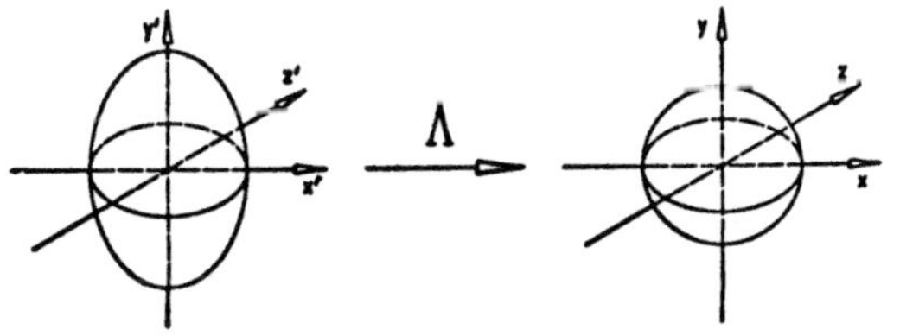

Fig. 6: Nonlinear scale transformation which transforms an ellipsoid in the primed coordinate system into a sphere in the unprimed system.

D < 1 corresponds to oblate and D > 1 to prolate deformation. To solve the Dirac equation is such a cavity Viollier et al[23] have proposed a very useful method which will be employed here.

First the ellipsoid is mapped onto a sphere by a nonlinear scale transformation from the primed to the unprimed coordinate system (fig. 6). In Cartesian coordinates this transformation has the representation

$$\Lambda = \begin{pmatrix} 1/\sqrt{D} & 0 & 0 \\ 0 & 1/\sqrt{D} & 0 \\ 0 & 0 & D \end{pmatrix} \tag{6.7}$$

Λ is volume conserving such that the bag energy $\mathcal{B}V^+$ is unchanged. In the primed system the boundary conditions (6.3a) and (6.3b) become spherical, however the quark and pion hamiltonian are now anisotropic. The Dirac equation (6.1a) becomes (primes are omitted for convenience)

$$-i\left\{A_q(D)\underset{\sim}{\alpha}\cdot\underset{\sim}{\nabla} + B_q(D)\left[\underset{\sim}{\alpha}\times\underset{\sim}{\nabla}\right]_{20}\right\}\psi(\underset{\sim}{r}) = \omega\psi(\underset{\sim}{r}) \tag{6.8a}$$

$$A_q(D) = 1/3(2\sqrt{D} + 1/D) \tag{6.8b}$$

$$B_q(D) = \sqrt{2/3}\,(D^{-1}-D^{1/2}) \tag{6.8c}$$

and the Helmholtz equation (6.1b) transforms to

$$\left\{A_\pi(D)\underset{\sim}{\nabla}^2 + B_\pi(D)\left[\underset{\sim}{\nabla}\times\underset{\sim}{\nabla}\right]_{20} + k^2\right\}\phi_k(\vec{r}) = 0 \tag{6.9a}$$

$$A_\pi(D) = 1/3\,(2D+D^{-2}) \tag{6.9b}$$

$$B_\pi(D) = \sqrt{2/3}\,(D^{-2}-D) \tag{6.9c}$$

Sphericity of the boundary conditions on the free fields implies that the solutions for the scalar parts of (6.8) and (6.9) are known. They are the MIT-solutions for positive and negative energy quark states[12)]

$$u^o_{\nu k\mu}(\underset{\sim}{r}) = \begin{Bmatrix} g_{\nu k}(r) \\ -if_{\nu k}(r)\underset{\sim}{\sigma}\cdot\hat{r} \end{Bmatrix} y^\mu_{j\ell 1/2}(\hat{r}) \tag{6.10a}$$

$$v^o_{\nu k\mu}(\underset{\sim}{r}) = \begin{Bmatrix} f_{\nu k}(r)\underset{\sim}{\sigma}\cdot\hat{r} \\ -ig_{\nu k}(r) \end{Bmatrix} y^\mu_{j\ell 1/2}(\hat{r}) \tag{6.10b}$$

with

$$g_{\nu k}(r) = \left\{1/2\ \frac{\Omega_{\nu k}}{\Omega_{\nu k}+k}\right\}^{1/2} \frac{j_\ell(\Omega_{\nu k} r/R)}{|j_\ell(\Omega_{\nu k}/R)|} R^{-3/2} \tag{6.10c}$$

$$f_{\nu k}(r) = \left\{1/2\ \frac{\Omega_{\nu k}}{\Omega_{\nu k}+k}\right\}^{1/2} \mathrm{sgn}(k)\ \frac{j_{\ell-\mathrm{sgn}(k)}(\Omega_{\nu k} r/R)}{|j_\ell(\Omega_{\nu k}/R)|} R^{-3/2} \tag{6.10d}$$

$y^{\mu}_{j\ell 1/2}$ = spin-spherical harmonics, $j = |k|-1/2$, $\ell = j-1/2\ \mathrm{sgn}(k)$.

The pion wavefunctions have been given by Chin[21])

$$\phi^o_k(\vec{r}) = \sum_{L,M} \phi^o_{kL}(r) Y_{LM}(\hat{r}) \tag{6.11a}$$

$$\phi^o_{kL}(r) = 1/2\left\{\exp(i\alpha_L)h_L(kr) + \exp(-i\alpha_L)h^*_L(kr)\right\} \tag{6.11b}$$

$$\exp(2i\alpha_L) = -\frac{h^{*\prime}_L(kR)}{h^{\prime}_L(kR)} \tag{6.11c}$$

The full solutions of (6.8) and (6.9) are now straightforwardly obtained by diagonalizing the tensorial pieces in the spherical basis. They do not conserve total angular momenta of quarks and pions since they couple to states of different angular momenta. However the projections μ and M as well as parity remain good quantum numbers. Because of parity conservation, i.e. reflection symmetry states with $\pm\mu$ and $\pm M$ are degenerate. After diagonalization the solutions of the Dirac equation which spheroidal boundary conditions are given by

$$u_{\alpha\mu\pi}(\underset{\sim}{r}) = \sum_{\nu k} C^{\alpha\mu\pi}_{\nu k}(D) u^o_{\nu k\mu}(\underset{\sim}{r}) + C'^{\alpha\mu\pi}_{\nu k}(D) v^o_{\nu k\mu}(\underset{\sim}{r}) \tag{6.12}$$

(α enumerates the states for given μ and π.) Note that U is a mixture of spherical positive and negative energy states which follows from the completeness of the basis. Since the strength of the tensorial part of the Hamiltonian depends on D the quark frequencies will also. The positive energy states are displayed in the left part of fig. 7 for positive and negative parity. The right part of fig. 7 shows the admixtures of spherical states in the lowest deformed orbit. The results agree with refs. 22 and 23. Fast convergence was found for both energies and wavefunctions even for large deformations.

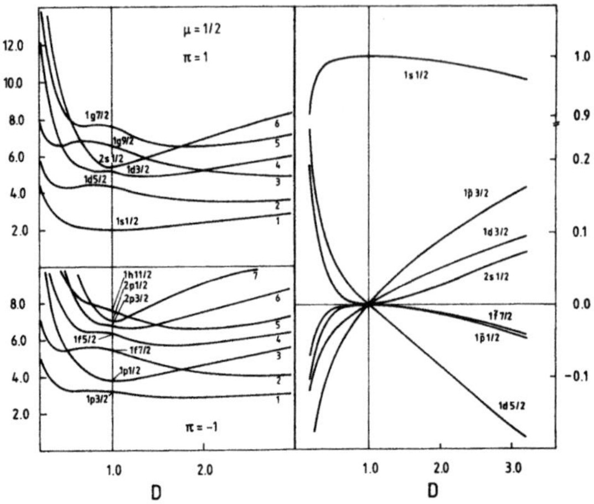

Fig. 7: Positive energy quark levels for $\mu=1/2$ and $\pi=\pm1$ (left part) and mixing amplitudes of spherical states in the lowest deformed state as a function of the deformation parameter D (the bar denotes negative energy spherical states).

In the case of the deformed pion field the diagonalization of (6.9) seems problematic. Because of the Neumann boundary condition the matrix elements are nondiagonal in the linear momentum k. This momentum space nonlocality goes to zero as $R\to0$ and will be ignored in the following. Furthermore the rank-2 tensor couples all odd or even partial waves which leads to an infinite tridiagonal matrix problem[24] with a continuous bound set of eigenvalues λ. One finds that the matrix elements tend to a constant independent of L as $L\to\infty$. In terms of λ and the spherical solutions ϕ^o one then finds

$$\phi_{\vec{k}}(\vec{r}) = \int d\lambda \sum_M N_{\hat{k}}^M(D) \sum_L C_L^{\lambda M}(D)\phi^o_{kL}(r)Y_{LM}(\hat{r}) \tag{6.13}$$

where N is a $\hat{k}$-dependent normalization constant. Using the eigenmode expansion (6.5) for the pion propagator Δ and the partial wave expansion

$$\Delta(\underset{\sim}{r},\underset{\sim}{r}',\omega) = \sum_{L,L',M} \Delta_{L,L',M}(r,r',\omega)Y^*_{L'M}(\hat{r})Y_{LM}(\hat{r}') \tag{6.14a}$$

one obtains

$$\Delta_{L,L',M}(r,r'\omega) = \frac{2}{\pi}\int_0^\infty k^2dk\ \phi^o_{kL}(r)\phi^o_{kL'}(r') \int d\lambda\ \frac{C_L^{\lambda M}(D)D_{L'}^{\lambda M}(D)}{\omega^2-k^2\lambda+i\eta} \tag{6.14b}$$

In practice (6.14b) is evaluated by truncating the tridiagonal matrix at finite order which yields a discrete set $\{\lambda_i\}$. Obviously this truncation amounts to a discretization of the λ-integral. Again convergence is reached rapidly with the number of integration points, i.e. the number of partial waves.

With the solutions for the free quark field and the free propagator Δ the second order correction (6.4) to the ground state energy from quark-pion coupling can be evaluated as a function of the shape parameters R and D. For convenience (6.4) is written as a matrix element of a frequency dependent quark-pion exchange interaction $V_{ij}(\omega)$ between quark i and j. Both i=j and i≠j are allowed. The first case corresponds to quark-selfenergy correction which diverges if all intermediate states are allowed[21). Because of this difficulty which originates from the point-coupling between quarks and pions i=j terms are excluded. It is also clear that they can not cause dynamic deformation. The ground state $|\phi_0>$ in (6.4) for a multi-quark state has to be constructed as a totally symmetric state in the quark quantum numbers α, μ, π and t_z ((6.12)). Defining

$$|\alpha\mu t_z> \equiv \sum_{k\nu} C^{\alpha\mu}_{\nu k}(D) \left(\frac{\Omega_{\nu k}}{2(\Omega_{\nu k}+k)}\right)^{1/2} \frac{j_L(\Omega_{k\nu})}{|j_L(\Omega_{k\nu})|} y^{\mu}_{j\;1/2}(\hat{r})\chi^{t_z}_{1/2} \tag{6.15}$$

(π is omitted for brevity) the wavefunctions for nucleon and isobar are

$$|N> \equiv \frac{1}{\sqrt{2}}\Big(\underset{|J_z|=1/2}{\begin{smallmatrix}\Box\Box\\ \Box^{+}\end{smallmatrix}} \times \underset{T=1/2}{\begin{smallmatrix}\Box\Box\\ \Box^{+}\end{smallmatrix}} + \underset{|J_z|=1/2}{\begin{smallmatrix}\Box\Box\\ \Box^{-}\end{smallmatrix}} \times \underset{T=1/2}{\begin{smallmatrix}\Box\Box\\ \Box^{-}\end{smallmatrix}}\Big) = |\underset{1/2\;1/2}{\Box\Box\Box}> \tag{6.16a}$$

$$|\Delta> \equiv \underset{|J_z|=1/2}{\Box\Box\Box} \times \underset{T=1/2}{\Box\Box\Box} = |\underset{3/2\;3/2}{\Box\Box\Box}> \tag{6.16b}$$

J_z denotes the projection of the total quark angular momentum on the axis of symmetry and T the total isospin.

While the nucleon is mixed symmetry in the labels (α,μ) and t_z respectively the isobar is symmetric in each one of them. With (6.15), (6.16a) and (6.16b) the second order correction to the ground state energy is given by

$$\Delta E_0^{(2)}(R,D) = <\Box\Box\Box|\sum_{i\neq j} V_{ij}(\omega)\,|\Box\Box\Box> \tag{6.17a}$$

with

$$V_{ij}(\omega) = \frac{1}{2f_\pi^2R^3}\sum_{L,L',M} \tilde{\Delta}_{L,L',M}(D,\omega)\underset{\sim}{\sigma}_i\hat{r}_i\underset{\sim}{\sigma}_j\hat{r}_j\, Y^*_{LM}(\Omega_i)Y_{L'M}(\Omega_j)\vec{t}_i\vec{t}_j \tag{6.17b}$$

It has been used that at the bag surface where we only need to know Δ

$$\Delta_{L,L',M}(R,R,\omega) = \tilde{\Delta}_{L,L',M}(D,\omega)/R \tag{6.17c}$$

In the spherical limit (D=1) one has for the static part (ω=0) of $\tilde{\Delta}$ the usual result[21)

$$\tilde{\Delta}_{L,L',M}(1,0) = 1/(L+1)\ . \tag{6.17d}$$

It should be stressed that if pions are expelled from the bag interior the Neumann boundary condition (6.6b) introduces a deformation dependence to the free pion propagator. One sees from (6.17b) that deformation does not prevent the bag from collapsing since ΔE_0 still goes like $1/R^3$. Representation (6.17) exhibits the operator structure of the quark-pion interaction nicely. V_{ij} contains both a central and a tensor term since

$$\underset{\sim}{\sigma}_i \hat{r} \underset{\sim}{\sigma}_j \hat{r} = S_{ij}^{12} - \frac{1}{3} \underset{\sim}{\sigma}_i \underset{\sim}{\sigma}_j \tag{6.18}$$

(As mentioned in ref. 20 the fact that we obtain V_{ij} depending on $\hat{r}_i$ and $\hat{r}_j$ rather than $r_i - r_j$ has to be attributed to spurious center of mass motion not projected out here.) The deformation is caused by the tensor term.

Expressions (6.17a) - (6.17c) together with (6.15), (6.16a) and (6.16b) form the central result. They give ΔE_0 rather easily once $|\phi_0\rangle$ is specified. First all quarks are in the lowest state ($\alpha=1, |\mu|=1/2$). For given R one finds a minimum with respect to D. (This also holds for the quark kinetic energies as well as discussed by Viollier et al.[23]). They show that without gluons or pions the ground state is spherical.) The second order results for nucleon and isobar are shown in the upper part of fig. 8. The isobar shows a minimum for D<1, the nucleon remains spherical however.

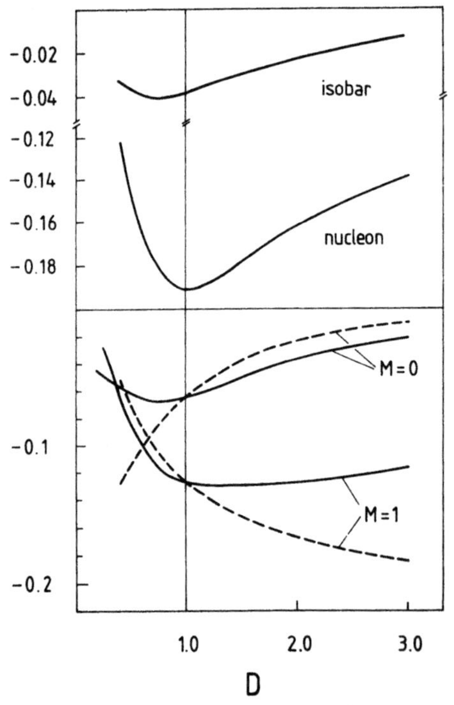

Fig. 8: Second order ground state energy correction ΔE_0 (in units $1/f_\pi^2 R^3$) as a function of D for nucleon and isobar (upper part). All quarks are in the lowest energy state. In the lower part the M-contributions to ΔE_0 (eq. (16b)) are shown separately. The dashed lines indicate results neglecting the deformation dependence of the pion propagator.

This can be understood from the permutation symmetry of $|\phi_0\rangle$. In the spherical limit one can switch to an LS-representation. (In (6.17a)-(6.17c) V_{ij} acts in the space of the large components only and we can use the nonrelativistic notations.) For the nucleon L=0 and S=1/2. Keeping all quarks in the lowest state this is not changed for $D \neq 1$ and therefore the tensor force can not operate ($\langle 1/2\ S_z |\ |\sigma \times \sigma|_{2M} | 1/2\ S_{z'} \rangle = 0$).

For the isobar the situation is different since in the spherical limit S=3/2 and the tensor force can act.

In the lower part of fig. 8 the D-dependence of ΔE_0 is analysed further by showing the M=0 and M=1 contributions (6.17a) and (6.17b) separately. For the isobar one only has M=0. For the nucleon the pion field can flip the quark μ and therefore both M=0 and M=1 contribute cancelling linear terms in D exactly. In the dashed lines deformation effects on the propagator are neglected replacing (6.17c) by (6.17d). Individually M=0 and M=1 no longer have a minimum with respect to D and deformation effects are much stronger. Expelling pions from a deformed bag costs energy and thus reduces the tendency to deform.

To obtain an intrinsic deformation for the nucleon one has to allow for configuration mixing by promoting one or more quarks to higher orbits. The effect is demonstrated with one quark in an $\alpha=3,|\mu|=1/2$-state which for D=1 corresponds to $d_{3/2}$. The construction of the corresponding three mixed symmetry states is somewhat involved. The matrix elements of V_{ij} are diagonalized in the space of these states together with the state in which all quarks are in the lowest orbits ($\alpha=1,|\mu|=1/2$). Since $\alpha=3$ and $\alpha=1$ are not degenerate one in principal needs the frequency dependence of V_{ij}. It turns out however that the static approximation $\omega=0$ is sufficient. The results of the diagonalization are plotted in fig. 9. In the dashed curve all quarks are in $\alpha=1,|\mu|=1/2$. As the mixing is allowed and R decreases minima on both the prolate and the oblate side evolve with the oblate one always deeper. The additional R-dependence, other than R^{-3} is induced by the R-dependence of quark energy gap $\omega_{\alpha=3}-\omega_{\alpha=1}$.

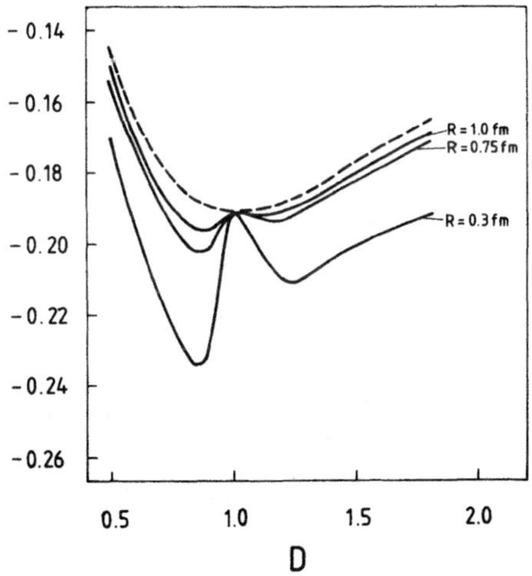

Fig. 9: ΔE_0 for the nucleon (in units $1/f_\pi^2 R^3$) as a function of D for various R after configuration mixing (see text). The dashed curve represents the result with all quarks in the lowest orbit.

To summarize the linearized CBM predicts bag deformations for both nucleon and isobar. If all quarks occupy the lowest deformed single particle state the nucleon stays spherical and only acquries deformation after genuine D-states are mixed in. The deformation dependence of the pion propagator acts like a surface tension and resists deformation.

I would like to thank the theory group at Stony Brook for many fruitful discussions on the subjects presented here and especially Z.Y. Ma for a wonderful collaboration on bag deformations.

References

1. J. Kogut et al., Phys. Rev. Letts. 50, 393 (1983).
2. G.E. Brown and M. Rho, Phys. Lett. 82B (1979) 177.
3. A.D. Jackson and M. Rho, Stony Brook preprint (1983).
4. M. Rho, G.E. Brown and A.S. Goldhaber, Stony Brook preprint (1983).
5. V. Vento et al., Nucl. Phys. A345, 413 (1980).
6. T.H.R. Skyrme, Proc. Roy. Soc. London, Ser. A260, 126 (1961); Nucl. Phys. 31, 556 (1962).
7. H.R. Fiebig and B. Schwesinger, Nucl. Phys. A393, 349 (1983).
8. S. Théberge, A.W. Thomas and G.A. Miller, Phys. Lett. 91B, 192 (1980); Phys. Rev. D22, 2838 (1980).
9. A. Chodos and C.B. Thorn, Phys. Rev. D12, 2733 (1975).
10. S. Weinberg, Phys. Rev. 166, 1568 (1968).
11. V. Vento and M. Rho, Stony Brook preprint (1983).
12. A. Chodos et al. Phys. Rev. D9, 3471 (1974).
13. L.D. Faddeev, Lett. Math. Physics 1, 289 (1976); N.K. Pak and H.Ch. Tze, Ann. Phys. (N.Y.) 117, 164 (1979).
14. A.P. Balachandran et al., Phys. Rev. Lett. 49, 1124 (1982); Phys. Rev. D27, 1153 (1983).
15. G.S. Adkins, C.R. Nappi and E. Witten, Princeton preprint (1983).
16. J. Wambach and I. Hulthage, Contributed paper, Workshop on Nuclear Physics with real and virtual photons 'From collective states to quarks in nuclei', Bologna (1980).
17. G.E. Brown, M. Rho and V. Vento, Phys. Lett. 97B 423 (1980).
18. R. Jackiw and C. Rebbi, Phys. Rev. D13 3398 (1976).
19. S.L. Glashow, Physika 96A, 27 (1979).
20. V. Vento, G. Baym and A.D. Jackson, Phys. Lett. 102B, 97, (1981).
21. S.A. Chin, Nucl. Phys. A382, (1982) 355.
22. C. De Tar, Phys. Rev. D17, 323 (1978); K. Hahn, R. Goldflam and L. Wilets, Phys. Rev. D27 (1983) 635; R. Shanker et al., J. Phys. 68 (1982) 133.
23. R.D. Viollier, S.A. Chin and A.K. Kerman, Nucl. Phys. A407 (1983) 269
24. J. Wambach and Z.Y. Ma, to be published in Phys. Lett.

CENTRE OF MASS CORRECTION TO BAG MODEL

A. Szymacha
Institute of Theoretical Physics,
Warsaw University,
Hoża 69, 00-681 Warszawa, Poland.

Introduction

In the MIT bag model [1], the calculation of any property of hadron is bound to the static sferical cavity approximation in which bag is well localised in space. By the uncertainity principle, it follows that such a state can not be an eigenstate of the total linear momentum. The expectation value of the total momentum squared in this state is different from zero, in other words, the centre of mass is in motion. Nevertheless in early applications of the bag model, the ground state of quarks in the cavity was treated as representing hadron at rest. The results for magnetic moments,axial coupling constant, etc, calculated in such approximation contained contributions from the centre of mass motion. There have been many attempts to improve the original model and to calculate appropriate corrections [2,3,4,5,6,7]. The results obtained by various authors are very different; not only magnitude, but also the signs of some corrections differ from paper to paper. This concerns particularly the corrections to magnetic moments, but other corrections are not free from ambiguity either.

The various approaches leading to contradictory results differ by their specific assumptions. In order to clarify the situation, and to understand the source of discrepancy I shall present an approach to the centre of mass (CM) corrections based essentially on only one approximate assumption, which is however common to all calculations we are talking about. I will restrain from making additional, more specific assumptions until I get formulae showing unambiguously sign and order of magnitude of the appropriate correction. Only then I shall be more specific and apply these results to making detailed fits to the data.

The plan of my talk is following: Chapter I will be devoted to the discussion of basic assumption that the bag state at rest represents, at a given time, the superposition of free physical particle states with different linear momenta. In Chapter II I shall present formulae for corrected static parameters of baryons. Chapter III contains comparison of these results with particle data.

The three chapters correspond to three papers [8,9,10] written in collaboration with J. Bartelski, Z. Ryzak from Warsaw University, and L. Mankiewicz, S. Tatur from N. Copernicus Astronomical Centre of Polish Academy of Sciences.

I. The wave packet approximation

The MIT bag model wave function has two characteristic features: 1) It is localised in space, and 2) It is a product of independent quark wave functions. The second property limits accuracy with which localised state may be represented as a wave packet of states of *the same internal motion*. It is well known that in nonrelativistic case, only for harmonic oscilator the product wave function $\Pi_i u_i(r_i)$ of independent particles may be cast in the "wave packet" form:

$$\prod_i u_i(r_i) = \int \Phi(P)\{e^{iPR_{CM}}\cdot\Psi(\text{int. coord.})\}d^3P = F(R_{CM})\cdot\Psi(\text{int.coord.}) \quad (1)$$

For two particles, for example, the following identity holds:

$$e^{-m_1 r_1^2}\cdot e^{-m_2 r_2^2} = e^{-MR^2}\cdot e^{-\ r^2} = N\int e^{-P^2/4M}\{e^{iPR}\cdot e^{-\mu r^2}\}d^3P \quad (2)$$

where μ is the rtduced mass, M is the total mass, r and R are the relative and CM positions respectively.

For any other functions, in particular for "bag-like" functions u_i localized within sharply defined boundary, representation (1) is not valid. However one may still wonder how good the representation (1) could be as an approximation. It is very difficult to answer this question for relativistic case, but we may get some feeling what the real situation is by investigating the simpler nonrelativistic case.

We choose an example of two nonrelativistic particles of equal masses in a one dimensional square well potential. It has the merit of being the simplest possible case, and at the same time the most vulnerable for discrepancy (the more degrees of freedom, the less severe the problem of CM motion). Sharp boundaries seem also to be as much contrasting with smoothness of the oscillator wave function as possible. Mathematically, we are looking for two functions F(R) and $\Psi(r)$ which minimize the deviation:

$$\delta\int|u(R+r/2)u(R-r/2) - F(R)\Psi(r)|^2 dR\,dr = 0 \quad (3)$$

for given $u(x)=N\sin kx$.

Solution of the above problem leads in a natural way to the expansion:

$$u(r_1)\,u(r_2) = \sum\sqrt{\lambda_i}F_i\Psi_i\,, \quad \sum \lambda_i = 1 \tag{4}$$

where $\{F_i\}$ and $\{\Psi_i\}$ are certain orthonormal sets defined by Eq(3). The left hand side of Eq(4) is well approximated by only one term if the largest eigenvalue λ_{max} is close to 1. In the opposite case, e.g. for $\lambda_{max} \cong 0.5$, we would say that the approximate representation of independent particles wave function $u_i(r_i)$ by the wave packet of fixed internal motion is completely unjustified.

In practice it turns out, that for the case under consideration

$$\lambda_{max} \cong 0.98$$

what is very close to 1. The obtained result is obviously not a proof of applicability of the formula analogous to Eq(1) for the relativistic bag model, but it certainly provides some credibility to such an approximate assumption.

In what follows we shall therefore assume:

$$|B,j_z\rangle = \int d^3p\,\Phi(|\vec{p}|)\,|\vec{p},s_z\rangle\ ; \qquad j_z = s_z \tag{5}$$

Letter B refers to bag, $|\vec{p},s_z\rangle$ is the state of physical particle with quantum numbers of the corresponding bag state. The wave packet on the right hand side of Eq(4) is assumed to have the same parity and angular momentum as the bag state on the left. This makes $\Phi(|\vec{p}|)$ a function of $|\vec{p}|$ only, what simplify calculations significantly. The Eq(5) is in fact *the most general one consistent with the neglect of particles mixing.*

II. Corrections for CM motion

The typical developpment from now, which seems to be one of the sources of discrepancies among published results, consists of the following:

Translating Eq(5) by $\vec{a}$ one obtains

$$e^{i\hat{\vec{P}}\vec{a}}|B\rangle_{\vec{0}} = |B\rangle_{\vec{a}} = \int d^3p\,\Phi(p)\,e^{i\vec{p}\vec{a}}|\vec{p}\rangle \tag{6}$$

which after integration over $\vec{a}$ gives:

$$|\vec{p}\rangle = \frac{1}{(2\pi)^3\Phi(p)}\int d^3a\,e^{-i\vec{p}\vec{a}}|B\rangle_{\vec{a}} \tag{7}$$

Taking appropriate scalar product one also obtains:

$$|\Phi(p)|^2 = \frac{1}{(2\pi)^3}\int d^3a \; {}_0\langle B|B\rangle_a \, e^{-ipa} \tag{8}$$

With help of Eq(7) we may now express any matrix element between physical particles through the matrix elements between bag states $|B\rangle$. In this way all expressions for physical matrix elements will contain in the integrand a quantity:

$$ {}_0\langle B|J|B\rangle_a \tag{9}$$

The matrix element (9) is then calculated by integrating quark wave functions over the overlaping part of two bags. Such procedure implicitely assume, that scalar product of two,partly overlaping, empty bags is 1:

$$ {}_0\langle EB|EB\rangle_a = 1$$

The above assumption introduces uncontrolable errors. It is impossible to calculate the empty bag matrix element because nonquark degrees of freedom (responsible for confinement) are not actually quantized in the bag model.

In these circumstances, for not to introduce uncontrolable assumptions I shall consider only equations containing matrix elements of currents (known to be built out of quark only) between states with *identical empty bags*. Contrary to a prejudice one may meet frequently, I claim that such a matrix element (without one of the bags being boosted or shifted with respect to the other) gives useful informations about physical matrix element, not only for momentum transfer equal zero, but also in some vicinity of this point, allowing calculations of magnetic moments, g_A, and electromagnetic (as well as weak) squared radii.

Making use of Eq(5), we may immediately write down:

$$\langle B,s_z'|J(\vec{x},0)|B,s_z\rangle = \int d^3p\,d^3p'\,\Phi^*(p')\Phi(p)e^{i(\vec{p}-\vec{p}\,')\vec{x}}\langle s_z',\vec{p}\,'|J(0,0)|\vec{p},s_z\rangle \tag{10}$$

Without boosting bags, Eq(10) contains already on the right hand side various fourmomentum transfers $(E(\vec{p}\,')-E(\vec{p}),\vec{p}\,'-\vec{p})$ between physical hadrons. This is enough for the appearence of physical formfactors which we shall try to extract from the above relation. In this sense Eq(10) takes into account both "CM correction" and "recoil correction" which need not be separated in any way. As long as one does not propose better connection between bag states and particle states than that expressed by Eq(5), one can not avoid Eq(10).

Multiplying Eq(10) by exp(ikx) and integrating over x we get

$$\int d^3x e^{ikx} \langle B|J(x)|B\rangle = \int d^3p \Phi^*(p+k)\Phi(p)\langle p+k|J(0)|p\rangle \tag{11}$$

(the spin indeces have been supressed). Formally the Eq(11) is an identity with respect to $\vec{k}$, but in practice it is as approximate as the basic assumption about wave packet representation given by Eq(5).

Differentiating Eq(11) with respect to k (once or twice) and putting k=0, we may generate relations involving measurable quantities (on the right hand side) and bag model quantities (on the other side).

For illustration let us consider this equation for $J = J^o_{em}$ taken between nucleon states with $s_z=s_z'$ and differentiated twice with respect to k. For k=0 on the left hand side we obtain:

$$-\int x^2 \langle B\ J^o_{em}(\vec{x},0)\ B\rangle d^3x \tag{12}$$

which may immediately be calculated ($=-0.53R_B^2$) and is called static charge squared radius. On the right hand side there appear several terms. One of them is connected with the argument of formfactors:

$$F_i(k_\mu k^\mu) = F_i\{(\sqrt{m^2+(\vec{p}+\vec{k})^2} - \sqrt{m^2+\vec{p}^2})^2 - \vec{k}^2\} \cong F_i\{-\vec{k}^2(1-\frac{1}{3}\frac{p^2}{E^2})\} \tag{13}$$

leading to:

$$\left.\frac{dF_i}{d\vec{k}^2}\right|_{\vec{k}^2=0} = -(1-\frac{1}{3}\frac{p^2}{E^2})F_i'(0) \tag{14}$$

F'(0) is directly related to physical electromagnetic (electric and magnetic, which are slightly different) squared radii, while factor $(1 - p^2/3E^2)$ averaged with $|\Phi(p)|^2$ is one of the typical corrections. It is of course, just the Lorentz contraction.

Another important term comes from differentiating $\Phi^*(p+k)$ in Eq(11) It is:

$$\langle R^2\rangle = -\int \Phi^*(p)\nabla^2\Phi(p)d^3p \tag{15}$$

With nonrelativistic profile, this would be just the average square of CM position; in more general case the physical interpretation of this quantity may be obscure (recall difficulties with the definition of position operator for relativistic particle), but this need not concerne us, because Eq(15) defines uniquely this quantity.

In this way we have obtained the following formulae:

$$g_A = g_A^{\text{st bag}} \frac{3}{1+2\langle\frac{M}{E}\rangle} \tag{16}$$

$$\mu = \mu^{\text{st bag}} \frac{3}{1+\langle\frac{M}{E}\rangle+\langle\frac{M^2}{E^2}\rangle} + \frac{M_p}{M} Q \frac{1-\langle\frac{M}{E}\rangle}{1+\langle\frac{M}{E}\rangle+\langle\frac{M^2}{E^2}\rangle} \tag{17}$$

$$\langle r^2\rangle = \frac{\langle r^2\rangle^{\text{st bag}} - Q\langle R^2\rangle}{\frac{2}{3}+\frac{1}{3}\langle\frac{M^2}{E^2}\rangle} + \begin{array}{l}\text{calculable but}\\ \text{numerically negligible}\\ \text{terms}\end{array} \tag{18}$$

All averages have to be calculated with profile $\Phi(p)$, e.g.

$$\langle\frac{M}{E}\rangle = \int|\Phi(|p|)|^2 \frac{M}{\sqrt{M^2+p^2}} d^3p \tag{19}$$

Till now we have made no assumption in addition to the basic one (Eq(5)), which can not be avoided however, if a given bag state is connected with one sort of physical particles. Having assumed so little, we are not able to calculate electroweak properties to the end; the profile (p) is not yet specified. Nevertheless, even without this specification Eqs (16), (17), (18) allow already to draw very definite conclusions. For example, whatever the profile $\Phi(|p|)$ is, we have always

$$\langle\frac{M}{E}\rangle < 1 , \qquad \langle\frac{M^2}{E^2}\rangle < 1$$

and consequently

$$\mu > \mu^{\text{static bag}} \tag{20}$$

what is not a trivial result in view of disagreement in the literature concerning the sign of $\mu-\mu^{\text{st bag}}$.

For numerical quantitative applications, Eqs (16), (17) and (18) are also very convenient, because they are in fact very unsensitive to the details of the distribution of the momentum defined by the profile $\Phi(\vec{p})$. this allows us to calculate reliably all corrections with few additional but very plausible assumptions.

First of all we shall assume, that averages like $<\frac{M}{E}>$ and $<M^2/E^2>$ are not independent. Formally, if $\Phi(\vec{p})$ is not specified, average $<M/E>$ and $<M^2/E^2>$ can take arbitrary, unrelated values. However for smooth distributions, containing no extreme velocities, functions like M/E or M^2/E^2 fulfill, with great accuracy, the naive relation:

$$f(<v>) = <f(v)> \tag{21}$$

The above observation reduces the information contained in $\Phi(\vec{p})$ to practically one parameter. More attention is needed in estimation of $<R^2>$. Because of the form of Eq(15), we have immediately the "uncertainity relation"

$$<R^2><p^2> \geqq \frac{9}{4} \tag{22}$$

In order to estimate $<R^2>$ we take the equality sign

$$<R^2> \cong \frac{9}{4<p^2>} \tag{23}$$

This result wold be exact for Gaussian profile. Introducing the average boost factor γ,

$$\frac{1}{\gamma} = <\frac{M}{E}> \tag{24}$$

we have

$$\begin{aligned} <\frac{M}{E}> &= \frac{1}{\gamma} \\ <\frac{M^2}{E^2}> &= \frac{1}{\gamma^2} \\ <R^2> &= \frac{9}{4M^2(\gamma^2-1)} \\ <p^2> &= M^2(\gamma^2-1) \end{aligned} \tag{25}$$

Finally one has to specify γ. The simplest choice is to take only contribution from quarks and calculate $<p^2>$.

$$\langle p^2\rangle = M^2(\gamma^2-1) = \frac{3x_0^2}{R^2} \qquad (26)$$

for nonstrange baryons, and similarly for other hadrons. This is the most arbitrry assumption in our approach. However, whatever the further developpment of the bag theory is, the order of magnitude of $\langle p^2\rangle$ will rather not change. And because our formulae (16), (17), (18) do not contain cancellation, and depend smoothly on $\langle p^2\rangle$, the arbitrariness of the assumption (26) is not very dangerous to the reliability of obtained results.

Before going to applications, let me quote one particular result, namely that for $\langle R^2\rangle$, in the case of massless quarks in the nucleon:

$$\langle R^2\rangle = \frac{9(x_0-1)}{2(2x_0^3-2x_0^2+4x_0-3)}\langle r^2\rangle^{\text{st bag}} = 0.34\langle r^2\rangle^{\text{st bag}} \qquad (27)$$

which is surprisingly close to the nonrelativistic, classical value $\frac{1}{3}$.

Applications

The MIT bag model contains several number of adjustable parameters. Inclusion of any correction,like CM correction, needs always refitting all of the parameters. The usual strategy is to choose certain particles and calculate parameters from their masses, and then to predict remaining masses and electroweak properties. The choice is of course arbitrary and the results somehow accidental. The most objective choice of parameters would consist in taking all calculable properties and fitting parameters by some kind of least square method. Because,however there is no obvious prescription how to weight various contributions of different physical dimensions,this will not be a very good procedure either.

We have decided to take all masses of the ground state multiplets and to look for model parameters which minimize the function

$$\chi^2(m_u,m_s,Z,B,\alpha_s) = \sum(M_{exp}-M_{bag})^2 \qquad (28)$$

The pion is not included in the above sum, because no stable solution could be found in this case (before CM corrections had been included). The minimalization procedure applied to the original, uncorrected version of the model, improves the overall agreement of masses which is already surprisingly good in the fit of de Grande et al.[1].The new parameters do not differ to much from the old one. They are compared in Table 1.

Table 1

orig. MIT	$m_u=0$	$m_s=279$ MeV	$\alpha_s=2.2$	Z=1.84	$B^{1/4}=146$ MeV	$\sqrt{\chi^2}=23$ MeV
χ^2 method CM uncorr.	12MeV	283 MeV	2.04	2.01	149 MeV	16 MeV
$M=\sqrt{E_B^2-\langle p^2\rangle}$	10MeV	286 MeV	1.68	0.86	144 MeV	16 MeV

The next step was to fit masses corrected for CM motion according to the formula

$$M = \sqrt{E_B^2 - \langle p^2\rangle} \tag{29}$$

The stability equation was taken as the minimum of E_B . Resulting parameters are shown in the third row of Table 1. The changes of Z and α_s go in theoretically desirable direction, but the accuracy stays at the same level ($\langle\delta M\rangle$=16 MeV). For electroweak properties however, the improvement is enormous! Results are shown in Table2

Table 2

		μ_{exp}	μ_B	μ_B (CM corrected)
magnetic moments	P	2.79	1.82	2.53
	N	-1.91	-1.21	-1.64
	Λ	-0.61	-0.50	-0.61
	Σ^+	2.37	1.88	2.39
	Σ^-	-1.18	-0.64	-0.93
	Ξ^0	-1.25	-1.08	-1.33
	Ξ^-	-0.69	-0.45	-0.58
	g_A	1.25	1.10	1.25
	$\langle r^2\rangle_P$	$(0.84\text{fm})^2$	$(0.69)^2$	$(0.67)^2$
	$\langle r^2\rangle_N$	$-(0.34\text{fm})^2$	0	0

The additional merit of taking into account the CM motion is,that the pion may now be treated at the same footing as other hadrons.

Repeating fit with pion contribution included into χ^2 we obtain the following results:

m_u=3MeV, m_s=283MeV, α_s=1.6, Z=0.75, $B^{\frac{1}{4}}$=142MeV, <δM>=19MeV. With the last value for average discrepancy, the mass of the pion comes out equal 125 MeV to be compared with experimental 138 MeV.

Summarizing the above results we may say, that simple MIT model supplemented by CM corrections in the way explained in Chapter II describes masses of hadrons (including pion) and their electroweak properties very well. One may even say, that it reproduces these properties (particularly masses) too well. The agreement is so good that it seems not to leave any room for corrections from the pion cloud, which must not be omitted if chiral symmetry is to be included into the bag model [11,12,13,14,15,16,17].

In view of the necessity of refitting everything each time something new is added to the model, this suspition has to be checked by explicit calculation. If the pion cloud contribution to the energy is taken in its simplest form

$$E_\pi = -\frac{1}{f_\pi^2 R^3} f(m_\pi R, m_u R) \sum_{\substack{nonstr\\quarks}} (\vec{\sigma}\tilde{\tau})^i (\vec{\sigma}\tilde{\tau})^j \tag{29}$$

no reasonable, stable fit to masses is possible. The main reason is the large pion cloud contribution to the pressure balance equation which typically exceeds the gluon one by a factor of three. One possible way out is to exclude pion cloud contribution from pressure equilibrium equation and retain it only in the expression for energy, as some authors do [18]. There is no convincing argument in favor of this procedure; we have tried another, purely phenomenological approach. It consists in replacing the bare value of the quark-pion coupling constant $g^o=1/2f_\pi$ in the expression (29) by an effective value g^{eff} which is then fitted together with B, α_s, etc. Such a replacement is motivated by a recent indication [19] that finite size of the pion may lower effectively the overall strength of the pion field in the bag model. The fitted value which emerges is $g^{eff}=0.57g^o$.

Conclusions

I want to conclude with three simple statements:

i) CM corrections are calculable and by themselves bring the MIT model to the very good agreement with measured static properties of hadrons

ii) $\bar{q}\,q$ pion state is at the same footing as other hadrons

iii) There is no room for full chiral pion cloud contribution to the energy. Effective coupling constant giving good fit is $0.57g^{o}$.

References

[1] A. Chodos, R.L. Jaffe, K. Johnson, C.B. Thorn, V.F. Weisskopf, Phys. Rev. D9 (1974) 3471; A. Chodos, R.L. Jaffe, K. Johnson, I. Kiskis, Phys. Rev. D12 (1975) 2060

[2] C.W. Wong, K.F. Liu, Phys. Rev. Lett. 41 (1978) 62.

[3] K.F. Liu, C.W. Wong, UCLA report (1980) unpublished.

[4] J.F. Donoghue, K. Johnson, Phys. Rev. D21 (1980) 1975.

[5] C.W. Womg, Phys. Rev. D24 (1981) 1416.

[6] C. Carlson, M. Chachkhunashwili, Nordita preprint, August 1981.

[7] K.F. Liu, C.W. Wong. Phys. Lett. 113B (1982) 1.

[8] Z. Ryzak, A. Szymacha, IFT/1/83, to be publ. in Nucl. Phys. A.

[9] J. Bartelski, A. Szymacha, L. Mankiewicz, S. Tatur IFT/4/83, to be publ. in Phys. Rev. D.

[10] J. Bartelski, A. Szymacha, Z. Ryzak, L. Mankiewicz, S. Tatur IFT/5/83.

[11] A. Chodos, C.B. Thorn, Phys. Rev. D12 (1975) 2733

[12] M.V. Barnhill, Phys. Rev. D20 (1979)

[13] G.E. Brown, M. Rho, Phys. Lett 82B (1979) 177

[14] A. Szymacha, S. Tatur, Zeit. fur Physik c7 (1981)311

[15] C. De Tar, Phys. Rev: D24 (1981) 762

[16] S. Théberge, G.A. Miller, A.W. Thomas, Can. J. Phys. 60 (1982) 59

[17] S. Théberge, A.W. Thomas, Nucl. Phys. A393 (1982) 252

[18] P.J. Mulders, A.W. Thomas, TH 3443- CERN (1982)

[19] J. De Kam, H.J. Pirner, Nucl Phys A389 (1982) 640.

DYNAMICS OF THE SOLITON BAG

L. Wilets and R. Goldflam*
Institute for Nuclear Theory, Department of Physics, FM-15
University of Washington
Seattle, Washington 98195

The MIT bag[1] was one of the earliest and most successful models of QCD, imposing confinement and including perturbative gluon interactions. An evolution of the MIT bag came with the introduction of the chiral[2] and cloudy[3] bags, which treat pions as elementary particles. (The pion is an "anomalously" light hadron, and in a chirally invariant QCD should emerge as the massless Goldstone boson.) Non-relativistic potential models also have had remarkable success, and have the advantage of being amenable to dynamic calculations.

As a model of QCD, the soliton model proposed by Friedberg and Lee[4,5] is particularly attractive. It is based on a covariant field theory and is sufficiently general so that, for certain limiting cases of the adjustable parameters, it can describe either the MIT or SLAC (string) bags. The confinement mechanism appears as a dynamic field. This allows non-static processes, such as bag oscillations and bag collisions, to be calculated utilizing the well-developed techniques of nuclear many-body theory.

In the soliton model, the (effective) Hamiltonian density is

$$\mathcal{H} = \mathcal{H}_q + \mathcal{H}_\sigma + \mathcal{H}_{q\sigma} + \mathcal{H}_G \quad \text{+(counter terms, Higgs fields,etc)}, \qquad (1)$$

where the individual terms have the following interpretation:

$\mathcal{H}_q = \Sigma_f\ \psi_f^+\ (\vec{\alpha}\cdot\vec{p} + \beta m_f)\ \psi_f$ describes the quarks as Dirac particles of mass m_f, where f is the flavor. We take $m_u = m_d = 0$.

$\mathcal{H}_\sigma = \frac{1}{2}\ (\pi_\sigma^2 + |\vec{\nabla}\sigma|^2) + U(\sigma)$ describes the scalar soliton field σ, which represents the complex structure of the vacuum, arising from virtual gluons and quark-pairs interacting among themselves. The non-linearity of the soliton field enters through the self-interaction function (see Fig. 1)

$$U(\sigma) = \frac{1}{2} a \sigma^2 + \frac{1}{3!} b \sigma^3 + \frac{1}{4!} c \sigma^4 + B. \tag{2}$$

The polynomial terminates in fourth order to ensure renormalizability. $U(0) = B$ is to be identified with the "bag constant" or volume energy density of a cavity. With a suitable adjustment of the constants, the function has two minima, one at $\sigma = 0$, and another, lower minimum at $\sigma = \sigma_v$. The physical vacuum corresponds to the second minimum, and the constant B is chosen so that $U(\sigma_v) = 0$.

The quarks interact with the soliton field through the term $\mathcal{H}_{q\sigma} = g\bar{\psi}\sigma\psi$. In the presence of (real) quarks, the sum $U(\sigma) + g\bar{\psi}\sigma\psi$ may have a minimum (depending on the parameters) near $\sigma = 0$ (the perturbative vacuum). This leads to a cavity in the σ-field, which is called the "bag."

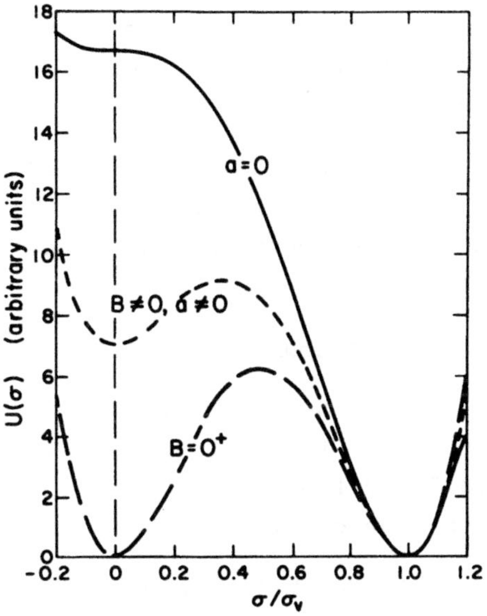

Fig. 1. Three forms for $U(\sigma)$.

Color gluon fields are introduced as in QCD, except that they interact with the soliton field through a dielectric function $\kappa(\sigma)$, chosen such that $\kappa(0) = 1$ and $\kappa(\sigma_v) = 0$. A convenient (but not unique) form is

$$\kappa(\sigma) = (\sigma/\sigma_v - 1)^2 . \tag{3}$$

The magnetic susceptibility is $\mu = \kappa^{-1}$. The requirements on κ yield color confinement. This can be seen easily if one keeps only terms linear in the gluon field. Then Gauss's law gives

$$\vec{\nabla}\cdot\vec{D}^c + \rho^c, \tag{4}$$

where c is the color index. If the total color charge does not vanish within some finite cavity, the D-field will fall off as r^{-2} as $r \rightarrow \infty$, and the color electric energy in medium

$$\frac{1}{2}\int d^3r\ D^2/\kappa(r) \tag{5}$$

will be infinite because $\kappa \to 0$ (exponentially in the model) as $r \to \infty$.

As long as one calculates diagrams only through order of one gluon exchange, there is no problem of double counting: the soliton field represents at least two-gluon structures. If higher order diagrams are calculated, the coefficients in the effective Lagrangian must be readjusted at each stage to compensate.

There are five parameters in the model: a, b, c, g and α_s. The first four involve only the soliton field and quarks; α_s is the quark-gluon coupling constant. The following key data may be used to help fix these parameters:

<u>Soliton-quark parameters</u>.

From the presentations of Drs. Engels and Schierholz earlier (see also refs. 6 and 7), it is encouraging to note that we now have available from lattice gauge theory (LGT) calculations the following pieces of data.

(1) The bag constant $B = (220+20)$ MeV/fm^3. This is considerably larger than the MIT value[8] of 57 MeV/fm^3.

(2) $m_{GB} = (720+40)$ MeV, the mass of the glueball state (0^{++}). We identify this state with an excitation of the pure soliton field, such that $U''(\sigma_V) = m_{GB}^2$.

The errors quoted in (1) and (2) above are presumably statistical only, and do not account for systematic effects of the lattice size or the omission of dynamical interactions with quarks.

From experiment we select the subset

(3) The mean of the nucleon and delta masses, $\bar{m} = (m_N + m_\Delta)/2 =$ 1.087 GeV.

(4) The proton size $\langle r_p^2 \rangle^{1/2} = 0.83$ fm.

(5) The proton magnetic moment, $\mu_p = 2.7925$.

(6) The ratio of the axial to vector coupling constants, $g_A/g_V =$ 1.25.

<u>The strong coupling constant α_s</u>. This is a "running" constant, and is evaluated for the regime of hadronic sizes. To fix it, we may utilize

(7) The delta-nucleon mass difference, $m_\Delta - m_N = 297$ MeV.

(8) The "string constant, which is the coefficient of the linear term in the potential between massive quarks, such as charmed quarks. This has been fit phenomenologically to the charmonium spectrum using a non-relativistic potential model[9] and has also been obtained in LGT calculations[7] ; the value is $t \cong 1$ GeV/fm. In the MIT model, $t = (32 \pi\alpha_s B/3)^{\frac{1}{2}}$. It is also calculable in the soliton model.

Before this workshop, we did not have data (1) and (2) available, so the results presented here are for fits of the parameters to the remaining data.

There is much more data to fit once the parameters are determined. These include hadronic spectra and resonances, decay widths, various reactions (especially nucleon-nucleon scattering), etc. There are many more experimental data of high and low quality, but the data from LGT calculations are still few and of uncertain accuracy.

In adjusting the model parameters, the fitting is dependent on the level of sophistication of the model calculations. Whenever more diagrams are included, one must readjust the model parameters. For example, if we were able to calculate all gluon interactions exactly, then presumably the effects of the soliton field would disappear completely.

We quote results here for the soliton model calculated through the mean field approximation plus realistic recoil corrections . The latter are calculable in the soliton model, since the "cavity" carries both energy and momentum. By measuring the proton form factor relative to its center-of-energy, the size is seen to be smaller than that measured with respect to the bag center; hence a larger bag is needed to fit experiment than would be the case for a static bag. In calculating hadronic masses, the total momentum term is subtracted according to the relativistic relation

$$m^2 = E^2 - P^2 . \tag{6}$$

The strong coupling constant has been fixed using one gluon exchange calculations for the $m_{\Delta} - m_N$ mass difference based on the MIT bag model (scaled for size).

Pion effects have not yet been included explicitly. Although these could be calculated in the manner of the chiral,[2] cloudy,[3] or chiral-soliton[12] bag models, our philosophy is that they should emerge in the soliton model as $q\bar{q}$ pairs interacting with the σ and gluon fields. The pion bag mass has been calculated in the MIT model[8] to be small. Recoil effects reduce it further. The pion cloud surrounding a nucleon should be describable as a surface oscillation of the nucleon σ-cavity accompanied by virtual $q\bar{q}$ pairs. Many reactions involving pions are certainly easier to handle mathematically in the chiral and cloudy bag models, however.

Through the order of calculations stated above, we find a "reasonable" parameter set which gives a moderately good fit to experimental data:

$$a = 43.84,\ b = -7\ 184,\ c= 180\ 000,\ g = 25,\ \alpha_s = 2.49. \qquad (7)$$

The structure of the soltion field and the quark wave function for the nucleon are shown in Fig. 2. Other hadronic properties are summarized in Table 1.

The method of generator coordinates (GCM) is being applied to large amplitude bag dynamics. Consider a parameter or set of parameters α which describe the static configuration of a system of quarks and the soliton field and let $|\alpha,n\rangle$ denote a set of basis states which is complete for any α. A method of obtaining these basis states is described below. The GCM state vector is written

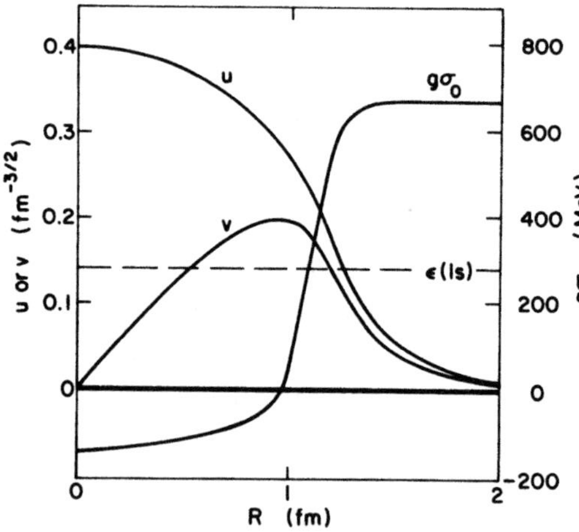

Fig. 2. Numerical results for the parameter set given in text above.

$$|\Psi> = \sum_n \int \phi_n(\alpha) \mid \alpha, n> d\alpha \tag{8}$$

Since the set is complete for each α, the expansion is overcomplete. In practice, this causes no problem since the sum is truncated to a small number of terms. In what follows, we consider only a single term and suppress n; actually, several configurations may be required. The generalization is straightforward.

The weight function $\phi(\alpha)$ is obtained by minimizing the expectation value of the Hamiltonian $<\Psi|H|\Psi>$ subject to the normalization constraint $<\Psi|\Psi> = 1$:

$$\delta<\Psi|H-E|\psi>/\delta\phi^*(\alpha) = 0 = \int d\alpha'<\alpha|H-E|\alpha'>\phi(\alpha') \; . \tag{9}$$

This is the basic GCM integral equation for $\phi(\alpha)$. Depending upon whether the spectrum is discrete or continuous, it is either an eigenvalue or a scattering equation. Although we can work with the integral equation, it is instructive to consider the approximate differential equation which it satisfies. For a system which has well developed collective-motion, we expect $<\alpha|H-E|\alpha'>$ to fall off rapidly as a function of $\alpha'-\alpha$. (This is a criterion for α to be a good collective coordinate.) Then we can expand $\phi(\alpha')$ in a Taylor series about $\alpha'=\alpha$, to obtain

$$\begin{aligned}&[\int d\alpha'<\alpha|H-E|\alpha'>]\;\phi(\alpha)\\ &+[\int d\alpha'<\alpha|H-E|\alpha'>(\alpha'-\alpha)]\;d\phi/d\alpha\\ &+[\int d\alpha'<\alpha|H-E|\alpha'>(\alpha'-\alpha)^2]\;\tfrac{1}{2}\;d^2\phi/d\alpha^2\\ &+\;....\;=0.\end{aligned} \tag{10}$$

With the definition $H_n(\alpha) = \int d\alpha'<\alpha|H|\alpha'>(\alpha'-\alpha)^n$

and $N_n(\alpha) = \int d\alpha'<\alpha|\alpha'>(\alpha'-\alpha)^n$, this assumes the simpler form

$$(N_0^{-1}H_0-E)\phi + N_0^{-1}(H_1-EN_1)\phi' + N_0^{-1}(H_2-EN_2)\tfrac{1}{2}\;\phi'' + \;....\; = 0. \tag{11}$$

If we terminate the Taylor series with ϕ'', we have the beginning of a Schroedinger-type equation. Note, however, that H_0/N_0 is not the conventional mean field term for the potential energy, which is usually taken to be $<\alpha|\,H\,|\alpha>$. There is a term in ϕ', and the coefficient of ϕ'' is α-dependent. By well-known techniques we can eliminate the ϕ' term or, better, recast the equation into the Hermitean form

$$[-\frac{d}{d\alpha}\frac{1}{2B(\alpha)}\frac{d}{d\alpha} + V(\alpha) - E]\tilde{\phi}(\alpha) = 0 . \tag{12}$$

It is an "elementary" exercise to relate $(\tilde{\phi},V,B)$ to (ϕ,H_n,N_n). Later we define α such that $\alpha \to r$ as $r \to \infty$, where r is (say) the separation of two bags. However, $V(\alpha)$ has no simple interpretation as a potential until $B(\alpha)$ is determined!

Following Mosel,[13] we now introduce a change of variable from α to x such that

$$x = X(\alpha) = \int^{\alpha}[B(\alpha')/\mu]^{\frac{1}{2}}d\alpha' , \tag{13a}$$

$$\tilde{\phi}(\alpha) = f(\alpha)\ \psi(x) , \tag{13b}$$

$$f(x) - \text{const.}\quad [\mu B(\alpha)]^{\frac{1}{4}} , \tag{13c}$$

where $\mu = M/2$ is the reduced nucleon mass. Then Eq. (12) becomes

$$(-\frac{1}{2\mu}\frac{d^2}{dx^2} + V + V_1 - E)\ \psi(x) = 0 , \tag{14}$$

with

$$V_1 = \frac{1}{8}\frac{B''}{B^2} - \frac{1}{3}\frac{(B')^2}{B^3} \tag{15}$$

Since we choose α so that for separated bags $\alpha \to r$ (the bag separation) this is consistent with $B \to \mu$, $x \to \alpha$.

We determine $|\alpha\rangle$ by minimizing the expectation value of the total Hamiltonian, $\langle\alpha|H|\alpha\rangle$ with respect to a variational mean field wave function for the quarks and a coherent state wave function for the soliton field subject to a constraint

$$\langle\alpha|Q|\alpha\rangle = Q_0(\alpha) , \tag{16}$$

where Q is some moment of the quark distribution

$$Q = \int\bar{\psi}\ q(\vec{r})\psi\ d^3r . \tag{17}$$

The constrained mean field equations now assume the form

$$(\vec{\alpha}\cdot\vec{p} + \beta[g\sigma_0(\vec{r}) - \lambda q(r)] - \varepsilon_k)\ \psi_k = 0 , \tag{18a}$$

$$-\nabla^2\sigma_0 + U'(\sigma_0) + g\sum_{k\text{-occ}}\bar{\psi}_k\psi_k = 0 , \tag{18b}$$

where λ is a Lagrange multiplier. Instead of specifying the constraint function q(r) explicitly and solving the pair of equations (18 a&b), it is more physical to specify the function in square brackets

$$[g\sigma_0(\vec{r}) - \lambda q(r)] \equiv \mathcal{V}(\vec{r}) \ , \tag{19}$$

then solve for the $\psi_k(\vec{r})$, and then for $\sigma_0(r)$; there is no iteration involved. The self consistency is implicit. The constraint can now be "discovered" by solving

$$\lambda q = g\sigma_0 - \mathcal{V} . \tag{20}$$

We parameterize $\mathcal{V}(\vec{r}) = \mathcal{V}(\alpha,\vec{r})$ by a folding procedure. We consider two spheres of radius R with centers separated by a distance α, as shown in Fig. 3. For $\alpha > 0$, we define a function $\Theta(\vec{r})$ equal to unity if $\vec{r}$ lies in the interior of either sphere and zero otherwise. If $\alpha < 0$, we choose $\Theta(\vec{r})$ to equal unity only in the intersecting, lens-shaped volume; this gives a natural continuation of α from positive values (prolate shapes) to negative values (oblate shapes). The radii of the spheres are chosen so that the enclosed volume is independent of α, and equals $2\times(4/3)\ \pi R_0^3$, where R_0 is the radius of a three quark bag. (The constant volume approximation, which is valid for a relativistic Fermi gas, can be relaxed to, say, minimize the energy as a function of R for each α, or to let R be another shape parameter. α can assume all values, $-\infty < \alpha < +\infty$, and is equal to the separation of isolated bags for $\alpha > 2R_0$.

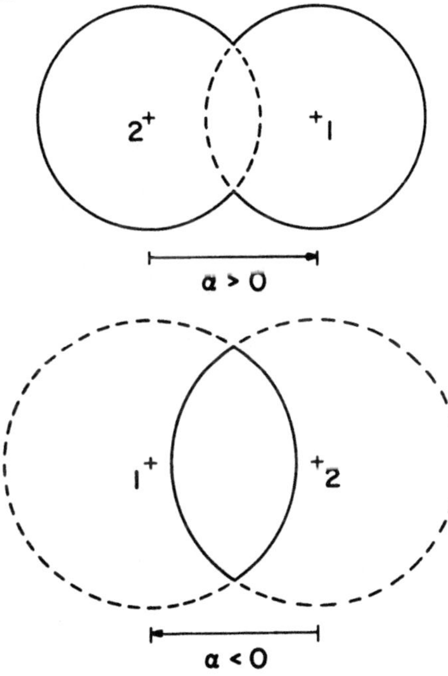

Fig. 3. Geometric shapes used to define $\Theta(\vec{r})$ and α.

Into this geometric shape is folded a Yukawa smoothing function, yielding

$$\mathcal{V}(\vec{r}) = g\sigma_v \left[1 - \frac{\mu^2}{4\pi} \int \frac{e^{-\mu|\vec{r}-\vec{r}\,'|}}{|\vec{r}-\vec{r}\,'|} \Theta(\vec{r}\,')\, d^3\vec{r}\,'\right] . \tag{21}$$

Note that $\mathcal{V} \to g\sigma_v$ as $r \to \infty$ and $\mathcal{V} \cong 0$ for $\vec{r}$ well inside the geometric volume. It is adjusted to approximate the self consistent, unconstrained spherical solution for isolated bags.

The method of solution of equations (18 a&b) is a generalization of that described in ref. (5). Here, however, both $\mathcal{V}(\vec{r})$ and $\sigma(\vec{r})$ are expanded in terms of even Legendre polynomials, and the quark functions are expanded in terms of Dirac spinors ot good quantum number k. An example of the shape of $\mathcal{V}(\vec{r})$ is shown in Fig. 4. Low lying eigenvalues of the Dirac equation as a function of α are shown in Fig. 5. Parity is a good quantum number for the quark functions, and we note that for well-separated bags the eigenvalues become doubly degenerate with respect to the two parities, corresponding to degenerate left and right states. All of the GCM calculations described here were performed by A. Schuh, and the work is continuing.

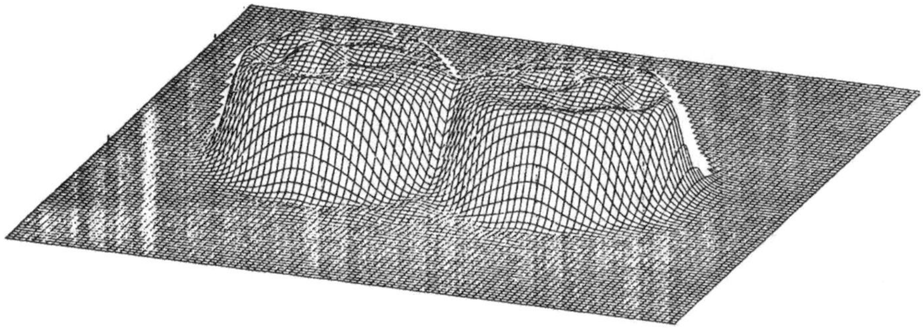

Fig. 4. The function $-\mathcal{V}(r)$ for $\alpha = 2$ fm and $R = 1$ fm.

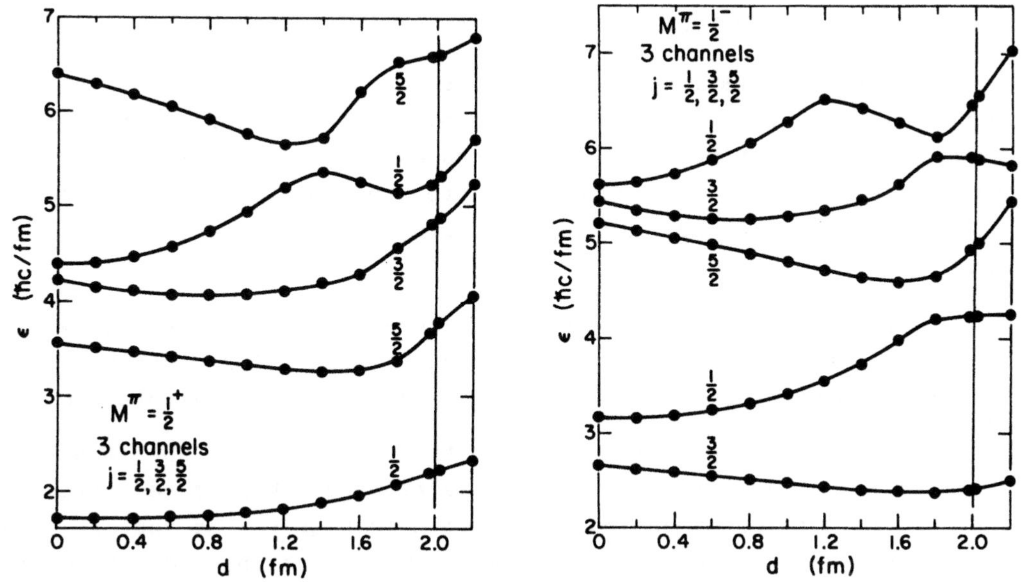

Fig. 5. Low lying quark eigenenergies calculated coupling three κ-states. Calculations coupling six states become flat, and exhibit degeneracy for + and - parities, for $\alpha > 2$ fm.

The GCM described to this point uses a static basis set, $|\alpha\rangle$. It is well known that for translational or rotational motion, where the GCM reduces to projection, the method fails to reproduce the proper mass or moment of inertia. The rectification of this problem in those cases has been addressed by many authors. We follow the self-consistent cranking method of Haff and Wilets[14], which builds dynamics into the wave function. In brief, we replace

$$|\alpha\rangle \rightarrow e^{i\beta(Q-\langle Q\rangle)}|\alpha\rangle \tag{22}$$

where β can be identified with the momentum conjugate to $\langle\alpha|\,Q\,|\alpha\rangle = Q_0(\alpha)$ in the resulting differential equation for $\phi(\alpha)$; that is,

$$\beta = -i\ \partial/\partial\langle Q\rangle = -i(dQ_0/d\alpha)^{-1}\ \partial/\partial\alpha. \tag{23}$$

In the translational case, where $Q = R_{GM}$(the center-of mass operator), $Q_\alpha = \langle Q\rangle = Z$ (the center of the bag), $\beta = P$ (the total momentum), $\phi = e^{iPZ}$, then the result is completely equivalent to projection into the zero momentum state followed by boosting to a state of some finite momentum:

$$e^{iPR_{cm}} \int_{-\infty}^{\infty} dZ\,|Z\rangle.$$

Further calculations are in progress, of which we mention a few: the structure of the virtual mesonic cloud around the nucleon; further corrections to hadronic spectra and the evalaluation of decay widths; transition from nucleonic matter to the quark-gluon plasma as function of density and temperature, using periodic (lattice) boundary conditions for the soliton field.

We have not begun, but would recommend to others who have the machinery, to do time-dependent mean field (TDMF) calculations, analogous to time-dependent relativistic Hartree-Fock calculations.

We conclude by re-emphasizing the utility of the model in calculating dynamical processes. Reactions which involve collision, fission, fusion, creation and oscillation of bags can be and are being calculated using the well-developed techniques of nuclear collective models. Such processes are difficult, at best, to calculate in other relativistic bag models.

We wish to thank our many collaborators in these projects, J. Achtzehnter, M. Betz, M. Bickeböller, J-L. Dethier, I. Duck, E. M. Henley, R. Horn, G. Lübeck, J. Rehr, A. Schuh, and E. Umland.

References

* Present address: Mission Research Co., P.O. Drawer 719, 735 State Street, Santa Barbara, CA 93102.

1. A. Chodos, R. L. Jaffe, K. Johnson and C. B. Thorn, Phys. Rev. D 10, 2599 (1974).

2. G. E. Brown and M. Rho, Phys. Lett. 82B, 177 (1979); G. E. Brown, M. Rho and V. Vento, ibid 84B, 383 (1979); A. Chodos and C. B. Thorn, Phys. Rev. D12, 2733 (1975); F. Myhrer, G. E. Brown and Z. Xu, Nucl. Phys. A362, 377 (1981).

3. G. A. Miller, A. W. Thomas and S. Theberge, Phys. Lett. 91B, 192 (1980); S. Theberge, A. W. Thomas and G. A. Miller, Phys. Rev. D 22, 2838 (1980); A. W. Thomas, S. Theberge and G. A. Miller, ibid 24, 216 (1981).

4. R. Friedberg and T. D. Lee, Phys. Rev. D 15, 1694 (1977); D 16, 1096 (1977); D 18, 2623 (1978); T. D. Lee, Particle Physics and Introduction to Field Theory (Harwood Academic, New York, 1981).

5. R. Goldflam and L. Wilets, Phys. Rev. D 25, 1951 (1982).

6. T. Celik, J. Engels and H. Satz, "The Latent Heat of Deconfinement in SU(3) Yang-Mills Theory" (Univ. of Bielefeld preprint BI-YP 83/07)

7. K. Ishikawa, G. Schierholtz and M. Teper, Phys. Lett. 116B, 429 (1982).

8. T. DeGrand, R. L. Jaffe, K. Johnson, and J. Kiskis, Phys. Rev. D 12, 2060 (1975).

9. E. Eichten *et al.*, Phys. Rev. D 21, 203 (1980).

10. J-L. Dethier, R. Goldflam, E. M. Henley and L. Wilets, Phys. Rev. D 27, 2191 (1983).

11. M. Betz and R. Goldflam, Phys. Rev. D (accepted for publication)

12. M. Gell-Mann and M. Levy, Nuovo Cimento 16, 705 (1960); M. C. Birse and M. K. Banerjee, "A Chiral Soliton Model of the Nucleon and Delta," U. of Maryland Preprint No. 83-201.

13. U. Mosel, Particles and Nuclei, 3, 297 (1972).

14. P. K. Haff and L. Wilets, Phys. Rev. C 10, 353 (1974)

QUARK DISTRIBUTIONS IN NUCLEI AND DEEP INELASTIC SCATTERING OF CHARGED LEPTONS

K. Rith
Fakultät für Physik
Universität Freiburg
Hermann-Herder-Str. 3
D-7800 Freiburg i. Br.

1. INTRODUCTION

Deep inelastic lepton (electron, muon, neutrino) scattering experiments are the basic tool to study the internal structure of hadronic matter. The results of these experiments have established our understanding of nucleons as a collection of quarks and gluons and have provided us with a lot of information about the momentum distribution of these constituents inside the nucleon and in general about their interactions as described by Quantum Chromodynamics (QCD). Recently the European Muon Collaboration (EMC) has discovered in a muon scattering experiment off deuterium and iron targets that these distributions are different for nucleons, which are embedded in nuclei, and for free nucleons {1}. These results caused a lot of excitement and they may help a lot for the understanding of nuclear physics in terms of quarks and gluons, which is the general topic of this conference. In this contribution I will briefly recall the phenomenology of deep inelastic scattering experiments, discuss the experimental data and their consequences and review the theoretical models which are available at present.

2. PHENOMENOLOGY OF DEEP INELASTIC SCATTERING OF CHARGED LEPTONS

In a deep inelastic scattering process the charged lepton ℓ (electron, muon) interacts with a nucleon via the exchange of a virtual photon γ or a neutral vector boson Z_o. (In the following discussion I will neglect the Z_o, because at present accelerator energies its contribution to the cross section is very small {2}). In the Quark Parton Model {3} the virtual photon couples to the charge of one of the pointlike, quasifree quarks which might be either one of the valence quarks (u_v, d_v) or a member of the quark-antiquark pairs contained in the 'sea'(u_s, $\bar{u}_s$, d_s, $\bar{d}_s$, s_s, $\bar{s}_s$, c_s, $\bar{c}_s$, ...). The process is shown in fig. 1.

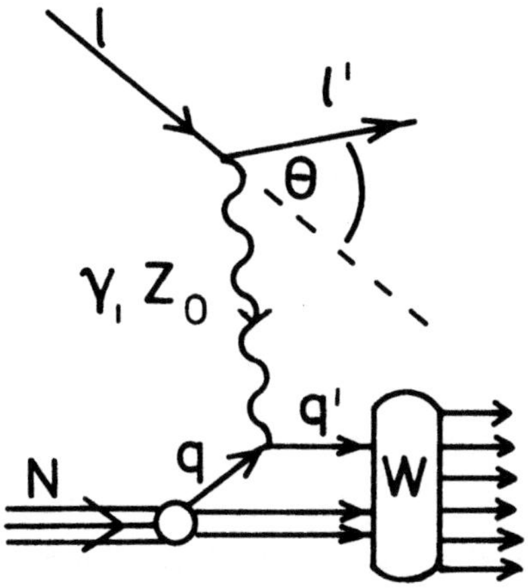

Fig. 1. Quark-Parton-Model picture of deep inelastic charged lepton nucleon scattering.

The two quantities which controll the interaction are Q^2 and x. Q^2 is the negative square of the fourmomentum carried by the virtual photon and is in the laboratory system given by $Q^2 = 2EE'(1-\cos\theta)$, E and E' being the energies of incoming and outgoing lepton and θ the lepton scattering angle. Q^2 is a measure of the spatial resolution $\lambda \sim 1/\sqrt{Q^2}$ at which the nucleon is probed. Present muon experiments reach values of Q^2 around 200 GeV^2 which correspond to a resolution of about 10^{-15}cm. The quantity x is defined as $x = Q^2/2M\nu$, where M is the nucleon mass and $\nu = E-E'$ is the energy transferred by the virtual photon from the lepton to the hadronic final state of mass W. In an infinite momentum frame {4} x can be interpreted as the fraction of the nucleon momentum carried by the interacting quark. It can be selected by a suitable choice of Q^2 and ν or E, E' and θ, respectively.

In the one photon exchange approximation the double differential cross section in terms of these two quantities can be written, using $y = \nu/E$, as

$$\frac{d^2\sigma}{dQ^2dx} = \frac{4\pi\alpha^2}{Q^4}\left\{ (1-y-\frac{Mxy}{2E})\frac{F_2(x,Q^2)}{x} + F_1(x,Q^2)y^2 \right\} \qquad (1)$$

Apart from kinematical factors this cross section depends on a combination of two structure functions $F_1(x,Q^2)$ and $F_2(x,Q^2)$. These two structure functions contain informations about the charge and spin of the quarks similar to the electromagnetic form factors in the case of elastic electron-nucleon and electron-nucleus scattering. The form factors can be related to the Fourier transforms of the spatial distribution of charge and magnetic moment of the nucleon or nucleus and show a strong Q^2 dependence because these are extended objects. In the case of scattering off point-like particles the Fourier transform is a constant and therefore no Q^2 dependence of the structure functions should be observed, what in first approximation is indeed the case.

F_1 and F_2 are linked by the ratio $R = \sigma_\ell/\sigma_t$ of the longitudinal to transverse virtual photon cross section via the expression

$$2\,x\,F_1 = F_2\,(1 + Q^2/\nu^2)/(1 + R). \qquad (2)$$

For spin $-\frac{1}{2}$ constituents $\sigma_\ell = 0$ and $R = 0$ and consequently

$$2\,x\,F_1 \approx F_2\,,$$

while for spin 1 constituents $\sigma_t = 0$, $R = \infty$ and $F_1 = 0$. Experimental results show {5} that at high Q^2 R is approximately equal to zero: quarks indeed carry spin $\frac{1}{2}$. The structure function F_2 is related to the quark charges and is given by

$$F_2(x) = \sum_f Q_f^2\,x\,(q_f(x)+\bar{q}_f(x)), \qquad (3)$$

where the sum runs over the different quark flavours, Q_f is the charge number of the quarks and $q_f(x)$ ($\bar{q}_f(x)$) is the probability that a quark (antiquark) of flavour f carries the fraction x of the nucleon momentum. The quantities $q_f(x)$ are also called quark distribution functions.

Let us apply this expression to the cases where the scattering takes place off a deuterium or a iron target.

If there are no binding effects, the structure function F_2^N per nucleon for the deuteron is given by the sum of the structure functions for a free proton and a free neutron

$$F_2^N\,(D) = \frac{1}{2}\,(F_2^p + F_2^n)$$

$$= \frac{5}{18}\,x\,\{u(x) + \bar{u}(x) + d(x) + \bar{d}(x) + s(x) + \bar{s}(x) + c(x) + \bar{c}(x)\}$$

$$- \frac{1}{6}\,x\,\{c(x) + \bar{c}(x) - s(x) - \bar{s}(x)\} \qquad (4)$$

The distributions u(x) and d(x) contain contributions from valence and sea quarks:

$u(x) = u_v(x) + u_s(x)$, $d(x) = d_v(x) + d_s(x)$, while the other distributions correspond to seaquarks only.

If there are no nuclear effects the structure functions F_2^N per nucleon for iron is given by

$$F_2^N\ (\mathrm{Fe}) = \frac{1}{56}\ (30\ F_2^n + 26\ F_2^p) = F_2^N\ (\mathrm{D})\ (1-K(x)), \tag{5}$$

with

$$K(x) = \frac{1}{14}\ \frac{F_2^p - F_2^n}{F_2^p + F_2^n} \tag{6}$$

K(x) is a correction factor which takes into account that iron is no isoscalar target and that proton and neutron structure functions are different. Approximating F_2^n by {6}:

$$F_2^n \approx (1-0.75x)\ F_2^p\ ,$$

one gets for x = 0.65, the highest x-value covered by the EMC data, a correction of $\sim$ 2.3%.

The x dependence of the structure function F_2 and of the contributions coming from the valence quarks and from the sea quarks is shown in fig. 2.

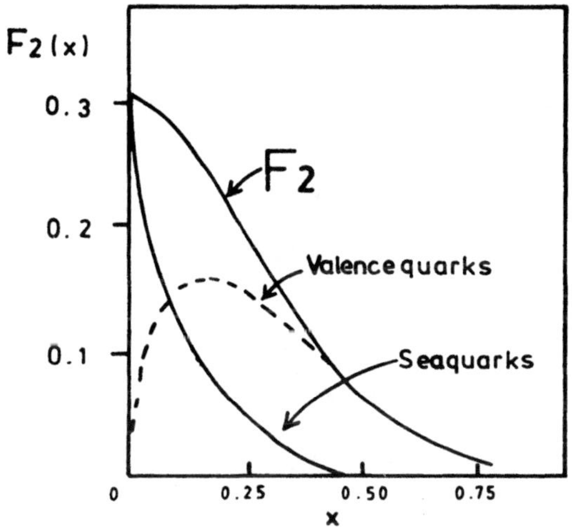

Fig. 2. The nucleon structure function F_2.

It must be mentioned here that the separation in valence and sea part can not be done in a direct way with electron or muon scattering experiments alone, for the virtual photon just couples to the charge of the quarks independent of their flavour. This is different in the case of neutrino scattering. Neutrinos and antineutrinos couple differently to different flavours and to quarks and antiquarks and the sea quark distribution can be extracted by a combination of antineutrino and neutrino cross sections {7}.

As can be seen from this figure sea quarks dominate at low x but their contribution has disappeared at values of x around 0.4 - 0.5.

Gluons do not contribute directly to the cross section obtained in deep inelastic scattering. But they show up indirectly. If one sums over all momentum fractions x and all probabilities q(x), this sum should yield 1 and the momentum integral $\int_0^1 F_2^N \, dx$ of equation (4) should give $\frac{5}{18}$. Experimentally this integral is only around 45% of the expected value, which shows that only 45% of the nucleon momentum is carried by charged quarks, the remaining 55% must be carried by neutral constituents, the gluons. Therefore the momentum sum rule has to be modified to

$$\int_0^1 (F_2^N + x\, g(x))\, dx = \frac{5}{18} \tag{7}$$

g(x) is the momentum distribution of the gluons inside the nucleon.
g(x) and q(x) are correlated by the elementary QCD processes of gluon emission by quarks and the creation of quark-antiquark pairs out of gluons.

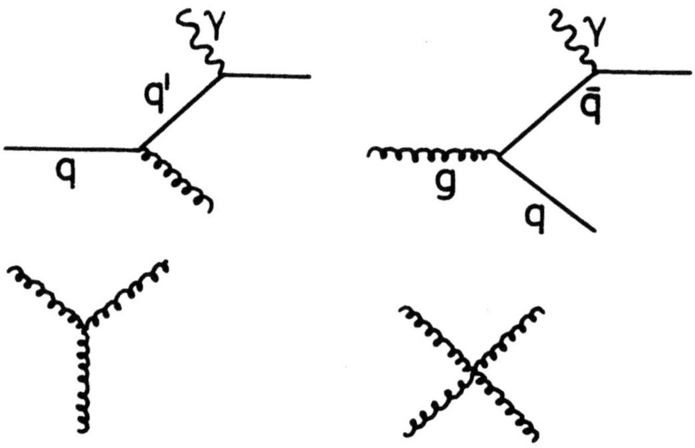

Fig. 3. First order QCD processes

Due to these processes and gluon-gluon interactions shown in fig. 3 the momentum distributions and the structure functions become Q^2-dependent: they are depleted with increasing Q^2 at high x and enhanced at low x as shown in fig. 4a as a function of x for different Q^2 and in fig. 4b as a function of Q^2 for fixed values of x.

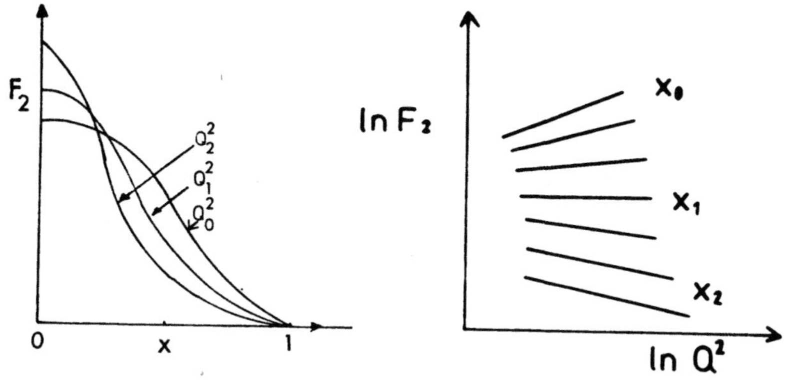

Fig. 4. The x-Q^2 dependence of F_2 due to QCD processes.
a) for fixed $Q^2 (Q_o^2<Q_1^2<Q_2^2)$, b) for fixed x $(x_o<x_1<x_2)$

The coupling of g(x) and q(x) and their logarithmic Q^2-dependence is expressed by the Altarelli-Parisi-equations {8}. Thus the analysis of the pattern of scale breaking of the structure functions with Q^2, as shown in fig. 4b, in terms of these equations allows a determination of the gluon distribution g(x) and its Q^2-dependence {9}.

Up to now we have only discussed the case of free nucleons and we have assumed that, apart from simple kinematical effects due to the Fermi motion of the nucleons, the quark and gluon distributions are not altered due to the neighbouring nucleons inside the nucleus. This might be a good approximation for the deuteron which is only a loosely bound system with a mean center to center nucleon distance of about 4 fm compared to the mean nucleon charge radius of $\sim$ 0.8 fm. But already in medium-weight nuclei the nucleons are packed much tighter together. Therefore there might be a lot of nuclear effects which can change the quark and gluon distributions compared to the free nucleon case.

3. THE EXPERIMENTAL DATA

Indeed nuclear effects in deep inelastic muon scattering at high Q^2 have been observed by the EMC by comparing structure functions for iron and deuterium. This group has measured the structure functions for hydrogen {10}, deuterium {11} and iron {12}. Apart from the different targets the identical apparatus has been used for these measurements and the data have been treated with the same analysis chain. Therefore a comparison can be made where the systematic errors are well understood. The apparatus is shown in fig. 5. It is fully described elsewhere {13}.

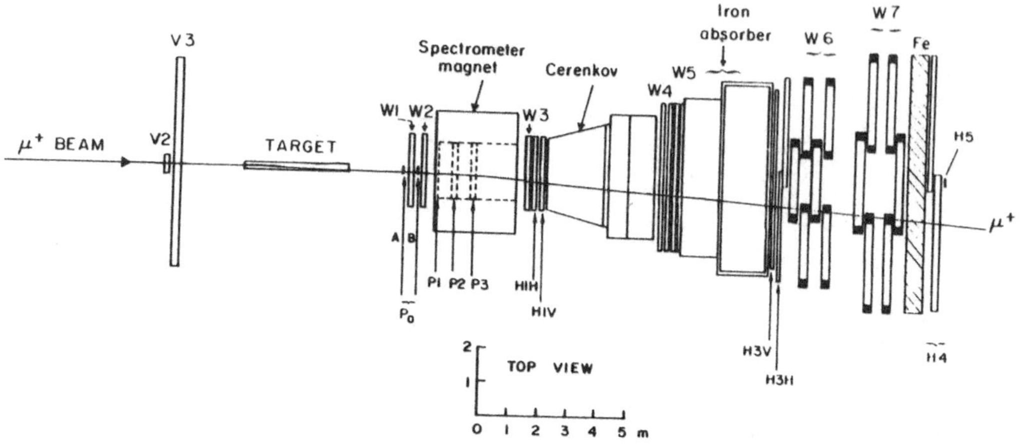

Fig. 5. The EMC Forward Spectrometer

The muon beam was incident either on a 6 m long target of liquid hydrogen or deuterium or on an iron scintillator target. Momentum and direction of the incoming muons were determined in a set of scintillator hodoscopes with high time and space resolution. Scattered muons were detected and momentum analyzed by a system of drift-and multi-wire-proportional-chambers W1 - W5, P1 - P3 and a large aperture (1 x 2 x 4 m^3) air gap dipole magnet and recorded in drift chambers W6, W7 after they had penetrated through a lead-iron-scintillator calorimeter and a 2 m thick iron hadron absorber.

The iron data have been obtained at four muon beam energies of 120, 200, 250 and 280 GeV and contain altogether one million events. The values of the nucleon structure function F_2^N (Fe) $= \frac{1}{56} F_2^{\mu Fe}$ are shown in fig. 6a as a function of Q^2 for fixed values of x. The data are corrected for the nonisoscalarity of iron according to equation (6) so that they correspond to a hypothetical target with the same number of protons and neutrons, no Fermi-motion corrections have been applied. The errors shown are statistical only. Apart from statistical fluctuations there is perfect agreement between these four data sets and the x and Q^2 dependence is very well defined. The rise with Q^2 at low x and the fall with Q^2 at high x, as predicted by QCD, shows up nicely. The solid lines represent a simple parametrization of the iron data, the dashed lines indicate where one would expect the deuterium data relative to the iron data according to Fermi motion model predictions a la Bodek and Ritchie {14}. Predictions of several nuclear Fermi motion models are shown in fig. 7 where the ratio of the structure function F_2^A for a nucleus, with mass number A, to the sum of the free nucleon structure functions for proton and neutron, weighted with the corresponding nucleon numbers, is plotted as a function of x for a iron nucleus. For the deuteron these corrections are expected to be much

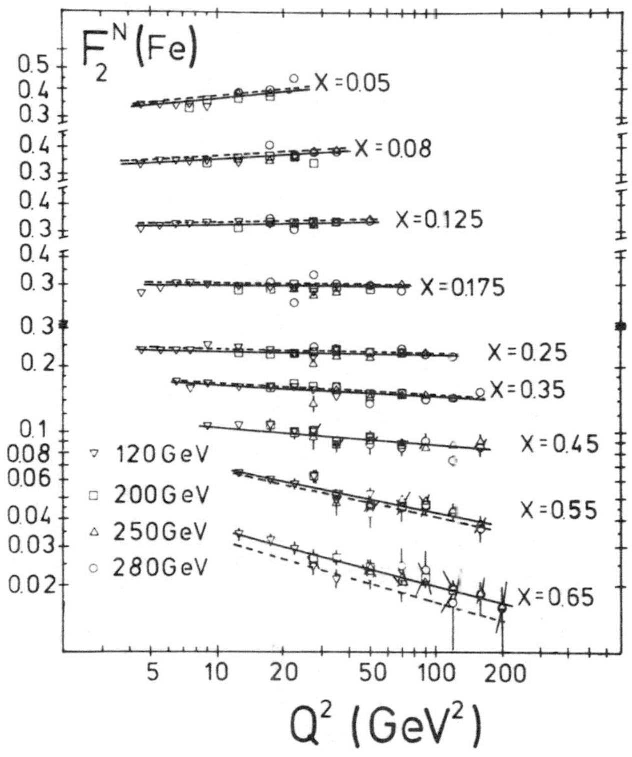

Fig. 6a. The nucleon structure function F_2^N(Fe) for iron as measured by the EMC.

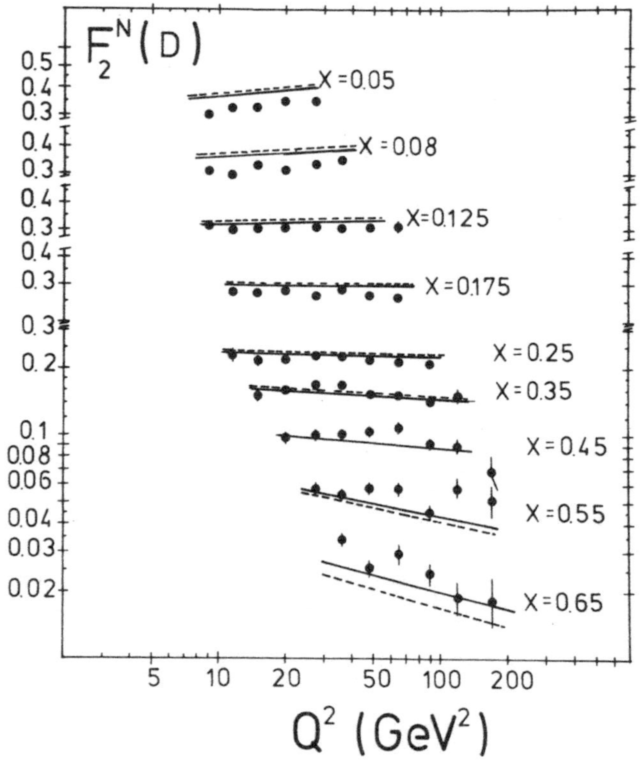

Fig. 6b. The nucleon structure function F_2^N(D) for deuterium as measured by the EMC (muon beam energy 230 GeV)

smaller as can be seen by a comparison with the curve labeled D which shows the deuterium corrections according to Bodek and Ritchie. For discussion of the different Fermi motion models {15} see reference {16}.

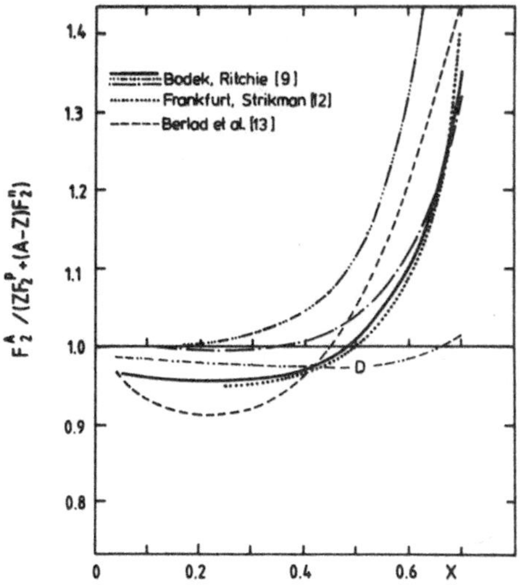

Fig. 7. Fermi motion model calculations for the ratio of the nucleon structure function F_2^A for iron and the weighted sum of the structure functions for proton and neutron.

Deuterium data have been obtained with three muon beam energies of 120, 200 and 280 GeV. The values of $F_2^N\,(D) = \frac{1}{2}\,F_2^{\mu D}$, not corrected for Fermi motion, are shown in fig. 6b for the data set taken at 280 GeV. There are negligible changes if the preliminary 200 GeV data are included, the 120 GeV data are still being analyzed. The curves are the same as in fig. 6a, which means that the deuterium data should well be represented by the dashed line if the simple kinematical Fermi-motion models would be correct which assume that the nucleus is a dilute gas of slowly moving nucleons. Obviously this is not the case. At low x the structure function for the iron should be 2-3% smaller than for deuterium but it is larger by about 15%, at high x the iron data should be larger than the deuterium data by about 20-25%, but they are smaller by about 15%. This means that at x around 0.65 there is a discrepancy of 35-40% between Fermi motion model predictions and experimental data. Within the limits of statistical and systematic errors there is no or little Q^2 dependence of this behaviour within the region of overlap of the datasets, which varies from $9 \leq Q^2 \leq 27$ GeV^2 for x = 0.05 over $11.5 \leq Q^2 \leq 90$ GeV^2 for x = 0.25 up to $36 \leq Q^2 \leq 170$ GeV^2 for x = 0.65.

The ratio of the measured nucleon structure functions $F_2^N(Fe)/F_2^N(D)$ has been calculated point by point and then averaged over Q^2 for each value of x. The x dependence is shown in fig. 8 where the error bars are statistical only. The ratio is falling from around 1.15 at x = 0.05 to a value around 0.89 at x = 0.65. A fit of the form

$$F_2^N(Fe)/F_2^N(D) = a + bx \tag{8}$$

yields a = 1.18 ± 0.01, and for the slope

b = − 0.52 ± 0.04 (stat) ± 0.21 (syst).

A conservative estimate of the possible effect of the systematic uncertainties on the slope is indicated by the shaded area in figure 8.

In terms of quark distributions this result means that a high x (valence quark region) a depletion of the quark momentum distribution in iron comparent to the momentum distribution in a free nucleon is observed, opposite to Fermi motion model predictions, while at low x (sea quark region) where only little difference is predicted, there is an increase.
After the EMC result became known, the SLAC-MIT-Rochester group has realized that the target walls used in SLAC electron-proton and electron deuterium experiments E49 and E87 {17} were constructed from aluminium and steel respectively. The target-wall contributions were measured using empty target replicas, subtracted from the full target rates and then forgotten. These data have been recovered and reanalyzed. The resulting ratios of the cross sections for aluminium to deuterium and steel to deuterium are shown in figures 9a and 9b {18,19}. They span a range of Q^2. The mean Q^2 for the steel data varies from 6.0 GeV^2 (x=0.25) to 19.3 GeV^2 (x=0.9) and for the aluminium data from 1.9 GeV^2 (x=0.075) over 5.2 GeV^2 (x=0.40) to 20.1 GeV^2 (x=0.863). The systematic errors for these data are calculated to be ± 2.3%. Also shown in figure 9b is the EMC result at larger values of Q^2. In the region of overlap the agreement between the two experiments is excellent. A straight line fit to the SLAC data in the region $0.2 \leq x \leq 0.6$ yields a=1.15 ± 0.04 ± 0.011 and b = −0.45 ± 0.05, values which are very similar to the EMC result. The SLAC data extend to higher values of x where the cross section ratio is rising again, indicating that the effects of Fermi motion become dominant there. Also plotted in this figure is the ratio for copper and deuterium from a low Q^2 experiment at SLAC {20} and photoproduction results {21}, which show that at low Q^2 the ratio is considerably reduced due to shadowing effects.
The aluminium results, which are shown in figure 9a together with data from a lower Q^2 experiment at SLAC and photoproduction show a similar trend as that observed for steel, the effect being somewhat weaker. The results for a straight line fit yield

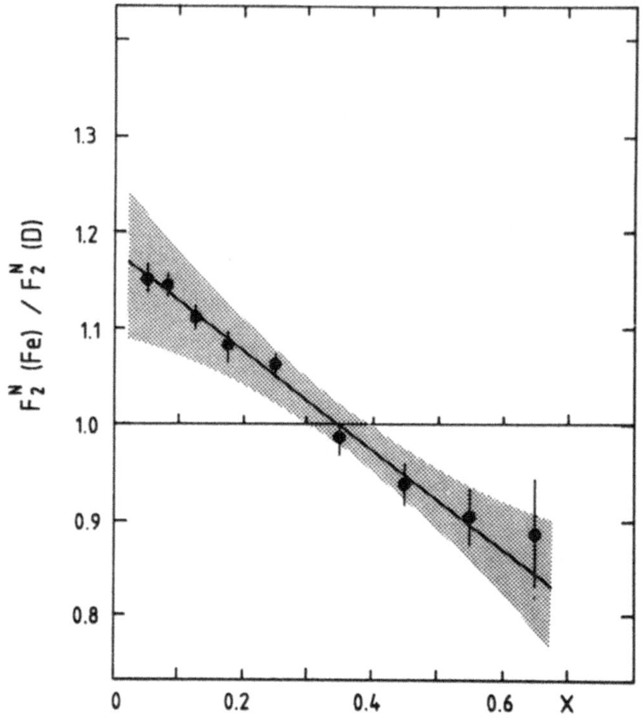

Fig. 8. The ratio of the nucleon structure function F_2^N for iron and deuterium measured by the EMC

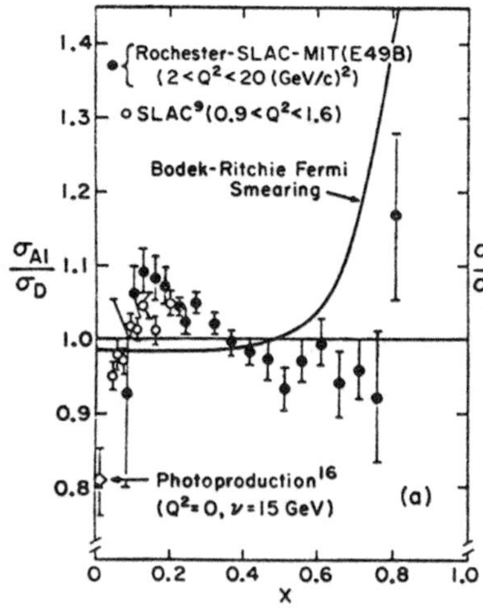

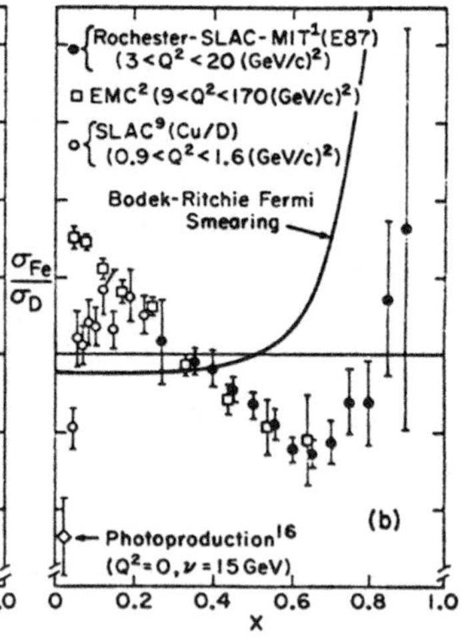

Fig. 9a. σ_{Al}/σ_D measured by SLAC

Fig. 9b. σ_{Fe}/σ_D measured by SLAC

$a = 1.11 \pm 0.02 \pm 0.23$ and $b = -0.30 \pm 0.06$, the slope being 1.5 standard deviations lower than for steel.

3. CONSEQUENCES

Taking all this informations together the following picture emerges, which is summarized in figure 10: quark distributions measured in deep inelastic scattering off nuclei $q_A(x)$ are far from being identical with those in free nucleons $q_{n,p}(x)$.
In the low x region up to x around 0.25-0.3, where sea quarks are important, nuclear effects cause a significant enhancement of $q_A(x)$ compared to the free nucleon case if Q^2 is large enough ($Q^2 \geq 10$ GeV^2?).

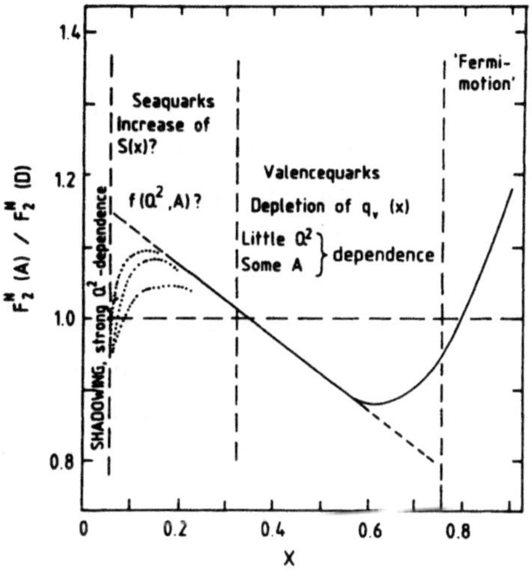

Fig. 10. Possible behaviour of nuclear effects in deep inelastic scattering.

At very low x values, which at present accelerator energies are only accessible at very low Q^2 and by photoproduction ($x=Q^2=0$), A dependent shadowing effects may play the dominant role. The combined effects of nuclear enhancement at low x and shadowing at low Q^2, which are due to completely different physics, may cause a strong Q^2 and A dependence of the observed effect. At medium x ($0.3 \leq x \leq 0.7$), the valence quark region, $q_A(x)$ is depleted opposite to what is expected from Fermi motion effects. The data indicate no or little Q^2 dependence of the effect for $Q^2 > 6$ GeV^2 while probably it is weakly increasing with mass number A. Finally, at even higher x(>0.7) the effects of Fermi motion become important, causing the effective quark distribution to extend far above x = 1, the kinematical limit being x = A. As a consequence also the gluon distribution in nuclei will differ from that in free nucleons as discussed in chapter 2. The best informations on the gluon and sea distributions have been extracted up to now from neutrino-iron and neutrino-marble structure functions {7} and multimuon data from iron targets {22}. Therefore we have little information on the gluon distribution in a free nucleon so far, but its knowledge is essential for input to calculations for other processes like jet productions in $p\bar{p}$ collisions which are dominated by gluon-gluon interactions.

Probably there is little effect on the QCD analysis of data obtained with nuclear targets {23,24}. But one should keep in mind that the Altarelli-Parisi-equations require an integration from x to 1. In the case of nuclear targets the integral has to be extended to x=A, the x dependence of the structure functions, even if made QCD compatible, being unknown. Fortunately the contribution of this region to the integral is suppressed by the splitting functions, therefore it may not influence the result very much. But obviously a QCD analysis combining datasets of different nuclear targets should no longer be done.

Of course there is some impact on the interpretation as well of nucleon-nucleus scattering data, where in the analysis the same quark distributions for incoming and target nucleon are being used, as of $p\bar{p}$ which are usually compared to quark distributions obtained from nuclear targets instead to those from hydrogen targets.

But there is also an exciting consequence. For the first time nuclear effects have been observed on the quark level. This measurement of the change of quark distributions due to the nuclear surrounding may be the starting point for a lot of other investigations and open the door for the understanding of nuclear physics in terms of quarks and gluons.

4. THEORETICAL INTERPRETATIONS

At the time when the EMC data became known the effect appeared quite surprising and many people argued that one could hardly think of any mechanism which could produce the observed behaviour. They obviously did not realize that there have been many publications {25} and a lot of conferences like this one {26} where 'quarks (mesons) in nuclei' have been the main topic and where several phenomena have been discussed which indeed can change the quark distributions compared to the free nucleon case. But to my knowledge there existed only one publication by Krzywicki {27} which, although somehow hidden in the context, predicted, how one of those effects could influence the nucleon structure functions obtained in deep inelastic lepton nucleus scattering.

In the meantime a lot of theoretical papers has been published (and the number is still rapidly increasing) in which several phenomenological models are being used to explain the effect. This audience has a lot of expertise concerning these models and we have heard several interesting presentations where they have been used for the description of other phenomena. Nevertheless let me (as an unexpert) try to summarize the main ideas and to put them into some order. For more details the reader is referred to the original references.

Before I start the discussion it is important to rise one basic question: is all effect due to iron or is it also caused by the deuterium? It is somehow funny that this question is not yet settled, although the deuteron is probably the nucleus which has been most extensively and carefully studied as well experimentally as theoretically. It is generally assumed that nuclear effects in the deuteron are small and well understood and that there are no other distortions to the quark distributions than those from Fermi motion effects. Experimental results indicate that this assumption is correct {28}. But recently there is again a lot of discussion going on and Frankfurt and Strikman {29} argue that nuclear effects inside the deuteron cause much larger corrections than the usual Fermi motion corrections {30,31}. Another example is the work by Glazek {32} which has been presented in a comment at this conference. He calculates that the assumption of the structure function being the sum of the structure functions for a free neutron and a free proton can be wrong on the 5-10% level already at low and medium x. At the moment this is still an open question and it might be that in the near future results concerning the ratio and the difference of neutron and proton structure functions extracted from deuterium data have to be revised.

Having this warning in mind let us turn to the existing theoretical explanations of the EMC effect which all assume that the observed effect is totally due to iron.

I have divided the models into different classes. Most of the model calculations can easily be sorted into one of them, but there are also a few which can not so easily be classified because they contain several ideas. If they are not as well presented as they would have served it, I apologize for this.

In the first approach part of the effect is explained by 'conventional nuclear physics'. It is based on the argument of Sullivan {33} that already the structure function for a free nucleon contains a contribution from the pion cloud associated with it: the virtual photon will either interact with a constituent of the nucleon or with a constituent of the surrounding pions. When several nucleons are brought together to form a nucleus there will be extra mesons associated with the exchange mechanisms responsible for the nuclear binding. It has been suggested by Llewellyn Smith {34} that these extra quasifree pions ($\sim$ 15% per nucleon) are responsible for the enhancement at low x. The measured structure function for iron contains an extra contribution which is a convolution of the probability $f_\pi(y)$ that a pion carries the fraction y of the nucleon momentum and the pion structure function F_2^{π}. Ericson and Thomas {35} have shown (thick full curve in fig. 13) that magnitude and x dependence of the effect are well reproduced (for x<0.3) with this assumption and reasonable nuclear physics parameters. Llewellyn Smith in addition argued that the momentum taken by these extra quasifree pions should show up in a depletion of the nucleons valence quark distribution at high x as it is observed in the data. The authors also conclude that in nucleons with large neutron excess π^- production should be favoured over π^+ production. In their model the enhancement should grow approximately as the density of the nucleus and saturate for heavy nuclei. The mean numbers of excess pions and the associated contribution of Δ-states as calculated by Friman et al.{36} offer a good gideline.

The influence of extra pions in nuclei on the nucleon structure function has also been studied in detail by Berger et al. {37}. In their approach they use a pion density $f_{\pi}(y)$ and a pion structure function F_2^{π} which is quite different from the one used by Ericson and Thomas. They find that the magnitude of the effect at very small $x(\leq 0.1)$ cannot be explained in the context of known constraints to F_2^{π} unless they postulate an enhanced pion density of about 0.4 extra pions per nucleon which is roughly three times the value suggested by nuclear models. Even with this large number they are unable to fit the full x dependence of the measured ratio. They therefore conclude that other effects have to be present in addition which cause an intrinsic distortion of the nucleons in a nuclear medium.

What are these intrinsic distortions? The models available in the literature can be roughly divided into three different classes: in the first class the individual nucleon bags stay intact and don't change their quark content, but the internal properties like mass, radius or spin might be modified due to the presence of other nucleons; in the second class some of the nucleon bags loose their identity and form multiquark clusters and in the third class finally quarks are able to flow through the whole nucleus which then in an extreme view can be understood as one big bag.

Change of internal properties. There are two extreme models for a change of the nucleon radius r_N inside nucleons. An increase of r_N by about 30% compared to the free nucleon radius r_{free} has been proposed {38} to explain inelastic electron-^{56}Fe data {39}, the other model {40} demands that nucleons inside a nucleus should be compressed to bags of very small radii by a very significant pion cloud. There is some doubt on the accuracy of both calculations {41,42}. Nevertheless one should keep in mind that there is no evidence to suggest that the nucleon should not change its size when put inside a nucleus. Furthermore a collective change of the structure of each individual nucleon is probably the most easiest way to produce the large effect seen in the data. A small increase of r_N (which corresponds to an inversely proportional decrease in mass) will reflect in a degradation of quark momenta to lower x.

Indeed Staszel et al. {43} get good agreement in the whole x range (dashed curve in fig. 11) with a decrease of effective nucleon mass (increase of nucleon radius) by about 13% and including Fermi motion corrections. Close et al. {44} conclude that an increase of confinement size of about 15% in iron compared to deuterium is needed to describe the data. I will come back to this model lateron in another context. Frankfurt and Strikman {29} argue that the observed behaviour is mainly a valence quark effect. In their model the increase at low x is due to antishadowing {45}, the relative enhancement at x around 0.1 for different nuclei should rise approximately like 1.8 log A. The behaviour at x>0.3 can be well explained by a swelling of nucleon size by only about 2.8% due to the nuclear surrounding. The predicted A-dependence of the effect at high x for mass numbers above around 12 is weaker than that at low x by about a factor of 2 while at lower mass numbers it is somewhat stronger. Also in another recent calculation by Levin and Ryskin {65} the EMC effect is explained by a swelling of nucleon size. To fit the experimental data they calculate an increase of the nucleon radius in iron of about 5.3 percent.

Although basically different the work of Fredriksson {46} somehow also fits into this class. In his model the nucleon is mostly in a bound quark-diquark state and the EMC effect is caused by an increase of radius of the dominating (ud) diquark in the nucleon surrounded by dense nuclear matter. This results in an change of Q^2-scale for two nuclei A_1 and A_2 by a multiplicative factor which is given by the ratio of the mean diquark radii squared for nuclei A_2 and A_1. Also Bjorken {47} interpretes the effect at high x as a change of the quark - diquark configuration due to the nuclear surrounding, without specifying in detail the mechanism for it.

A fraction of nucleons may transform into excited baryon states like Δ's. Szwed {48} claims that this fraction can be large in heavy nuclei. He calculates that a softening of the valence distribution can be explained if there is a 9-15% probability per nucleon of finding a Δ-isobar in a iron nucleus (dashed-dotted curve in fig. 11: 15% Δ-admixture; full curve: 9% admixture plus 3.3% decrease in gluonmomentum). With this fraction the data are well reproduced,but it seems unreasonably high compared to measurements and theoretical calculations {36,49}.

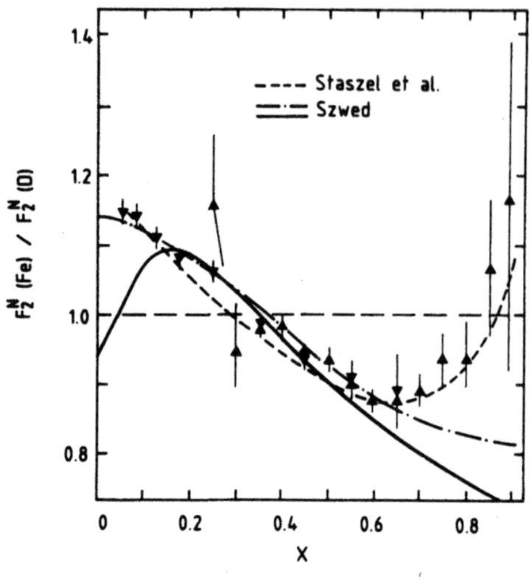

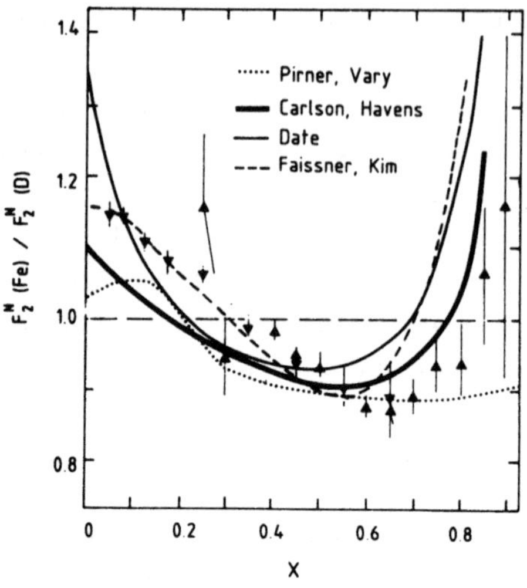

Fig. 11. Model calculations for the EMC effect-change of internal nucleon properties

Fig. 12. Model calculations for the EMC effect-multiquark states

Multiquark states. In the second class of models the quarks are no longer confined in their original nucleon bags but there is a certain probability that these bags can overlap and multiquark states are formed which contain six valence quarks ($|6q>$), nine valence quarks ($|9q>$), etc. This might be not so unlikely even at normal densities, for the mean nucleon distance in a high A nucleus is around 1.8 fermi compared to the free nucleon bag radius of around 1 fermi in the MIT bag model. Jaffee {50} was the first one to react on the experimental data and his paper contains several ideas which then have been picked up by other authors. He supposes that two nucleons may coalesce in $|6q>$ bags, having a bigger radius than $|3q>$ bags, the longer range being responsible for an enhancement of the quark distributions at low x. He calculates an increase of the sea in iron compared to deuterium of 60% and a softening as well of the valence as the gluon distribution. In figure 12 the results of several other calculations using this concept of multiquark states are shown. Pirner and Vary {51} (dotted curve) use a 16% probability for a nucleon being in a $|6q>$ cluster, the same number they have evaluated to fit inelastic ^3He data {52}. Carlson and Havens {53} (thick full curve) choose a fraction of 30% of nucleons to become part of $|6q>$ clusters which from counting rule arguments have much steeper x distributions for valence and sea quarks than for free nucleons. Date {54} (thin full curve) also includes clusters with more than

six valence quarks. He finds that sea quarks in multiquark states have to be enhanced while the gluons are rather suppressed compared to the free nucleon case. Faissner and Kim {55} use a very special class of multiquark clusters. They started from the observation that α-clusters play a dominant role in many nuclei. They assume that these α-clusters are bags of 12 valence quarks, the 6u and 6d quarks, forming a singlet in spin, isospin and colour, moving freely in it. This assumption, however, is in contradiction to other multiquark state calculations which predict only a very small fraction of 12-quark states in ^{4}He {56}. Including Fermi motion correction without high momentum tail for the nucleons and having an effective quark mass as additional parameter they get good agreement (dashed curve) with the data if (20 ± 5)% of the nucleons are contained in these $|12q\rangle$ clusters. They predict a very strong effect for ^{4}He and also for ^{12}C, ^{16}O and similar nuclei, but apart from this a roughly constant EMC effect throughout the periodic table, for the fraction of α clusters in normal nuclei is roughly constant.
Also another interpretation of the EMC effect by Cohen-Tannoudji and Navelet {57} is based on the idea of multinucleon clusters together with ingredients from QCD.

Quark percolation - the nucleus as one big bag. The last class of models goes even one step further and treats the nucleus as one big bag. If the baryon density inside a nucleus becomes big enough many nucleon bags may overlap in such a way that finally quarks will be able to move freely through the whole nucleus. This free flow of quarks (colour) is known as percolation {58}. In the extreme view the nucleus then can be treated no longer as an assembly of n nucleons or some nucleons plus multiquark bags but as one single bag inside which the 3n valence quarks and the sea quarks are completely delocalized. Calculations using this approach in another context have been performed by Bleuler et al. {59} and presented to this conference by Petry {60}. They show that characteristic nuclear properties connected with conventional shell structure are obtained in a natural way by starting directly from this enlarged scheme.
In the model of Nachtmann and Pirner {61} the quarks can extend over the whole transverse size of the nucleus with radius R_A. They conclude that in this case no longer Q^2 is the relevant parameter but $Q^2 \cdot R_A^2$. This means that for fixed x the structure function for deuterium measured at $Q'^2 = Q^2 \cdot (R_{Fe}^2/R_D^2)$ should be the same as for iron at Q^2. Thus the observed difference of Fe and D is purely a scale change similar to the model of Fredriksson {46} and of Close et al. {44} mentioned above where it is caused by the increase of diquark or of nucleon size, respectively. The latter authors have empirically determined a Q^2 scale factor of around 2 while Nachtmann and Pirner get a factor of $R^2_{Fe}/R_D^2 = 4.5$ which is clearly an overestimation of the effect. It follows automatically from these models that the difference of the structurefunction for all nuclei should disappear at the same x as the scaling violations do. In the Nachtmann-Pirner approach the deviation from unity for the ratio of the structurefunction for two different nuclei A_1 and A_2 at fixed x should in

first order be directly proportional to ln $(R_{A_1}^2/R_{A_2}^2)$ times the slope of ln $F_2(x,Q^2)$ versus ln Q^2 at this value of x. In my opinion the deuteron nucleus is too untypical to fit really in this 'big bag' picture and data from other nuclei (when being available) should be used to test the validity of this model. The picture of the nucleus as one big bag is also the basis of the work of Krzywicki {27}. He argued that in a snapshot of a nucleus one could not distinguish between virtual particles emitted and eventually reabsorbed by the same nucleon and those exchanged between different nucleons. He therefore treated the nucleus as one single bag of partons. The clue of his model is that this single bag contains an additional amount of sea quarks which then is responsible for the change of the nucleon structure function. A paper by Furmanski and Krzywicki {62} extends this work (dashed-dotted -curve in fig. 13). In their model the effect saturates for A>20, while in a similar approach by Godbole and Sarma {63} (thin full curve in fig. 13) it rises approximately with mass number A.

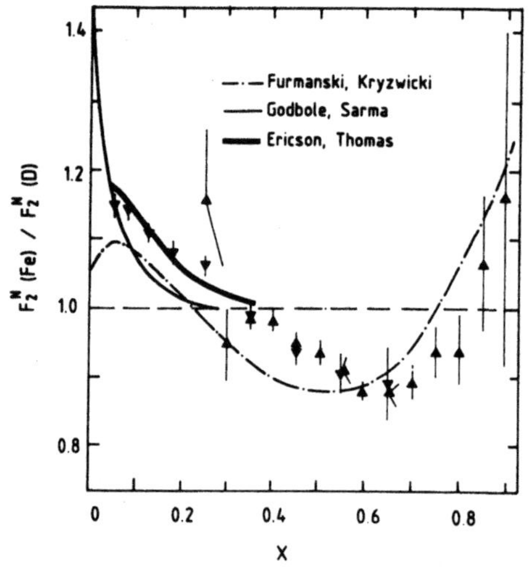

Fig. 13. Model calculations for the EMC effect-sea enhancement.

Finally there is another interesting paper by Barshay {64}. He predicts a strong effect in the tightly bound systems of He^3 and He^4 which is due to extra gluons and their self coupling. A part of the momentum is taken from the valence quarks to these extra gluons and a fraction of it again reappears in the sea quark distribution. The effect on medium-weight and heavy nuclei then would originate from the heliumlike clusters in these systems as it is proposed by Faissner and Kim {55}.

5. CONCLUSION

Deep inelastic scattering experiments with charged leptons have shown that the quark and gluon distributions in nuclei are different from those in free nucleons. The global features of the observed effect can be described by many 'simple' models but obviously more experimental information is needed to decide definitely which explanation is correct. As a first step a measurement of the A dependence has already been performed at low Q^2 in a SLAC experiment {66}. This has taken data with several targets ranging from deuterium and helium to gold. Preliminary results indicate a mild increase of the effect at high x with A up to the highest mass numbers, final results are expected shortly. A low angle experiment will be performed this autumn by the EMC to study for several nuclei the region at low x and low Q^2 where shadowing and nuclear effects seem to counteract each other. Measurements on several targets at higher Q^2 will be done by the two muon experiments EMC and BCDMS at CERN and data from different targets are also being compared by several neutrino experiments {67}. The EMC has already performed some studies of hadronic final states from nuclear targets and intends to continue this investigations with an upgraded apparatus. This group also tries to get out more information on the gluon distribution in different nuclei by comparing J/ψ events from hydrogen, deuterium and iron.

So there is a lot of experimental and theoretical work going on and all of this will hopefully help to develop a common understanding of nuclear and particle physics in terms of quarks and gluons.

REFERENCES

1. EMC, J.J. Aubert et al., Phys. Lett. 123B (1983) 275.
2. BCDMS, A. Argento et al., Phys. Lett. 120B (1982) 245.
3. For details see: F.E. Close, An Introduction to Quarks and Partons, Academic Press 1979.
4. S.D. Drell, T.M. Yan, Ann. Phys. (N.Y.) 66, (1971) 578.
5. EMC, J.J. Aubert et al., Phys. Lett. 121B (1983) 87.
6. A. Bodek et al., Phys. Rev. D20 (1979) 1471.
7. See for instance: CDHS, H. Abramowicz et al., Z. Phys. C 17 (1983) 283; CHARM, F. Bergsma et al., Phys. Lett. 123B (1983) 269.
8. G. Altarelli and G. Parisi, Nucl. Phys. B126 (1977) 298.
9. For a review see for instance: J. Drees, Proceedings of the 1981 Int. Symposium and Lepton and Photon Interactions at High Energies, ed. W. Pfeil, Bonn 1981, P. 474; F. Eisele, Proceedings of the 21st Int. Conference on High Energy Physics, ed. P. Petian and M. Porneuf, Paris 1982, P. C3-337.

10. EMC, J.J. Aubert et al., Phys. Lett. 105B (1981) 315.
11. EMC, J.J. Aubert et al., Phys. Lett. 123B (1983) 123.
12. EMC, J.J. Aubert et al., Phys. Lett. 105B (1981) 322.
13. EMC, J.J. Aubert et al., Nucl. Inst. Meth. 179 (1981) 445.
14. A. Bodek and J.L. Ritchie, Phys. Rev. D23 (1981) 1070; D24 (1981) 1400.
15. See ref. 14 and L.L. Frankfurt and M.I. Strikman, Nucl. Phys. B181 (1981) 22. G. Berlad et al., Phys. Rev. D22 (1980) 1547.
16. K. Rith, preprint Freiburg university THEP 83/4 (1983) (to appear in the Proceedings of the XVIII Rencontre de Moriond 1983).
17. A. Bodek et al., Phys. Rev. Lett. 30 (1973) 1087; A. Bodek et al., Phys. Lett. 51B (1983) 417; A. Bodek et al., Phys. Rev. D20 (1979) 1471.
18. A. Bodek et al., Phys. Rev. Lett. 50 (1983) 1431.
19. A. Bodek et al., Phys. Rev. Lett. 51 (1983) 534.
20. S. Stein et al., Phys. Rev. D12 (1975) 1884.
21. D.O. Caldwell et al., Phys. Rev. Lett. 42 (1979) 553; Phys. Rev. D7 (1973) 1362; A.M. Eisner in "Proceedings of the 1979 International Symposium on Lepton and Photon Interactions, Fermilab, Aug. 1979, p. 448; T.B.W. Kirk and H.D.I. Abarbanel ed.
22. EMC, J.J. Aubert et al., Nucl. Phys. B213 (1983) 1; BFP, A.R. Clark et al., Phys. Rev. Lett. 45 (1980) 682.
23. C.H. Llewellyn Smith, Phys. Lett. 128B (1983) 107.
24. C.H. Llewellyn Smith, Oxford University preprint 37/83 (1983).
25. For references see for instance: A.W. Thomas, CERN-TH 3368 (1982) (to appear in Advances in Nuclear Physics, Vol. 13 (1983).
26. See for instance: Nucl. Phys. A335 (1980); Nucl. Phys. A358 (1981); Nucl. Phys. A374 (1982); Progress in nuclear and particle physics, ed. D. Wilkinson, Vol 1 (1978), Vol 8 (1982); 'Mesons in nuclei', ed. M. Rho and D. Wilkinson, North Holl. 1979.
27. A. Krzywicki, Phys. Rev. D14 (1976) 132.
28. A. Bodek, University of Rochester, preprint UR 858, COO-3065-365 (1983) to appear in the Proceedings of the XIV International Multi Particle Symposium, Lake Tanoe, 1983.
29. L.L. Frankfurt and M.I. Strikman, Leningrad preprint 886 (1983).
30. G. West, Phys. Lett. 18 (1971) 509.
31. W.B. Atwood and G.B. West, Phys. Rev. D7 (1973) 773; L.L. Frankfurt and M.I. Strikman, Nucl. Phys. B148 (1979) 107.
32. S. Glazek, University of Warsaw Preprint 1983 and contribution to the International Europhysics conference of High Energy Physics, Brighton (UK), 1983, paper 27.
33. J.D. Sullivan, Phys. Rev. D5 (1972) 1732.
34. C.H. Llewellyn Smith, Phys. Lett. 128B (1983) 107.
35. M. Ericson and A.W. Thomas, Phys. Lett. 128B (1983) 112.
36. B.L. Friman et al., Phys. Rev. Lett. 51 (1983) 763.
37. E.L. Berger et al., Argonne preprint ANL-HEP-PR-83-24 (1983).
38. J.V. Noble, Phys. Rev. Lett. 46 (1981) 412.
39. R. Altemus et al., Phys. Rev. Lett. 44 (1980) 965.
40. G.E. Brown and M. Rho, Phys. Lett. 82B (1979) 177; G.E. Brown et al., Phys. Lett. 94B (1979) 383.
41. P.D. Zimmermann, Phys. Rev. C26 (1982) 265.
42. G.A. Miller, CERN TH-3516 (1983).
43. M. Staszel et al., University of Warsaw preprint IFT/9/83 (1983).
44. F.E. Close et al., Phys. Lett. 129B (83) 346.
45. N.N. Nikolaev and Z.I. Zakharov, Phys. Lett. 55B (1975) 397.
46. S. Fredriksson, Stockholm preprint TRITA-TFY-83-15 (1983).
47. J.D. Bjorken, unpublished note.
48. J. Szwed, Phys. Lett. 128B (1983) 245.
49. V. Bakken et al., Physica Scripta 19 (1979) 491; L.L. Frankfurt and M.I. Strikman, Phys. Rep. 76C (1981) 215.
M. Harvey, Nucl. Phys. A352 (1981) 326; M.H. Storm and A. Watt, Nucl. Phys. A to be published.
50. R.L. Jaffe, Phys. Lett. 50 (1983) 228.
51. M.J. Pirner and J.P. Vary, University of Heidelberg preprint UNI-HD-83-02 (1983).

52. H.J. Pirner and J.P. Vary, Phys. Rev. Lett. 46 (1981) 1376.
53. C.E. Carlson and T.J. Havens, Phys. Rev. Lett. 51 (1983) 281.
54. S. Date, Waseda University preprint WV-HEP-83-4 (1983) (submitted to Phys. Rev. Lett.).
55. H. Faissner and B.R. Kim, Phys. Lett. 130B (1983) 321.
56. M. Namiki et al., Phys. Rev. C25 (1982) 2157.
57. G. Cohen-Tannoudji and H. Navalet, Saclay preprint SPhT-83-97 (1983).
58. G. Baym, Physica 96A (1979) 131.
59. K. Bleuler et al., University of Bonn Preprint (1983).
60. H.R. Petry, these proceedings.
61. O. Nachtmann and H.J. Pirner, University of Heidelberg preprint HD-THEP-83-8 (1983).
62. W. Furmanski and A. Krzywicki, Orsay preprint LPTHE 83/11 (1983).
63. R.M. Godbole and K.V.L. Sarma, Phys. Rev. D25 (1982) 120, and private communication.
64. S. Barshay, TH. Aachen preprint (1983).
65. E.M. Levin and M.G. Ryskin, Leningrad preprint 883 (1983).
66. R.G. Arnold et al., SLAC experiment E139.
67. For a compilation of these results see:
K. Rith, Freiburg preprint THEP 83/5, to appear in the proceedings of the International Europhysics Conference on High Energy Physics, Brighton (UK), (1983).

NUCLEAR SHELLS FROM A QUARK-THEORETICAL VIEWPOINT

H.R. Petry
Institut für Theoretische Kernphysik der Universität Bonn
Nußallee 14-16, D-5300 Bonn, West Germany

Part I: A quark shell model of the nucleus

1. Introduction

Modern particle theory considers all hadrons as being made out of quarks which interact via gluon exchange (QCD). In particular, the nucleon is a 3-quark system of considerable size. This size is so large that the basic postulates of the nuclear shell model, assuming approximately point-like nucleons which move quasi-freely in a common average potential, become highly questionable. In fact the measured rms charge radius of the proton (0.9 fm) indicates that the nucleon radius is 1.2 fm (for a uniform charge distribution) whereas the average nucleon-nucleon distance in a nucleus is 1.8 fm. Thus we have to expect that the quark wave functions of the individual nucleons largely overlap and that the quarks develop a tendency to spread out over the whole nucleus. This effect seems to be observable in the recent μ-Fe scattering experiment which is discussed by Rith in this volume. Under these circumstances the question arises whether the overall nuclear structure might be considered from the outset as a system of quarks whose lowest energy states have the characteristic properties predicted by the standard shell model. This report reviews a first joint attempt[1)] in this direction. Part I concentrates on the physical aspects, part II on the mathematical techniques which we need.

2. Application of the MIT model to nuclei

The main ingredients of our new picture of the nucleus are taken from the MIT bag model of hadrons[2)]. This model is believed to constitute an intermediate step towards a true QCD calculation of hadronic properties and we just enlarge its application by treating now the nucleus as a system of 3N non-strange quarks. Hence the quarks are described by the Dirac equation

$$(|x|<R) \qquad (\vec{\alpha}\vec{p}+\beta m)\psi = \varepsilon\psi \ ,$$

together with the boundary condition

$$(|x| = R) \quad \vec{\alpha}\cdot\vec{x}/|x|\psi = i\beta\psi \ ;$$

i.e. the quarks move in a spherical, infinite potential well with a radius R, which we have to identify with the nuclear radius. It follows that quark states form shells of good angular momentum and parity and that we can label these states by quantum numbers (n,j,m,τ,i) representing radial excitations (n), total angular momentum (j), angular 3-momentum (m), isospin (τ) and colour (i). If we order these states with respect to increasing energy ε we find a sequence of shells with the same quantum numbers as in the Mayer-Jensen shell model, namely 1s $\frac{1}{2}$, 1p $\frac{3}{2}$, 1p $\frac{1}{2}$, 1d $\frac{5}{2}$ etc. Note that the spin-orbit splitting is already enforced here by the Dirac equation without any additional parameter.

Nuclear states are now formed by putting 3N quarks in these shells. In addition, we have to fulfill the requirement that the total wave function has colour zero. This requirement is trivially satisfied when a shell is completely filled. Remarkably enough one obtains in this simple way already the analogues of the magic nuclei within the quark model.

The major difficulty arises when a shell is not completely filled. We come to this problem later and discuss first the total energy E (the nuclear mass) which is up to now equal to the sum of the occupied single-particle energies ε. These energies depend on the quark mass m and the bag radius R. In particular, E is radius-dependent, too. In the MIT-model the hadronic system is stabilized by a universal outer pressure B which is supposed to be an effect of the degenerate gluonic vacuum. The equilibrium equation reads:

$$-\frac{\partial E}{\partial R} = 4\pi B R^2 \ ,$$

and can be used to determine the nuclear radius R, once the values of m and B are fixed. Choosing m = 270 [MeV] and $B^{1/4}$ = 127 [MeV], ($B^{1/4}$ = 100 [MeV] is the MIT value),we find the remarkable result that the nuclear radius computed in this way satisfies the well-known law $R = 1.3\ N^{1/3}$ [fm] valid all over the periodic table. Thus in our model the gluonic pressure B is directly related to the saturation properties of nuclei.

It should be mentioned that the last equation is slightly too complicated for practical computations. Using the approximate relation

$$-\frac{\partial E}{\partial R} = \frac{1}{R} E$$

(strictly valid for massless quarks), it can be converted into the simpler form

$$E = 4\pi B \cdot R^3$$

which states that the energy and the volume of nuclei are strictly proportional.

3. Non-filled quark shells

We now turn to the discussion of non-filled quark shells. For simplicity we consider the case of a core $|0\rangle$ of completely filled quark shells and M quarks in a shell with fixed quantum numbers n and j. It is convenient to lump state labels together, $(n,j,m,\tau,i) \rightarrow (\alpha,i)$, and to introduce the corresponding creation and annihilation operators $a^+_{i\alpha}$, $a_{i\alpha}$. Clearly the vector space of all M-quark states described above is then given by the vectors:

$$\prod_{r=1}^{M} a^+_{i_r \alpha_r} |0\rangle \ .$$

Most of these states are of no physical interest because they carry colour. The vector space of colour-free states alone is spanned by vectors of the form:

$$\prod_{r=1}^{N} B^+(\alpha_r,\beta_r,\gamma_r)|0\rangle \ ,$$

where $M = 3N$ and (using from now on the convention of summing over repeated indices):

$$B^+(\alpha,\beta,\gamma) = \varepsilon_{ikl} a^+_{i\alpha} a^+_{k\beta} a^+_{l\gamma} \ .$$

The proof of this statement has already been given by H. Weyl[3)] who applied it in his discussion of spinless electron systems (Russell--Saunders coupling). We denote the space of colour-free states from now on by V_N and proceed by characterizing it in more detail by certain group-theoretical properties. Let $d_{\alpha\beta}$ $(\alpha,\beta = 1,\ldots,n=2(2j+1))$, be an antihermitean trace-free matrix, i.e.

$$d_{\alpha\beta} = -\bar{d}_{\beta\alpha} \ .$$

Such matrices form the Lie-algebra of SU(n) and the assignment

$$d_{\alpha\beta} \rightarrow \hat{d} = d_{\alpha\beta} a^+_{i\alpha} a_{i\beta}$$

yields a representation of SU(n) in the 3N-quark space. This representation commutes with colour rotations and, therefore, leaves V_N invariant. Hence we can restrict it to V_N, and this restriction turns out to be irreducible[3]. Fortunately, all such irreducible representations are classified, either by Young diagrams[4] or by maximal weights[5]. Our representation belongs to the Young diagram with 3 columns and N rows, or, equivalently to the maximal weight $3f_N$[5].

From this maximal weight we can compute[5] the dimension of V_N (see part II). It turns out that

$$\dim V_N = 2\binom{n}{N}\binom{n+1}{N}\binom{n+2}{N}/(N+1)^2(N+2) \ .$$

A moment's thought will show us that this leads us into trouble. In fact, the situation with 3N quarks in one shell should correspond to N nucleons in a shell with the same quantum numbers. Hence the dimension of V_N should be equal to $\binom{n}{N}$ which is in general much smaller than the number given before. We come to the conclusion that the nuclear states we are looking for, can at best be a subspace of V_N, which must contain in addition other hadronic states. Such new states are, of course, expected because the nucleon itself has a lot of excited states arising from different quark configurations, and our description of the nucleus will include such excitations, too. The simplest new excitation will, for example, occur when a nucleon is changed into a Δ. Note, however, that in our model such a change is described by an excitation of the whole nucleus.

A Δ-excitation is energetically separated from the normal nuclear states by 300 MeV, but up to now, it has in our model the same energy as the ground state. We conclude that we have to include an additional part into our Hamiltonian, and the natural choice is a quark-quark interaction. From the discussion above it is clear what this force should do: It should yield as the lowest energy states a subspace of V_N which can be identified with the space of N nucleons in a shell of angular momentum j; this requirement fixes in particular the total spin and isospin content of this subspace completely. Largely separated in energy (300 MeV) from these states we should then find new excited states, which we can identify with Δ-excitations.

4. The quark-quark interaction

Unfortunately, we are working on a low energy problem far away from asymptotic freedom; hence QCD does not provide us with a reliable expression for our quark-quark interaction. We may assume, however, that it is short-ranged because long-range effects are already built into the bag boundary conditions. Since we are interested in an effective force P, which acts only between quarks in one shell the short-range character leads us to the assumption that our force will be of the pairing type, i.e. it will act only between two quarks coupled to angular momentum zero. It is well-known that the pairing part is always dominant when a force is short-ranged; thus our assumption yields at least a reasonable approximation. But then SU(3)--invariance and the Pauli-principle allow only two possibilities: Either the quarks are in a colour antitriplett, isospin zero state, or they form a colour sextett, isospin one state. The latter possibility can be disregarded because such a force would favour energetically high isospin states. We are left with the first possibility which can be written down in the form:

$$P = -gA_i^+A_i/4 \quad ;$$

$$A_i = \varepsilon_{ikl}g_{\alpha\beta}a_{k\alpha}a_{l\beta} \; .$$

In this formula ε_{ikl} represents the totally skew Kronecker symbol and

$$g_{\alpha\beta} = (-1)^{j_\alpha - m_\alpha - \tau_\alpha + \frac{1}{2}} \cdot \delta_{m_\alpha, -m_\beta} \delta_{\tau_\alpha, -\tau_\beta}$$

is essentially the Clebsch-Gordon coefficient achieving the coupling to spin and isospin zero. g is a coupling constant which we discuss later.

Our total Hamiltonian reads as follows:

$$H = \varepsilon N_{op} + P + E_{core} .$$

E_{core} is the energy contribution of the filled core which can be treated as a pure number in this context; ε and N_{op} are the energy and the number operator for the non-filled shell. This Hamiltonian commutes with colour rotations and, therefore, leaves the physical subspace V_N (the colourfree states) invariant. We have to diagonalize it in this subspace in order to determine the new mass values which will be shifted by the interaction P. The diagonalization can be done exactly with the help of group-theoretical methods as will be shown now.

First note that $g_{\alpha\beta}$ can be interpreted as symmetric bilinear metric. Consider the antihermitean trace-free matrices $d_{\alpha\beta}$ which leave in addition the metric $g_{\alpha\beta}$ invariant, i.e. the equation

$$d_{\alpha\gamma}g_{\gamma\beta} + d_{\beta\gamma}g_{\alpha\gamma} = 0$$

holds. These matrices form a Lie subalgebra of SU(n), which is naturally isomorphic to the Lie algebra of SO(n). The representation of SU(n) described in section 3 does not have the property to commute with H, but when we restrict it to the subalgebra SO(n) we can easily check that all the generators $\hat{d} = d_{\alpha\beta}a^{+}_{i\alpha}\alpha_{i\beta}$ commute with P provided that the last equation holds.

5. The diagonalization of the Hamiltonian

The invariance of our Hamiltonian under the group SO(n) can be made even more explicit; in fact P can be expressed entirely by the Casimir operators C_1 and C_2 of the Lie algebras SU(n) and SO(n), respectively. More precisely, the following formula holds (see part II):

$$P = -g[(n-2)C_2-nC_1 + \frac{1}{2N} N_{op}^2 - \frac{1}{2} N_{op}] .$$

(We use the same normalization of the Casimir operators as in ref. 5)).

In an irreducible representation the values of these Casimir operators are constant all over the representation space and completely determined by the maximal weights[5]. Since we know already that SU(n) acts irreducibly in V_N and has the maximal weight $3f_N$ in this space, we can already compute the value of C_1. To compute the values of C_2 we note that our irreducible representation of SU(n) in V_N will yield a <u>reducible</u> representation of SO(n) (the invariance group of our Hamiltonian) which splits into irreducible components. We can determine the maximal weights of these irreducible subspaces and hence compute the values of the Casimir operators C_2. Alternatively we may solve our problem by using Young diagram techniques[4]. The first method is described in full detail in part II; here for the moment we just quote the result which is remarkably simple: Set $k = \min(N, n-N)$; then the space V_N splits into k+1 components which are irreducible under SO(n). Moreover, we can label these subspaces by $l = 0, 1, \ldots, k$ such that the eigenvalue of H in each subspace takes the form

$$E_l = E_{core} + 3N\cdot\varepsilon - g[N(n+3-N) - l(n+3-l)] .$$

The lowest eigenvalue occurs for $l = 0$. The states in the corresponding subspace should be the normal nuclear shell model states corresponding to N nucleons in a shell with angular momentum. The first excitation occurs for $l = 1$; those states should describe Δ-excitations of the nucleus. (We will prove this in the next section). Hence $E_1 - E_o$ must equal the Δ-N mass difference which fixes the value of g to be $g = 120/(2j+1)$ [MeV].

The parameters of our model are now completely determined and we can compute nuclear masses. We do this by starting with 1s $\frac{1}{2}$ quarks and $E_{core} = 0$. Having filled the 1s $\frac{1}{2}$ shell we begin to fill the p-shells with the 1s $\frac{1}{2}$ core energy taken from the preceding computation. This process can obviously be continued as far as we wish, but we stop at Ca^{40} (N = 40) for physical reasons because afterwards the Coulomb energy begins to destroy the isospin invariance which, up to now, is a strict symmetry of our model. The numerical values for the nuclear masses then turn out to satisfy approximately the relation $E = \mu N$ where μ is the nucleon mass, as it should be. The deviations from this formula are for an individual nucleus, on the average ± 1.5 percent and become only slightly higher when a shell is completely filled. This result can be combined with the relation $E = 4\pi BR^3$ and leads to the radius law $R = 1.3\ N^{1/3}$ [fm] anticipated in section 2.

6. The structure of the SO(n)-irreducible subspaces

As we have stated in the previous section, we have to split the SU(n)-irreducible space V_N into SO(n)-irreducible components. This problem can be solved in the following way: Recall that V_N is spanned by vectors of the form

$$\psi = \prod_{r=1}^{N} B^{+}(\alpha_r,\beta_r,\gamma_r)\,|0\rangle \ ,$$

where

$$B^{+}(\alpha,\beta,\gamma) = \varepsilon_{ikl}a^{+}_{i\alpha}a^{+}_{k\beta}a^{+}_{l\gamma} \ .$$

Next note that any contraction of the indices carried by the vector ψ defines a linear map in V_N which commutes with the SO(n)-action. This invariance property of the contraction operator has already been used by H. Weyl[6] in order to determine all irreducible representations of SO(n). We will see that a similar technique works also here. Let us define the operator

$$C^{+}_{\alpha} = B^{+}(\alpha,\beta,\gamma)\cdot g_{\beta\gamma} \ ,$$

and let F_l denote the vector space spanned by all vectors φ_l of the form

$$\varphi_l = \prod_{r=1}^{l} B^{+}(\alpha_r,\beta_r,\gamma_r)\cdot \prod_{s=1}^{N-l} C^{+}_{\alpha_s}\,|0\rangle \ .$$

In particular F_N equals V_N and $F_o \subset F_1 \subset \ldots \subset F_{N-1} \subset F_N$. Moreover, due to the SO(n)-invariance of the contraction operator, each space F_l is SO(n)-invariant. Next we define the new spaces G_l by $G_o = F_o$ and

$$G_l \oplus F_{l-1} = F_l \ , \qquad (l = 1,\ldots,N) \ ;$$

i.e. G_l is the orthogonal complement of F_{l-1} in F_l. Clearly the new spaces G_l are also SO(n)-invariant and, by construction,

$$V_N = \bigoplus_{l=0}^{N} G_l \ .$$

Hence we have achieved an orthogonal decomposition of V_N into SO(n)-invariant subspaces. (In part II we will show that $G_l = 0$ if $N > n/2$ and $l > n-N$).

In order to show that G_1 has indeed the properties stated in the last section, we must establish SO(n)-irreducibility and then compute the maximal weight. The lengthy details will be described in part II. Here we want to discuss the states in G_o and G_1, i.e. the ground-states and the first excited states of our model. Evidently G_o is spanned by the vectors

$$\varphi_o = \prod_{r=1}^{N} C^+_{\alpha_r} |0> .$$

φ_o is antisymmetric in the indices α_r because the operators C_{α_r} anti-commute. Obviously, such states are in one-to-one correspondence with Slaterdeterminants build from nucleonic states in a shell with the same angular momentum and parity as our quarks. Thus we have shown that the quark-quark interaction P has indeed energetically singled out a nucleonic subspace with the correct physical properties and hence fulfills all the requirements of section 3. The fact that the states in this subspace are still degenerate can be accepted as a reasonable approximation and should not disturb us; in reality the corresponding energy splittings are of the order of a few MeV, and hence indeed small in comparison with the first genuine excitation in our model at 300 MeV.

These first excitations form the subspace G_1 and can be described as follows: Let $d_{\alpha\beta\gamma}$ be a symmetric tensor with the property that contractions with the metric $g_{\alpha\beta}$ vanish. Assign to any such $d_{\alpha\beta\gamma}$ the operator

$$D^+_d = B^+(\alpha,\beta,\gamma)\, d_{\alpha\beta\gamma} .$$

From its very definition, it then is obvious that G_1 is spanned by vectors of the form

$$\varphi_1 = D^+_d \cdot \prod_{r=1}^{N-1} C^+_{\alpha_r} |0> .$$

We have seen before that $C^+_{\alpha_r}$ is the creation operator of a nucleon with the quantum numbers α_r. On the other hand, a straightforward comparison with the standard quark model calculations of baryons shows that the operator D^+_d creates an excited nucleon, i.e. a Δ-particle. (The known physical quantum numbers arise when d is appropriately specialized).

It has now been rigorously established that the ground states and the first excited states of our model have the right properties. It remains to analyse the higher excitations. The most obvious interpretation would be to relate them to states with more Δ-particles on even higher resonances of the nucleon, but we have only made an incomplete analysis so far. An interesting point which we do not want to omit here, is the phenomenon that, e.g., a naive study of the Δ-excitations of the nucleus treating the Δ as an elementary particle, yields much more states than our quark model. This effect is due to Pauli principle alone and we illustrate it here with 6 quarks in the 1s $\frac{1}{2}$ shell. The two-nucleon states have in our model the spin and isospin quantum numbers (1,0) and (0,1), respectively; they stand for a rough description of the deuteron and the nucleon-nucleon antibound state and are degenerate as they should. The first excitation consists of a (2,1) and a (1,2) system. Note that N-Δ system with Δ elementary yields in addition states with the quantum numbers (2,2) and (1,1). The next higher excitations which then exhaust all possibilities of our model have the quantum numbers (3,0) and (0,3), whereas two elementary Δ-particles also give the combinations (3,2), (1,0), (1,2), (0,1), (2,3), (2,1). Hence, the use of the quark model reduces the number of states tremendously, and we speculate that this effect should be directly observable.

Conclusion

Our model description of nuclear states in terms of quarks can be summarised as a sequence of correlations which are built into the multi-quark wave functions step by step. The first correlation arises from the requirement of vanishing colour which enforces a grouping of quark creation operators into triplett clusters which are individually colourfree. This is achieved by contracting the colour indices with the ε-symbol. The second correlation then is introduced by coupling two creation operators in each cluster to angular momentum and isospin zero and amounts to contraction with the metric tensor $g_{\alpha\beta}$. The main result of our work consists in the proof that such a configuration is energetically favoured by our pairing force. Bearing in mind that this force can at best be regarded as an approximation to the full effective quark-quark interaction, it is to be expected that additional correlations must occur when a more realistic expression is used. Nevertheless we believe that our model will be of practical interest also for such a more refined theory, provided that the pairing part of the force remains dominant. In this case one could diagonalize the pairing part exactly by our methods and treat the rest perturbatively.

That more correlations are actually needed can indeed be checked by computing certain nuclear properties like magnetic moments, weak decay constants, etc. In particular, we have found that magnetic moments become too large for high angular momentum states. This behaviour is, in fact, due to the specific couplings of angular momenta in a nucleonic cluster which we use in our model, and it can be cured by admixing different configurations probably due to a more complicated quark force.(As long as the quark-quark interaction remains theoretically undetermined such calculations can be used to gain some phenomenological insight into its physical properties).

The basically new feature of our quark model of the nucleus consists in a radical reinterpretation of the nuclear saturation properties. These are usually determined by the following procedure (see R. Machleidt's contribution to this volume): First an expression for the nuclear force is derived partly from theoretical models (boson exchange) but still containing phenomenological ingredients (form factors). In the next step a certain approximation of the many-body problem, mostly based on Brueckner's theory, is applied to compute the density and energy of nuclear matter. Such a program is notoriously difficult and never gave completely satisfactory values for both of these data. Instead we apply now the MIT model prescription and obtain nuclear saturation directly as the result of the universal gluonic pressure on the walls of the nuclear bag.

Our model is obviously not invariant under translations. Hence our total energies must be corrected by subtracting the energy ΔE of the center of mass motion. We have no reliable method to compute ΔE but we can assume that ΔE will be proportional to the inverse of the quark number. Hence we make the phenomenological ansatz $\Delta E = \gamma/3N$, and treat the constant γ as an additional fit parameter. We find $\gamma = 370$ [MeV]. This purely numerical correction has already been incorporated into the results which we have described before.

Part II: The group theory of the model

1. Introduction

This part describes our mathematical techniques. They are based on the representation theory of compact Lie algebras. In section 2 we summarize the main theorems which we need and which we have all taken from ref. 5). In sections 3 and 4 we specialize these theorems to the groups SU(n) and SO(n) and we prove the basic mathematical properties of our model.

2. Compact Lie algebras

Let A be a real Lie algebra of finite dimension k with Lie product [,]. The adjoint representation ad:A → L(A), (the algebra of linear transformations of A), is defined by the formula

$$ad(x)y = [x,y], \quad (x,y \in A) \; . \tag{1}$$

Setting

$$\langle x,y\rangle_A = -\mathrm{tr}\; ad(x)ad(y) \; , \tag{2}$$

one obtains a bilinear symmetric form $\langle , \rangle_A$ on A, called the Killing form. A is called compact, if $\langle , \rangle_A$ is positive definite. From now on we assume that A has this property. (In particular, this implies that A is semi-simple).

Let H be an abelian subalgebra of A. H is called a Cartan algebra of A, if its dimension r is maximal which means that any abelian subalgebra of A has a dimension less than or equal to r. We fix H and denote by H^* its dual, i.e. the space of linear functions on H. Let $\{C_i\}$ be a basis of H. Hence $f \in H^*$ is completely specified by the values $f(C_i)$. Let γ_{ij} denote the inverse of the matrix $\langle C_i, C_j\rangle_A$. For arbitrary $f,g \in H^*$ we may set

$$\langle f,g\rangle_{H^*} = \sum_{i,j} f(C_i)g(C_j)\gamma_{ij} \tag{3}$$

to obtain a positive definite scalar product $<,>_{H^*}$ on H^*. With these definitions we are now able to recall the basic theorems on compact Lie algebras.

Theorem I (structure of compact Lie algebras): There is a unique set ϕ of linear functions $\omega \in H^*$ with associated vectors A_ω, B_ω which form an orthonormal basis of $H^\perp$, (the space orthogonal to H in A), and fulfill the equations

$$[h,A_\omega] = \omega(h)B_\omega \quad ,$$

$$[h,B_\omega] = -\omega(h)A_\omega \quad , \tag{4}$$

for all $h \in H$. Moreover, ϕ contains a subset $\{\omega_i\}$ which forms a basis of H^* and has the property that each $\omega \in \phi$ can be written in the form

$$\omega = \sum_{i=1}^{r} k_i \omega_i \tag{5}$$

with integers $k_i \geq 0$.

$\omega \in \phi$ is called a positive root of A, ϕ is the positive root space of A and $\{\omega_i\}$ is called a root basis.

Theorem I is needed in order to classify the unitary representations of compact Lie algebras. Let V be a complex Hilbert space of finite dimension. A linear map $D:A \rightarrow L(V)$, (the algebra of linear transformations on V), is called a representation if

$$D([x,y]) = D(x)D(y) - D(y)D(x), \quad (x,y) \in A .$$

D is called unitary if, in addition, $D(x) = -D(x)^*$. D is irreducible if it contains no invariant subspace. Two unitary representations D and D'are equivalent if there exists an isometry $U:V \rightarrow V'$ such that $UD(x) = D'(x)U$ for all $x \in A$.

Theorem II: Let D be unitary. Then V splits into orthogonal subspaces V_i, $V = \bigoplus_i V_i$ such that D restricted to V_i is irreducible.

This shows that D is equivalent to a direct sum of irreducible representations.

Theorem III (classification of unitary representations): Let D be a unitary irreducible representation in V. There is a nonzero maximal vector $x \in V$, which is uniquely determined up to scalar multiplication, together with a unique linear function $\Lambda \in H^*$, called the maximal weight of D, such that the equations

$$(D(A_\omega) - iD(B_\omega))x = 0 \tag{6}$$

$$D(h)x = i\Lambda(h)x \tag{7}$$

hold for all $\omega \in \phi$, $h \in H$. Two irreducible representations are equivalent if and only if their maximal weights coincide.

All unitary representations are, therefore, uniquely classified by maximal weights $\Lambda \in H^*$. The following theorem states how the maximal weights look like:

Theorem IV (maximal weights): Set r = dim H. Define the so-called fundamental weights f_i, i = 1,...,r by the formula

$$2\langle f_i, \omega_j\rangle_{H^*} / \langle \omega_j, \omega_j\rangle_{H^*} = \delta_{ij} \ . \tag{8}$$

Then each maximal weight Λ can be written in the form

$$\Lambda = \sum_{i=1}^{r} m_i f_i \tag{9}$$

with integers $m_i \geq 0$. Moreover, to each r-tuple of non-negative integers m_i belongs a unitary irreducible representation of A with a maximal weight given by the last formula.

The last two theorems classify the unitary, irreducible representations D of A completely up to equivalence in terms of their maximal weights. Hence it is to be expected that any property of D which is invariant under equivalence transformation can be expressed in terms of the maximal weight. One such quantity is certainly the dimension of V. Indeed the following equation holds:

$$\dim V = \prod_{\omega \in \phi} \langle \Lambda + \delta, \omega\rangle_{H^*} / \langle \delta, \omega\rangle_{H^*} \tag{10}$$

with $\delta = \frac{1}{2} \sum_{\omega \in \phi} \omega$.

Another quantity of this type is the eigenvalue $\lambda(D)$ of the Casimir-operator C(D) of D. This operator can be defined for any unitary representation D by the formula:

$$C(D) = \sum_{i,j=1} \gamma_{ij} D(C_i) D(C_j) + \sum_{\omega\in\phi} (D(A_\omega)^2 + D(B_\omega)^2) \tag{11}$$

It commutes with all element D(x), x∈A; hence, by Schur's lemma, it must be proportional to the identity if D is irreducible, i.e.

$$C(D) = \lambda(D) 1 .$$

In terms of Λ we have the formula

$$\lambda(D) = -\langle \Lambda + 2\delta, \Lambda \rangle_{H^*} . \tag{12}$$

3. The Lie algebra of SU(n)

In this section we specify A to be the Lie algebra of SU(n), i.e. the Lie algebra of antihermitean, trace-free matrices d. In this case the Killing form <,> can be expressed in terms of an ordinary matrix trace; more precisely if d,d' are antihermitean matrices, then

$$\langle d, d' \rangle_A = -2n \operatorname{tr} dd' . \tag{13}$$

This formula shows that $\langle , \rangle_A$ is indeed positive definite, whence A is compact. As a basis for the Cartan algebra H we can choose the diagonal matrices C_k, (k = 1,...,n-1), whose matrix elements $C_{k;\alpha\beta}$ read as follows:

$$C_{k;\alpha\beta} = i(\delta_{k\alpha}\delta_{k\beta} - \frac{1}{n}\delta_{\alpha\beta}) . \tag{14}$$

The matrix γ_{ij} in equ. (3) can then be computed:

$$\gamma_{ij} = \frac{1}{2n}(\delta_{ij} + 1) , \tag{15}$$

whence, according to equ. (3)

$$\langle f, g \rangle_{H^*} = \frac{1}{2n} \sum_{i=1}^{n-1} f(C_i) g(C_i) + \frac{1}{2n} \left(\sum_{i=1}^{n-1} f(C_i) \right) \left(\sum_{j=1}^{n-1} g(C_j) \right) \tag{16}$$

for all $f,g \in H^*$.

Next we consider the antihermitean, trace-free matrices A_{kl}, B_{kl}; $(1\leq k<l\leq n)$, whose matrix elements $A_{kl;\alpha\beta}$, $B_{kl;\alpha\beta}$ are defined as follows:

$$A_{kl;\alpha\beta} = \frac{1}{2\sqrt{n}} (\delta_{k\alpha}\delta_{l\beta}-\delta_{l\alpha}\delta_{k\beta}) , \tag{17}$$

$$B_{kl;\alpha\beta} = \frac{i}{2\sqrt{n}} (\delta_{k\alpha}\delta_{l\beta}+\delta_{l\alpha}\delta_{k\beta}) . \tag{18}$$

The matrices A_{kl} and B_{kl} form indeed an orthonormal basis of H . Moreover, if we define the linear functions $\omega_{kl}\in H^*$ by the equation

$$\omega_{kl}(C_m) = \delta_{km} - \delta_{lm} , \quad (1\leq k<l\leq n,\ 1\leq m<n) \tag{19}$$

then it is easily checked that the quantities A_{kl}, B_{kl}, ω_{kl} satisfy equ. (4). It follows that the set ω_{kl} is the space of positive roots for the Lie algebra of SU(n). We now set

$$\omega_i = \omega_{i,i+1} \tag{20}$$

Evidently, for $1\leq k<l\leq n$,

$$\omega_{kl} = \sum_{i=k}^{l-1} \omega_{i,i+1} = \sum_{i=k}^{l-1} \omega_i ,$$

which proves that ω_i is in fact a root basis. We can now insert these ω_i and the expression for $\langle,\rangle_{H^*}$ given by equ. (16) into the equation (8) in order to determine the fundamental weights f_k. The result is

$$f_i(C_k) = \begin{cases} -i/n & i<k , \\ (n-i)/n & i\geq k . \end{cases} \tag{21}$$

Next let D be a unitary, irreducible representation of the Lie algebra of SU(n). Let the maximal weight be

$$\Lambda = \sum_{i=1}^{n-1} k_i f_i .$$

It follows from equ. (21) that

$$\Lambda(C_k) = -\frac{1}{n}\sum_{i=1}^{n-1} ik_i + \sum_{i=k}^{n-1} k_i . \tag{22}$$

The equations (19) and (22) together with the expression (16) for $\langle,\rangle_{H^*}$ can now be inserted into the general formulas (10) and (11) for dim V and the eigenvalue $\lambda(D)$ of the Casimir operator. We find the results

$$\dim V = \prod_{1\leq i<j\leq n} (j-i + \sum_{m=i}^{j-1} k_m)/(j-i) \; . \tag{23}$$

$$\lambda(D) = -\frac{1}{2n} [\sum_{i=1}^{n-1} i(n-i)k_i(1+k_i/n) + \frac{2}{n} \sum_{1\leq i<j\leq n-1} k_i k_j i(n-j)] \tag{24}$$

These formulas can now be applied to the representation introduced in section 3 of part I. Recall that this representation was defined by assigning to each trace-free, antihermitean matrix d with matrix elements $d_{\alpha\beta}$ the operator $\sum_{,\beta=1}^{n} \sum_{r=1}^{3} d_{\alpha\beta} a^+_{r\alpha} a_{r\beta}$. Hence we have to set $D(d) = \hat{d}$. Using the formulas (15), (17) and (18) we find the identities:

$$D(C_\mu) = i(\sum_{r=1}^{3} a^+_{r\mu} a_{r\mu} - \frac{1}{n} N_{op}), \quad (\mu = 1,\ldots,n-1) \tag{25}$$

$$D(A_{\mu\nu}) = \sum_{r=1}^{3} (a^+_{r\mu} a_{r\nu} - a^+_{r\nu} a_{r\mu})/2\sqrt{n} \; , \quad (1\leq\mu<\nu\leq n) \tag{26}$$

$$D(B_{\mu\nu}) = i\sum_{r=1}^{3} (a^+_{r\mu} a_{r\nu} + a^+_{r\nu} a_{r\mu})/2\sqrt{n}, \quad (1\leq\mu<\nu\leq n) \tag{27}$$

A vector x is maximal, if according to equation (6),

$$(D(A_{\mu\nu}) - iD(B_{\mu\nu}))x = 0 \; .$$

From equations (26) and (2) we find that

$$D(A_{\mu\nu}) - iD(B_{\mu\nu}) = \frac{1}{\sqrt{n}} \sum_{r=1}^{3} a^+_{r\mu} a_{r\nu} \; . \tag{28}$$

Consider now the vector (compare section 3 of part I for the notation),

$$x = \prod_{\alpha=1}^{N} B^+(\alpha,\alpha,\alpha)|0\rangle = \prod_{\alpha=1}^{N} (\varepsilon_{rst} a^+_{r\alpha} a^+_{s\alpha} a^+_{t\alpha})|0\rangle \; ,$$

which is certainly in the colourfree 3N-quark space V_N considered in section 3 of part I. With the help of (28) it is easy to check that x is indeed maximal. Since we know that the representation D acts irreducibly on the colour-free states, theorem III tells us that it must be the only maximal vector (up to a scalar factor) in that space. With the help of (25) we next compute that

$$D(C_\mu)x = i\lambda_\mu x$$

with

$$\lambda_\mu = \begin{cases} 3(1-N/n), & \mu \leq N\ , \\ -3N/n & , \mu > N\ : \end{cases}$$

Comparing this with equation (22) we find that

$$\lambda_\mu = 3f_N(C_\mu)\ ;$$

hence $3f_N$ is the maximal weight of the SU(n)-representation in the space V_N. In order to compute the dimension of V_N, we can now use formula (23) with $k_N = 3$ and $k_m = 0$ otherwise. The result turns out to be

$$\dim V = 2\binom{n}{N}\binom{n+1}{N}\binom{n+2}{N}/(N+1)^2(N+2)\ . \tag{29}$$

We consider now the Casimir operator of D. After inserting the formulas (15), (25), (26) and (27) into equation (11) we find the simple formula

$$C(D) = -\frac{1}{2n}\left(\sum_{\alpha,\beta=1}^{n}\sum_{r,s=1}^{3} a^+_{r\alpha}a_{r\beta}a^+_{s\beta}a_{s\alpha} - \frac{1}{n}N^2_{op}\right) \tag{30}$$

On the other hand, the eigenvalue $\lambda(D)$ in the space V_N of colourfree 3N-quark states can now be computed by using formula (24) with $k_N = 3$ and $k_m = 0$ otherwise. It follows that

$$\lambda(D) - -3(1+3/n)N(n-N)/2n \tag{31}$$

4. The Lie algebra of SO(n)

Throughout this section we consider only the case that n is even i.e. n = 2m. The Lie algebra of SO(n) can be defined as the set of trace-free antihermitean matrices d whose matrix elements $d_{\alpha\beta}$ satisfy the additional relation

$$d_{\alpha\gamma}g_{\gamma\beta} + d_{\beta\gamma}g_{\alpha\gamma} = 0\ , \tag{32}$$

where $g_{\alpha\beta}$ is a symmetric matrix with

$$g_{\alpha\beta} = \pm\delta_{n+1-\alpha,\beta} \, . \tag{33}$$

The Killing form can be written as follows:

$$\langle d,d'\rangle_A = -(n-2)\,\mathrm{tr}\,dd' \, .$$

A basis $\tilde{C}_i$ for the Cartan algebra H is given by (see equ. (14)),

$$\tilde{C}_i = C_i - C_{n+1-i} , \quad (i = 1,\ldots,m) \, . \tag{34}$$

It follows that

$$\gamma_{ij} = \delta_{ij}/2(n-2) \, . \tag{35}$$

Hence the formula

$$\langle f,g\rangle_{H^*} = \frac{1}{2(n-2)} \sum_{i=1}^{m} f(\tilde{C}_i)g(\tilde{C}_i) \tag{36}$$

yields the scalar product for arbitrary linear functions $f,g \in H^*$. Next we define an orthonormal basis of H by the elements, $(1 \leq k < m, k < l < n+1-k)$,

$$\tilde{A}_{kl} = (A_{kl} - \sum_{p,q} g_{lp}g_{kq}A_{pq}) \cdot (\tfrac{n}{n-2})^{1/2}$$

$$\tilde{B}_{kl} = (B_{kl} - \sum_{p,q} g_{lp}g_{kq}B_{pq}) \cdot (\tfrac{n}{n-2})^{1/2} , \tag{37}$$

and check that these quantities, together with the linear functions $\tilde{\omega}_{kl}$, $(1 \leq k < m, k < l < n+1-k)$, given by

$$\tilde{\omega}_{kl}(\tilde{C}_i) = \delta_{ki} - \delta_{li} + \delta_{l,n+1-i} , \tag{38}$$

satisfy equ. (4). Therefore, the set $\{\tilde{\omega}_{kl}\}$ defines the positive roots of the Lie algebra of SO(n). A root basis $\tilde{\omega}_i$ is given by

$$\tilde{\omega}_i = \tilde{\omega}_{i,i+1} , \qquad i = 1,\ldots,m-1$$

$$\tilde{\omega}_m = \tilde{\omega}_{m-1,m+1} , \tag{39}$$

and the fundamental weights $\tilde{f}_i$ can now be computed by inserting the expressions (39), (38) and (36) into formula (8). We obtain in this way the coefficients

$$\tilde{f}_i(\tilde{C}_k) = \theta_{ik} - \frac{1}{2}(\delta_{i,m} + \delta_{i,m-1}) , \tag{40}$$

with

$$\theta_{ik} = \begin{cases} 0 & i<k \\ 1 & i\geq k \end{cases} . \tag{41}$$

Next let D be an irreducible representation of SO(n) with maximal weight

$$\Lambda = \sum_{i=1}^{m} k_i \tilde{f}_i \tag{42}$$

From the last equations we conclude that

$$\Lambda(\tilde{C}_k) = \sum_{i\geq k}^{m} k_i - (k_m+k_{m-1})/2 \tag{43}$$

The eigenvalue $\lambda(D)$ of the Casimiroperator and the dimension of the representation D can now be computed by inserting (43), (38) and (36) into the general formulas (10) and (12). We obtain in this way the formulas:

$$\dim V = \prod_{1\leq i<j\leq m} (j-i + \sum_{r=i}^{j-1} k_r)/(j-i) \tag{44}$$

$$\cdot \prod_{\substack{1\leq i<m \\ m<j<n+1-i}} (j-i-1 + \sum_{r=i}^{m} k_r + \sum_{r=n+1-j}^{m} k_r - k_m - k_{m-1})/(j-i-1) ,$$

$$\lambda(D) = -\frac{1}{2(n-2)} [\sum_{i=1}^{m} ik_i(k_i-k_m-k_{m-1}+n-i-1) + 2\sum_{1\leq i<j\leq m} ik_ik_j \tag{45}$$

$$+ m(k_m+k_{m-1})(k_m+k_{m-1}-2m+2)/4] .$$

We now apply these general formulas to the particular representation D of the Lie algebra of SU(n) in the space V_N of colourfree N quark--states. This representation restricts to a representation of SO(n), which is reducible but, according to theorem II splits into irreducible components. Our aim is to determine these components together with their maximal weights. In order to make contact with the physical application, we note that the metric $g_{\alpha\beta}$ in formula (32) has to be identified with the expression given in section 4 of part I. We begin by writing down the formulas for $D(\tilde{C}_\alpha)$, $D(\tilde{A}_{\alpha\beta})$, $D(\tilde{B}_{\alpha\beta})$:

$$D(\tilde{C}_\alpha) = \sum_{r=1}^{3} a^+_{r\alpha}a_{r\alpha} - \sum_{r=1}^{3} \sum_{\mu,\nu} g_{\alpha\mu}g_{\alpha\nu}a^+_{r\mu}a_{r\nu} \ , \quad (\alpha = 1,\dots,m) \tag{46}$$

$$D(\tilde{A}_{\alpha\beta}) = [\sum_{r=1}^{3} (a^+_{r\alpha}a_{r\beta}-a^+_{r\beta}a_{r\alpha} - \sum_{r=1}^{3} \sum_{\mu,\nu} g_{\beta\mu}g_{\alpha\nu}(a^+_{r\mu}a_{r\nu}-a^+_{r\nu}a_{r\mu})]/2\sqrt{n-2}$$

$$D(\tilde{B}_{\alpha\beta}) = i[\sum_{r=1}^{3} a^+_{r\alpha}a_{r\beta}+a^+_{r\beta}a_{r\alpha} - \sum_{r=1}^{3} \sum_{\mu,\nu} g_{\beta\mu}g_{\alpha\nu}(a^+_{r\mu}a_{r\nu}+a^+_{r\nu}a_{r\mu})]/2\sqrt{n-2}$$

$(1\leq\alpha<m;\ \alpha<\beta<n+1-\alpha)$.

According to equation (6) a vector x is maximal, if for all these indices α,β.

$$(D(\tilde{A}_{\alpha\beta})-iD(\tilde{B}_{\alpha\beta}))x = 0 \ .$$

From the equations (46) we conclude that in our example:

$$D(\tilde{A}_{\alpha\beta})-iD(\tilde{B}_{\alpha\beta}) = (\sum_{r=1}^{3} a^+_{r\alpha}a_{r\beta} - \sum_{r=1}^{3} g_{\beta\mu}g_{\alpha\nu}a^+_{r\mu}a_{r\nu})/\sqrt{n-2} \tag{47}$$

Consider now the vectors $x_l \in V_N$ defined by $(l = 0,\dots,\ k = \min(N,n-N))$

$$x_l = (\prod_{\alpha=N-l+1}^{N} B^+(\alpha,\alpha,\alpha) \prod_{B=1}^{N-l} C^+_\beta |0\rangle. \tag{48}$$

(Compare sections 3 and 6 of part I for the notation; note that $x_l = 0$ for all $l>n-N$ if $N>m$.) By construction $x_l \in G_l$, the SO(n)-invariant subspace introduced in section 6 of part I. Moreover, it is easily checked with the help of equ. (47) that x_l is in fact maximal. If N = m an additional set of maximal vectors $x_l \in G_l$ is found:

$$\tilde{x}_l = B^+(N+1,N+1,N+1) \prod_{\alpha=N-l+1}^{N-1} B^+(\alpha,\alpha,\alpha) \prod_{\beta=1}^{N-l} C^+_\beta |0\rangle \tag{49}$$

for $l\geq 1$ and

$$\tilde{x}_0 = C^+_{N+1} \prod_{\beta=1}^{N-1} C^+_\beta |0\rangle \tag{50}$$

By applying $D(\tilde{C}_\alpha)$, as given by equ. (4), to each of these vectors we find that the maximal weights Λ_l corresponding to the vectors x_l are given by the list:

$$\Lambda_l = \tilde{f}_k + 2\tilde{f}_l \quad , \quad (l \leq k < m-1) \; ,$$

$$\Lambda_l = \tilde{f}_{m-1} + \tilde{f}_m + 2\tilde{f}_l \quad , \quad (l < k = m-1) \; ,$$

$$\Lambda_l = 3(\tilde{f}_{m-1} + \tilde{f}_m) \quad , \quad (l = k = m-1) \; ,$$

$$\Lambda_l = 2(\tilde{f}_m + \tilde{f}_l) \quad , \quad (l < m-1, \; k = m) \; ,$$

$$\Lambda_l = 4\tilde{f}_m + 2\tilde{f}_{m-1} \quad , \quad (l = m-1, \; k = m) \; ,$$

$$\Lambda_l = 6\tilde{f}_m \quad , \quad (l = k = m) \; , \tag{51}$$

whereas the maximal vectors $\tilde{x}_l$ given by (49) and (50) correspond to

$$\tilde{\Lambda}_l = 2\tilde{f}_{m-1} + 2\tilde{f}_l \quad , \quad (l < m-1) \; ,$$

$$\tilde{\Lambda}_l = 4\tilde{f}_{m-1} + 2\tilde{f}_m \quad , \quad (l = m-1) \; ,$$

$$\tilde{\Lambda}_l = 6\tilde{f}_{m-1} \quad , \quad (l = m) \; . \tag{52}$$

From this list we can immediately read off the coefficients k_i entering into the general formulas (44) and (45) for the eigenvalues λ_l of the Casimiroperator and the dimension d_l of the irreducible subspace, to which x_l belongs. The actual computation of these numbers gives the following result:

$$\lambda_l = - \; [l(3+n-l) + N(n-N)/2]/(n-2) \tag{53}$$

$$d_l = \binom{n+2}{N}\binom{n+4}{l}\binom{n+4}{l+1} \frac{(N+1-l)(N+2-l)(n+1-N-l)(n+2-N-l)(n+3-2l)}{(N+1)(N+2)(n+1)(n+2)(n+3)(n+4)} \tag{54}$$

These formulas are also valid for the weights $\tilde{\Lambda}_l$ in the exceptional case N = m. The Casimiroperator itself can be rewritten in the form:

$$C(D) = - \frac{1}{2(n-2)} \Big(\sum_{r,s=1}^{3} \sum_{\alpha,\beta=1}^{n} a^{+}_{r\alpha} a_{r\beta} a^{+}_{s\beta} a_{s\alpha} - \sum_{r,s=1}^{3} \sum_{\alpha,\beta,\mu,\nu=1}^{n} a^{+}_{r\alpha} a_{r\beta} a^{+}_{s\mu} a_{s\nu} \, g_{\mu\alpha} \cdot g_{\nu\beta} \Big) \tag{55}$$

which is obtained after inserting the expressions (46) and (35) into formula (11).

Up to now we are not sure that all maximal weights and vectors are indeed found. We have by section 6 of part I an orthogonal, SO(n)--invariant decomposition $V_N = \bigoplus_{l=0}^{N} G_l$. In each subspace G_l we have found one maximal vector x_l when N<m, two maximal vectors x_l and $\tilde{x}_l$ when N = m. When N>m, we have found one maximal vector in G_l, if $l \leq N-m$ and no maximal vector when $l > N-m$. Let us now assume that, in the last case, G_l is in fact zero for $l > N-m$, that each G_l is in fact irreducible for $N \neq m$ and splits into two irreducible components for N = m. This conjecture is equivalent to the statement that the formulas (48), (49) and (40) give all the maximal vectors in V_N, because theorem III forbids any more candidates. On the other hand the conjecture is true if and only if the sum of the dimensions d_l associated to the maximal weights λ_l equals the dimension of V_N, i.e. we must check that

$$\dim V_N = \sum_{l=0}^{k} d_l$$

for $N \neq m$, and

$$\dim V_N = 2 \sum_{l=0}^{k} d_l$$

for N = m. By using the formulas (54) and (29) for d_l and $\dim V_N$, these identities can indeed be verified.

We conclude this section by rewriting the pairing force P of part I as a sum of Casimir operators for SU(n) and SO(n). From section 4 of part I we have the expression

$$P = -\frac{g}{4} \left(\sum_{i,k,l,k',l'=1}^{3} \sum_{\alpha,\beta,\mu,\nu=1}^{n} \varepsilon_{ikl}\varepsilon_{ik'l'} g_{\alpha\beta} g_{\mu\nu} a^{+}_{l'\nu} a^{+}_{k'\mu} a_{k\alpha} a_{l\beta} \right)$$

By using the identity $\sum_i \varepsilon_{ikl}\varepsilon_{ik'l'} = \delta_{kk'}\delta_{ll'} - \delta_{kl'}\delta_{lk'}$, we rewrite P in the form

$$P = -\frac{g}{2} \left(\sum_{k,l=1}^{3} \sum_{\alpha,\beta,\mu,\nu=1}^{n} g_{\alpha\beta} g_{\mu\nu} a^{+}_{l\nu} a_{l\beta} a^{+}_{k\mu} a_{k\alpha} - N_{op} \right)$$

By comparing this with the expressions (55) and (30) for the Casimir-operators C_1 and C_2 of SU(n) and SO(n), respectively, we see that in fact

$$P = -g[(n-2)C_2 - nC_1 + \frac{1}{2n} N_{op}^2 - \frac{1}{2} N_{op}] \tag{56}$$

It follows that all vectors in $G_l \subset V_N$ are eigenvectors of P; the eigenvalue $\lambda(P)$ is obtained by replacing in (56) N_{op} by N and C_1 and C_2 by their eigenvalues (31) and (53), respectively. It follows that

$$\lambda(P) = -g[N(n+3-N) - l(l+3-l)] . \tag{57}$$

This is the result which we have anticipated in part I, when diagonalizing our quark Hamiltonian. The mathematical aspects of our model are now completely described.

References

1) K. Bleuler, H. Hofestädt, S. Merk and H.R. Petry, Z. Naturforsch. 38a, (1983) 705
2) A. Chodos, R.L. Jaffee, K. Johnson, C.B. Thorn and V.F. Weisskopf, Phys. Rev. D9, (1974), 3471
3) H. Weyl, "The theory of groups and quantum mechanics" Dover Publications (1931)
4) D.E. Littlewood, "The theory of group characters", Oxford University Press (1970)
5) N. Jacobson, "Lie algebras", J. Wiley and Sons (1965)
6) H. Weyl, "The classical groups", Princeton University Press (1945)

EFFECTIVE QCD-LAGRANGIAN FOR NUCLEAR PHYSICS

E.R.Hilf and L.Polley
Institut fuer Kernphysik, T.H.Darmstadt, D-6loo Darmstadt, FRG

SUMMARY

A semi-classical approach to Nuclear Physics is proposed, which makes use of the one-loop effective thermal action for the Yang-Mills field, and the c-number colour source approximation to quarks.

NUCLEI AS COMPLEX SYSTEMS

Atomic Nuclei are complex systems which can be either disintegrated and fragmented into free nucleons,-or, in principle, be crunched to a plasma,-which both are objects, formed by the interaction of quarks by virtue of the strong interaction mediated by the gluon field $A_\mu^a(x)$.

Nuclei show some similarities to Solid Grains which also can either be disintegrated and fragmented into free molecules, or be crunched to a Plasma. Both are complex objects, formed by the interaction of charged Nuclei and Electrons, by virtue of the electro- magnetic interaction mediated by the photon field $A_\mu(x)$.

Theoretically the exact theory to treat a Solid Grain is known, the QED(or at least the QM). With it one may try to calculate and thus understand the properties of the individual free molecules.

On the other hand classical limits of the theory may be sufficient to treat the ionic plasma. However it is hopeless to try to calculate the properties of the complex system, the Grain, by the full theory. But it is also not necessary. As is wellknown it is sufficient to use semiclassical concepts derived from the exact theory such as the WKB-method, or the selfconsistent Thomas-Fermi. These methods which contain no additional parameters are well suited to study the slight disturbance of the outer part of the atomic Electron shell, the atomic polarisation, which results in the saturation binding,- covering a very small but important sector of the average density.

Analogously, the nuclear structure may ultimately be successfully described in terms of semiclassical concepts, derived from QCD, which contain no additional parameters, whereas it is hopeless to treat such extended systems (the Pb^{208} nucleus covers about $10^3 fm^3$, even a single nucleon about $6fm^3$) with full QCD, say on a lattice, due to the semiinfinite necessary computing time (sum over all links at present possible for fractions of a fm^3 only).

The wellstudied bag-type models although calculable, contain parameters fitted to free hadrons and thus are open to some arbitrariness in their semiempiric describing the polarisation of nucleons in nuclei. The soliton model does offer an however (parametrised but consistent) simplified scalar microscopic field description.

The to be developed Semiclassical Concepts may be useful for treating other complex systems composed of complex QCD-objects, such as Glue-Crystals of Glueballs, pion condensates of pions, neutron- and hadron-star matter.

Some experimental facts on nuclei to be understood by a semiclassical concept are the increased formfactor of nucleons in nuclei (see Rith, this volume), and the nucleon-nucleon force in nuclei. The latter can be formulated in terms of the semidivergence of the action-expansion, which is then normally stated in terms of a (fitted) density-dependence of the coefficients of the Hartree-Fock (or extended Thomas-Fermi)Hamiltonian. Fit data are experimental data[1] for the nuclear ground state such as the saturation density $\rho_o = (4\pi r_o^3)^{-1}$ with r_o=1.17fm, the saturation energy density of about 16MeV, the compressibility $K_o \approx 250$MeV, the surface tension

$a_s \approx 20MeV$, the asymmetry energy of 32MeV, and on the other hand e.g. recent experimental data[2,3] for warm nuclei, yielding a critical temperature of $T_c = 12MeV$ and a critical exponent of $\kappa = 1.8$, which are not in contradiction to some density-dependent Hartree-Fock scheme[4], but different to what one had expected from an earlier simple quadratic velocity dependence of a selfconsistent Thomas-Fermi approach[5]. A full-scale evaluation of all these nuclear informations will be the topic of a future conference[6].

THE SEMICLASSICAL METHOD

Wishes to a semiclassical approximation of QCD are that it should reproduce asymptotic freedom, and confinement.

The quarks, assumed to be confined in the nucleons, even if their distribution might be slightly polarised if embedded in a nucleus, are dealt with by treating them as external c-number sources $j_\mu^a(x)$, and we will discuss in this paper only static sources.

The gluons are represented by the expectation values of their field operators,

$$A_\mu^a(x) = \langle \hat{A}_\mu^a(x) \rangle \ .$$

The system is assumed here to be in its groundstate, thus we look for the limit $T \to 0$. But still there are contributions from quantum fluctuations of the field.

Then we can make use of the thermodynamic method[7,8]. For a given Hamilton operator as a functional of the classical sources $H[j]$ the partition function is given by

$$Z[j] = \text{tr}\, \exp(-\beta H[j]) \ . \tag{1}$$

This partition function is non-zero, for we do not impose unphysical boundary condition. as in some early lattice calculations, where periodic b.c.

were imposed, resulting as could be proved[9] in an empty Hilbert-space and a trivial partition function. If then formally Z=0 was assumed and inserted into (2), then the numerical results were interpreted as a first order phase transition, although this just reflects the fact[10] that the screening length exceeds the chosen lattice size.

For nonzero Z the Free Energy is

$$F[j] = -(1/\beta)\log Z[j] \tag{2}$$

From this thermodynamic potential we obtain the new effective action, supposed to be dependent on the conjugate variable of $j^a_\mu(x)$,

$$A^a_\mu(x)=-\delta F[j]/\delta j^{a\mu}(x) \tag{3}$$

by the contact-transformation

$$\Gamma[A] = F[j[A]]+\int d^3x\ j^{a\mu}A^a_\mu \ . \tag{4}$$

The functional derivative of this action yields the field equations

$$\delta\Gamma[A]/\delta A^a_\mu(x)=j^{a\mu}(x) \tag{5}$$

its local density is the effective Lagrangian L[A],

$$\Gamma[A] = \int d^3x\ L_{eff}[A(x)] \ . \tag{6}$$

In alliteration to Classical Mechanics terms (5) is the Euler-Lagrange equation. Only formal manipulations have been involved so far, exact in principle.

In the past a somewhat different approach was proposed[7]. From L[A] a local classical Hamiltonian was gained by a contact-transformation,

$$H[j,A] = jA - L[A]$$

yielding a global function

$$F[j,A] = \int d^3x\ H[j,A],$$

which is not a thermodynamic potential, in contrast to (2). Still its functional derivative

$$\delta F[j,A]/\delta A|_{j=const} = j$$

is one of the equations of motion of the Hamilton-Jacobi formalism analogue. For static sources it coincides with (3). Sometimes -H[j,A] was called a Lagrangian[11], which was tried to be cured by a new variational principle, but yielding a spurious sign, which disagrees with the respective QED limit.

THE EFFECTIVE LAGRANGIAN

Before introducing the effective gluonic Lagrangian we remind of the photonic analogue.

The Hamiltonian for photons and external electromagnetic sources j, is known,

$$H[j]=\sum_{\lambda=1}^{2} \int d^3k\ \ a^*(k,\lambda)\ \omega a(k,\lambda)+\sum_{\lambda=1}^{2} \int d^3k/\sqrt{(2\omega)}\ \times(j(k,\lambda)a^*(k,\lambda)+h.c.)$$
$$+ \iint d^3xd^3x'\ \rho(x)\rho(x')/(8\pi|x-x'|)\ . \tag{7}$$

so the partition function and the Free Energy can be calculated. The partition function factorises into a free photon part and a source-dependent part. In the zero temperature limit, only the latter contributes to the Free Energy,

$$F[j]= \iint d^3xd^3x'(\rho\rho'-jj')\ /(8\pi|x-x'|)\ . \tag{8}$$

Executing the thermodynamic method (section 2) yields

$$L_{eff} = \tfrac{1}{2}(B^2-E^2) = F_{\mu\nu}F^{\mu\nu} = F^2 \qquad (9)$$

This coincides with the classical Lagrangian of the electromagnetic field.

In classical Yang-Mills Theory, the purely classical action is of similar form,

$$L = g^{-2}F^2$$

where

$$F^a{}_{\mu\nu} = \delta^{ac}(\partial_\mu A^c_\nu - \partial_\nu A^c_\mu) + f^{abc} A^b_\mu A^c_\nu . \qquad (10)$$

The field equations are

$$(\delta^{ac}\partial^\mu + f^{abc}A^{b\mu})\, F^c_{\mu\nu} = j^a_\nu , \quad \nu=0,1,2,3. \qquad (10a)$$

The field equations have been solved numerically for different classical colour-charge configurations, such as the one charge shell[12], two charges[13], periodic plates[14]. Neither confinement nor asymptotic freedom was found.

One loop constant background-field calculations[15], as well as renormalisation group arguments[16], have suggested to consider an effective action that is obtained by replacing the bare coupling constant of the classical action with a running coupling which depends on the local colour field strength in units of a renormalisation group invariant mass.

This approach may also be plausible in that the running coupling constant of perturbative QCD has a relative momentum q^2 under the logarithm which in a thermodynamic framework would be proportional to the temperature or to F^2.

The effective QCD-Lagrangian is obtained by inserting the one-loop running coupling constant into the Lagrangian, yielding

$$L_{eff} = (1/4g^2)\, F^a_{\mu\nu} F^{a\mu\nu} \log[\tfrac{1}{2} F^a_{\mu\nu} F^{a\mu\nu}/e\kappa^2]. \quad (11)$$

The properties of this ansatz will be evaluated in section 4.

A quite different approach[18], starting from the Dyson equations in the limit $q_o=0$, $q\to 0$, resulted in $L=FD^2F/\mu^2$.

It seems that the discrepancy between this and the effective action used in this paper is due to different choices of renormalisation conditions[19]. Roughly speaking, the running coupling of Baker[18], $\bar{g}^{-2}=D^2/\mu^2$, coincides with the bare coupling at a particular momentum $D^2 = p^2 = \mu^2$, whereas the running coupling of refs. 13,14 does so for a particular value of the field strengths, $F^2 = \mu^2$.

PROPERTIES OF THE EFFECTIVE LAGRANGIAN

Qualitatively, L_{eff} is displayed as a function of B^2-E^2 in fig.1. Its prominent feature is the local minimum at $B^2-E^2 = \kappa^2$, which is even an absolute minimum for configurations with a real action (complex action configurations which have $B^2<E^2$ somewhere, are regarded as unstable.)[16].

For vacuum configurations, i.e. if $j=0$, the free energy equals the action by (4), and also the energy for $T=0$. Thus a thermodynamically stable zero temperature vacuum solution has to fulfill the extremum condition

$$\delta\Gamma/\delta A = 0 \quad \text{or} \quad dL/dF^2 = 0 \,. \quad (12)$$

This indeed is an Energy minimum and thus thermodynamically stable. Because this minimum for L of the vacuum occurs at

$$B^2 - E^2 = \kappa^2 > 0 \,. \quad (13)$$

the stable vacuum has nonvanishing colour-magnetic fields which are stronger than the electric fields.

The dielectric function

$$\varepsilon = dL/dF^2 = \log F^2/\kappa^2. \tag{14}$$

is displayed in fig.2. For the vacuum we have $\varepsilon=0$.

The effective Lagrangian, although developed from the one-loop approximation, is proposed and taken to be a useful approximation for all configurations of static sources.

The reason is, that for the vacuum $\varepsilon=0$, as to be expected[15], while with larger $|\varepsilon|$, signalling the presence of colour-charges, the running coupling $\bar{g}^2 = 1/\log(F^2/\kappa^2)$ goes to zero.

For saturation properties of nuclear matter, where we are not interested to study the deep interior of the nucleons, but their slight gluon-cloud polarisation due to the presence of neighbouring nucleons, the application of our pure gluonic L_{eff} seems to be reasonable.

Although qualitatively we do expect, because of the correct limits of either ε or g^2, that the L_{eff} will describe confinement, we have to study explicitely how this comes about.

CONFINEMENT WITH L_{eff}

Confinement in this rather simplified approach should mean that for a given charge-distribution, restricted to a finite volume in space, the expectation values of the Gluon fields should asymptotically, that is sufficiently far from the sources, tend to fulfill the vacuum condition $B^2-E^2=\kappa^2$, and that at least solutions for non-zero total charges Q should energetically be infinitely unfavourable. We shall see, that in our context it will be nontrivial to find solutions for the fields at all, and that we were successfull only in SU_{3c}, and that it is possible only if the vacuum condition holds identically outside a finite volume, which contains a colour charge distribution of total colour charge zero. Thus L_{eff} seems to enforce a rather strict version of confinement.

To obtain some quantitative results by analytical considerations we proceed by assuming a reasonable function $\varepsilon(r)$, namely $\varepsilon\approx 1$ inside the colour-charge distribution and $\varepsilon\approx 0$ far from sources. Next we try to find a set of solutions to the Yang-Mills analogue of the Biot-Savart law (spatial components of Maxwell's equations) for zero current density. Then we pick the solution which fullfills $\varepsilon=\log F^2/\kappa^2$, and if there is one, we can calculate the actual colour-charge distribution from $\rho=D_i\varepsilon F_{io}$.

Let us start with an Abelian exercise. With only one relevant colour direction assumed, all selfcoupling terms vanish, yielding $\nabla(\varepsilon E)=\rho$, $\nabla x(eB)=0$ with $E=\nabla A_o$ and $B=\nabla xA$.

For the vacuum abelian solutions with $B^2\geq\kappa^2$ do exist with $\varepsilon=0$ and $\rho=0$ everywhere.

Let us prescribe $\rho\neq 0$ in a small region. Then in the vicinity $\varepsilon\neq 0$, but $\varepsilon\approx 0$ far away from the source region. Let us assume for simplicity $\varepsilon\equiv 0$ outside a surface $S=\partial V$ more general proof with $\varepsilon\approx\exp(-\lambda r^2)$ yields the same result). Close to S but inside, the smooth function ε will be small and the field equation yields $\nabla xB+Bx\nabla\ln\varepsilon=0$ with the second term getting very large near S, while ∇xB is assumed to be continuous. Thus at S we have B to be parallel to $\nabla\ln\varepsilon$, that is parallel to the surface normal n, and $B^2\geq\kappa^2$ because $\varepsilon=0$. Since the field configuration would thus form a colour magnetic monopole in contradiction to the starting point $B=\nabla xA$, the result is that in this simple Abelian exercise no solutions exist,- the vacuum cannot be perturbed by introducing localised charges.

To obtain a simple nonabelian field configuration we consider spherically symmetric fields only. The spherically symmetric ansatz by Wu and Yang[20], which is a simple special case of Witten's ansatz, is applicable for our semi-classical model too. For the SU_{2c} gauge subgroup, it reads

$$A_o^a = r^{-1}\, G(r)r_a \quad , \; A_i^a = r^{-2}(H(r)-1)\varepsilon_{ian}r_n \quad . \tag{15}$$

$$\rho^a(r) = r^a\rho(r)/r^3 \tag{16}$$

with the boundary conditions

$$H(0) = 1, \; H'(0) = 0, \; G(0) = 0 \;, \tag{17}$$

which yields

$$B_i^a B_i^a = 2(H')^2/r^2 + (H^2-1)^2/r^4 \tag{18}$$

$$E_i^a E_i^a = (G')^2 + 2G^2H^2/r^2 \quad , \tag{19}$$

so that the equations of motion take the form of two coupled ordinary differential equations of second order

$$-H'(\ln|\varepsilon|)' = (H'' + H(1-H^2)/r^2 + G^2H) \tag{20}$$

$$\rho(r) = (r^2\varepsilon G'(r))' - \varepsilon G H^2 \text{ with } \varepsilon = \log F^2/\kappa^4. \tag{21}$$

For 'soft' bags, indicated by ε going smoothly to zero as $r\to\infty$, but with $\varepsilon\neq 0$ everywhere, it has been proven[17] that there are no continuous smooth solutions for the fields, since $H(r)$ runs into a singularity at some finite r.

For 'hard' bags, with $\varepsilon\equiv 0$ beyond some small r_1 outside V, a somewhat lengthy argument[17] dealing with inequalities leads to $H'(r_o)\equiv 0$ as a further boundary condition, and to $H(r_o)^2<1$.

Then there exist solutions of the sourceless field equation $D_\mu \varepsilon F_{\mu k}=0$ if

$$r_1 \le r_o \le 1/\sqrt{\kappa}, \tag{22}$$

which with the adopted values of κ of 2oo MeV is about $r_o \le 1\text{fm}$.

In this context, let us say, a proton would be described as a classical colour-charge distribution of, say, o.8fm radius, surrounded by a polarised gluon cloud, which at r_o slides into a gluon-vacuum solution with $B^2-E^2=\kappa^2$.

A special solution of $D\varepsilon F=0$ can be gained by choosing $-1 \leq H(r_o) \leq 0$ and $G(r)\equiv 0$, then the ansatz $H(r) \approx H_1(r)=[1-\kappa(1+\delta)r^2]^{\frac{1}{2}}$ for $r \leq r_1 << r_o$, $H(r) \approx H_2(r)=H(r_o)+\alpha(r-r_o)^3$. for $r \leq r_o$ with some smooth interpolation yields a solution with the features

$$|\varepsilon| \approx \exp(-c/(r-r_o)) \quad \text{for } r<r_o, \tag{23}$$

$$B^2(r) \geq B^2(r_o) \geq \mu^4(1+\delta)^2. \tag{24}$$

These solutions do not yet satisfy the consistency requirement that $\varepsilon(r)=\log F^2(r)/\kappa^2$.

In a larger gauge group, SU(3), the 8-component of the charge density and the colour electric field can be used to adjust the total magnitude of the field strength to the value required by the dielectric function $\varepsilon(r)$ fixed in advance. We only obtain, as an additional equation,

$$\nabla (\varepsilon E^8) = \rho^8 \tag{25}$$

which is easy to solve.

The total colour charge is the integral of div(εE) over the bag volume, which can be evaluated as a surface integral in a region where ε vanishes identically. Thus for the Wu-Yang radial symmetric SU_2 fields plus an ρ^8 source a bag-type colour-singlet solution is found and constructed.

Of course, $r_o=1/\sqrt{\kappa}$ is not to be compared with the bag-radius of bag models. Instead this bag-radius would be somehow defined by the distribution of the colour charge distributions. Then at $r_B \leq r_o$ the undisturbed gluon vacuum is reached.

L_{eff} AND NUCLEAR PHYSICS

A first consequence of the results would be that the interaction of free nucleons vanish at latest when their half-distance is larger than $1/\sqrt{\kappa}$.

Comparing this to the average nucleon half-distance[1] inside nuclei of about 1.17±0.01 fm, we see that in this frame the wellknown nuclear saturation density sets an upper limit for the renormalisation group invariant mass, for

$$1.17 \simeq r_{n.m.} < r_o \simeq 1/\sqrt{\kappa}. \tag{26}$$

Thus $\sqrt{\kappa}<169$ MeV.

Since $\sqrt{\kappa}$ is defined using field strengths as a renormalisation point, it cannot directly be compared to the more widely used invariant mass Λ, defined by using a momentum space renormalisation point. However, an estimate[21] within the perturbation limit, using the bag model estimate for the running coupling constant $\alpha_c(p_f) \simeq 2.2$ yields for the known nuclear matter density $n \simeq 0.15\ \text{fm}^{-3}$ the value of $\Lambda \simeq 107$ MeV. If this value would be inserted into (28) a zero force nucleon-nucleon half-distance of 1.82 fm would result which is not in contradiction to free n-n scattering data.

Numerical studies of solutions starting from L_{eff} for various given external source configurations are under way.

For application to the interior configurations of strongly interacting systems in equilibrium such as a free nucleon first the approach has to be improved by treating the quarks as Dirac particles, with triplet colour charges.

Finally, for the interior of quark systems, the colour-dielectric function ε may depend not only on the field strength F^2 but also on other invariant quantities.

The complex spatial geometry of multinucleon systems such as the nucleus may,- for studying the polarised nucleon formfactor, the density dependence of the nucleon force etc.,- well be approximated by studying the same problem of a nucleon in a nuclear surrounding but in the simpler spherical geometry.

This method is wellknown in Solid State Physics, named after Wigner and Seitz, where the calculation of the electron cloud of an atom in a spherical cell, and with the boundary conditions of zero Coulomb potential gradient at the surface and total charge zero yields the binding energy per atom of solids to better than 2%. A similar approach was used for nuclei in a thin neutron gas occuring in the neutron star skin[22]. Thus already in solids the gain in binding due to specific crystalline configurations but with the same total density is small. For nucleons in nuclei this approximation should be even better due to that the shell effects for (heavy) nuclear masses are only of order of less than 10 MeV[1] which is e.g. for Pb^{208} only 0.6%, reflecting the large ratio of kinetic to binding energy per nucleon.

At present, this method is applied in various approaches to semi- classical QCD.

In the Soliton Model[23], treating the nucleon as three Dirac- particles coupled to a colour-singlet scalar field σ, the soliton field, the selfconsistent solutions for ψ_q and $\sigma_o(r)$ are known. Thus for the application to a nucleon within a nucleus only the boundary conditions for the σ-field have to be changed to

$$\sigma'(r_1) = 0 , \tag{27}$$

where r_1 is given by the nuclear density

$$\rho_o = 1/(4\pi r_1{}^3/3). \tag{28}$$

Because $\sigma(r)$ will be only slightly different from $\sigma_o(r)$ in the region where the quark-density is non-zero, the soliton model equations can be solved[24] perturbatively to yield selfconsistent solutions for the slightly polarised quark density and the slight binding increase.

In the effective Lagrangian approach, the respective boundary condition is the vanishing of the radial derivatives of $A^a_\mu(x)$ at r_1. However for the respective minimisation either an additional constraint is needed, say,the square of the total charge of the external colour-charge density to be kept constant, or better the concept is to be improved by quantising the sources. Work in that respect is in progress.

ACKNOWLEDGEMENTS

One of us (E.H.) gratefully acknowledges stimulating discussions and the hospitality of the Nuclear Theory Division, University of Washington, Seattle, during the Summer Institute of Theoretical Physics, the attendence of which was supported by the Gesellschaft fuer Schwerionenforschung.

REFERENCES

[1] H.v.Groote,E.R.Hilf,and K.Takahashi, Atomic Data and Nuclear Data Tables 17(1976)418
[2] X. Panagiotou,D.Scott et al.Preprint MSUCL,Sept.(1983)
[3] D.Scott, preprint MSUCL(1983)434
[4] G.Sauer,H.Chandra,U.Mosel, Nucl.Phys.A264(1976)221
[5] W.A.Kueppers,G.Wegmann,E.R.Hilf,Ann.of Physics,88(1974)454
[6] Int.Conf.Atomic Masses and Fundamental Constants, VII, Bad Seeheim, FRG; September (1984)
[7] S.L.Adler, Phys.Rev.D23(1981)2905
[8] S.Coleman and E.Weinberg, Phys.Rev.D7(1973)1888
[9] E.R.Hilf and L.Polley, Phys.Lett.(1983) in press;
[10] B.Svetitsky,priv.comm.(1983)
[11] S.L.Adler and T.Piran, preprint(1983)and refs.therein
[12] M.Magg,Nucl.Phys.B158(1979)154
[13] R.Freedman,L.Wilets,S.D.Ellis,E.M.Henley,Phys.Rev.D22 (1980)3128
[14] L.Carson, and E.R.Hilf, unpublished(1980);
[15] S.K.Savvidy, Phys.Lett.71B(1977)133; H.Leutwyler, Nucl.Phys.B179(1981)129
[16] H.Pagels and E.Tomboulis, Nucl.Phys.B143(1978)485
[17] E.R.Hilf and L.Polley,(1982), submitted to Nucl.Phys.
[18] M.Baker and F.Zachariasen,Phys.Lett.108B(1982)206 and preprint40048-16 P3 May(1983)Univ.of Washington
[19] M.Baker and D.Pottinger, priv. comm.
[20] For a review, see A.Actor, Rev.Mod.Phys.51(1979)461
[21] G.Baym, in"Proc.Statist.Mech.of Quarks and Hadrons", ed.H.Satz, North-Holland Publ.Comp.(1981)461
[22] E.R.Hilf,'Early Neutron Star Matter and Hadrons', in "Int.Conf.Mesonic Effects in Nuclear Structure, Wiss.Verlag Bibl.Inst.(1975)17 and refs. therein;
[23] L.Wilets, this volume;
[24] A.Schuh,L.Wilets,and E.R.Hilf, in preparation

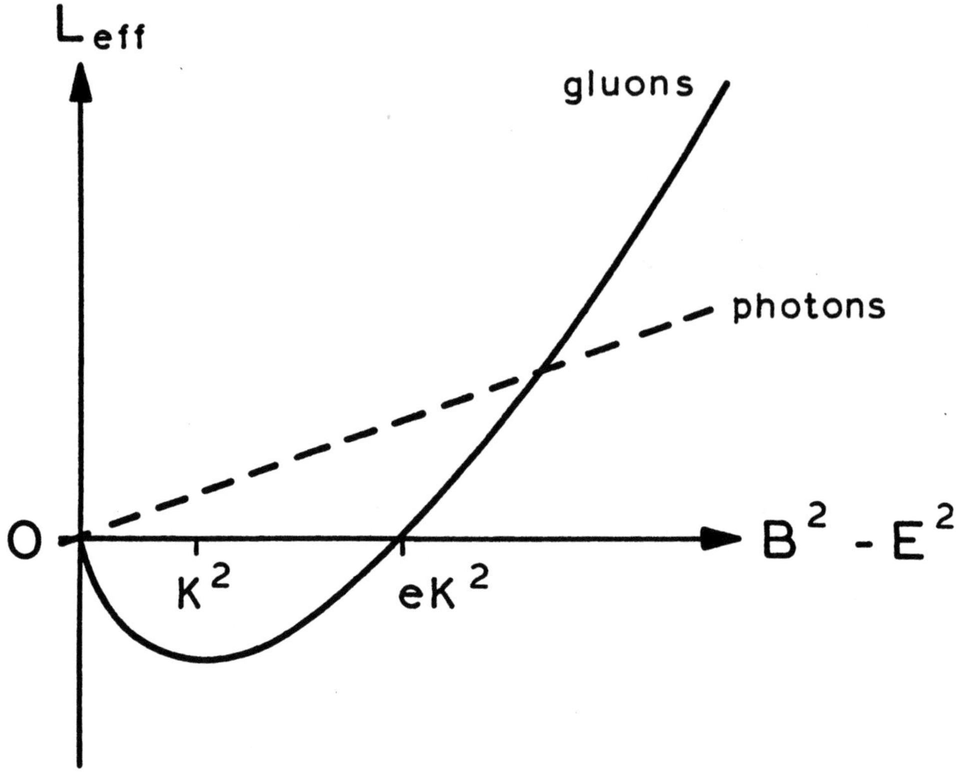

Fig.1

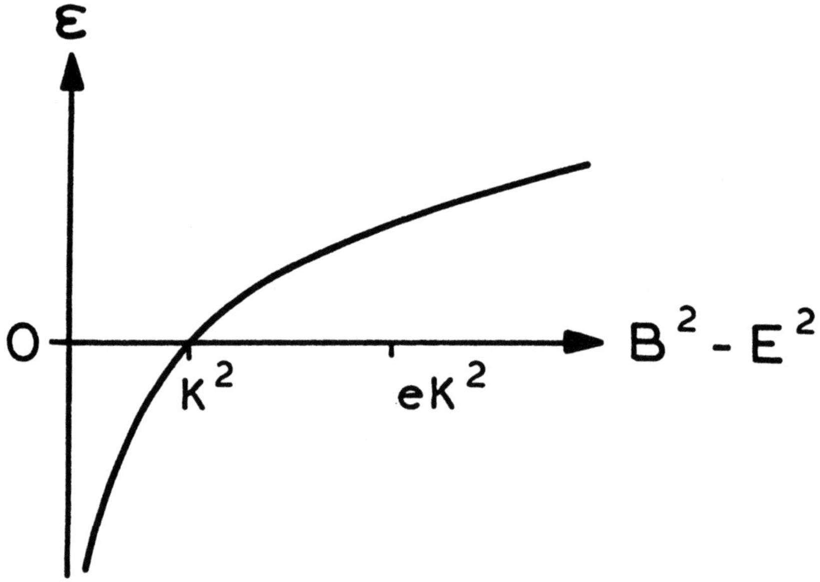

Fig.2

STRANGE QUARKS IN LOW ENERGY HADRONIC INTERACTIONS

H.J. Pirner
Institut für Theoretische Physik, Universität Heidelberg
and
Max-Planck-Institut für Kernphysik, Heidelberg, FRG

The dynamics of the K^+N-interaction is calculated in the constituent quark model with QCD-residual forces. The obtained S-wave interaction agrees rather well with the experimental data. We also discuss the status of spin-orbit forces in KN, Λ-nucleus and Σ-nucleus interactions.

I. Introduction

Strong interactions are independent of flavor. The forces mediating the strong interaction are colored vector-gluons which carry no flavor. The masses of elementary particles and their interactions nevertheless show strong flavor-dependence: e.g., the K-meson with $m_K \approx 500$ MeV is much heavier than the π-meson with $m_\pi = 137$ MeV. The Lagrangian density of QCD includes this flavor asymmetry by a nontrivial mass matrix M^{ij}, i.e.

$$L(x) = -1/4\, F^a_{\mu\nu}(x) F^{a\mu\nu}(x) + \bar{\psi}^{i,\alpha}(x)\,(i\gamma^u\, D^{\alpha\beta}_\mu\, \delta^{ij} - M^{ij})\psi^{j,\beta}$$

$$D_\mu^{\alpha\beta} = \partial\mu - ig\,\frac{\lambda_a^{\alpha\beta}}{2}\,A^a_\mu\,.$$

For only three flavors the mass matrix M^{ij} has the form:

$$M^{ij} = \begin{pmatrix} m_u & 0 & 0 \\ 0 & m_d & 0 \\ 0 & 0 & m_s \end{pmatrix} \qquad \text{with } m_s \approx \frac{m_u+m_d}{2} + 0.2\ \text{GeV}$$

In this picture the chiral SU3 x SU3 symmetry is broken due to the large strange quark mass. We therefore believe that chiral dynamics plays a less important role in strange hadron-hadron interactions. If we loosely identify chiral dynamics with pion or pion-cloud effects, we would expect that long-range meson exchanges are not distorting the short-range QCD-picture in strange quark interactions. Indeed the situation is very clear for K^+N-scattering which has clear advantages to NN-scattering. For the K^+N-interaction

(i) one-pion exchange is not allowed, because the coupling of three pseudo-scalars has to vanish;

(ii) two-pion exchange is suppressed, since the $KK^*\pi$ coupling is smaller than the $N\Delta\pi$ coupling.

The K^* has half the width of the Δ in spite of the larger phase space in $K^* \to K\pi$ compared to $\Delta \to N\pi$. In addition the t-channel $NN \to NN$ amplitude has the square of ($N\bar{N} \to \pi\pi$) amplitudes as weight functions, whereas in $K^+N \to K^+N$ the $\bar{K}K \to \pi\pi$ and $\pi\pi \to N\bar{N}$ amplitudes may have different signs. Hence the quark model alone may give rather realistic phase shifts for KN-scattering. Another advantage of the quark description in the KN-system is the isospin selectivity. Different isospins show complementary features in the KN-cluster model. The $I = 0$ channel is determined by quark-gluon exchange only, whereas the $I = 1$ channel has contributions to the scattering from quark rearrangement and quark-gluon exchange. Therefore, in the KN-system we are able to investigate the relative importance of the QCD-residual interaction and of quark antisymmetrization.

It is sometimes incorrectly stated that the repulsive NN-interaction is a consequence of quark antisymmetry. This is not true. Quark rearrangement alone leads to a vanishing interaction for NN-scattering. Only antisymmetrization together with the QCD-residual force, i.e. quark-gluon exchange, gives a repulsive short-range NN-interaction. Historically this work on strange quarks has been triggered by the experimental observation of Λ-hypernuclear spectra. In the meantime also Σ-hypernuclei have been seen. I will discuss their preliminary interpretation in the context of quark models. Let me start, however, with the KN-interaction. The understanding of this system would represent an important step forward towards a theory of hypernuclei. For a detailed discussion I would like to refer to ref. [1].

II. K^+N-scattering

In the constituent quark model all the complications and subtleties of non-perturbative QCD are parametrized by the constituent quark mass and a confining potential. The Hamiltonian has the form

$$H = \sum_{i=1}^{4} \frac{p_i^2}{2m} + \frac{p_5^2}{2m_s} + \sum_{i<j} \mathcal{V}_{ij}^{conf} + \sum_{i<j} \mathcal{V}_{ij}^{QCD} \tag{1}$$

where the confinement is included via the two-body interaction

$$\mathcal{V}_{ij}^{conf} = -\frac{D}{2} \lambda_i \lambda_j \ |\underset{\sim}{x}_i - \underset{\sim}{x}_j|^2 \tag{2}$$

and the QCD-residual interaction is the non-relativistic limit of the one-gluon exchange potential (without LS and tensor terms for the S-wave interaction).

$$\mathcal{V}_{ij}^{QCD} = \frac{\alpha_s}{4} \lambda_i \lambda_j \ \left\{\frac{1}{|\underset{\sim}{x}_i - \underset{\sim}{x}_j|} - \frac{2\pi}{3} \frac{1}{m_i m_j} \underset{\sim}{\sigma}_i \underset{\sim}{\sigma}_j \ \delta(\underset{\sim}{x}_i - \underset{\sim}{x}_j) - \frac{\pi}{2} \left(\frac{1}{m_i^2} + \frac{1}{m_j^2}\right) \delta(\underset{\sim}{x}_i - \underset{\sim}{x}_j)\right\} \tag{3}$$

We use the following parameters: the constituent quark mass is m = 0.3 GeV, the oscillator parameter D is adjusted to reproduce the p-wave nucleon excitation at 1.535 GeV. The QCD-coupling constant $\alpha_s = 0.7$ is obtained from the ΔN-splitting. The strange quark mass is related to the Λ, Σ and N mass difference.

The Hamiltonian of eq. (1) has as asymptotic states the kaon and the nucleon. Color confinement does not act asymptotically because of the color neutrality of the hadronic clusters. It is natural (but not unique) to choose a trial 5-quark wavefunction which takes into account the hadronic cluster states ϕ_N and ϕ_K.

$$\psi = A \int c(\underset{\sim}{R}) \ \ \psi_{\vec{R}} \ \ d^3\underset{\sim}{R} \tag{4}$$

with

$$\psi_{\vec{R}} = \phi_N(\xi_1 \xi_2) \ \phi_K(\xi_3) \ \chi_{\vec{R}}(r) \ \phi_{cm}(\underset{\sim}{R}_G) \ \chi_{KN}^{SIC} \tag{5}$$

and $A = 1 - P_{14} - P_{24} - P_{34}$.

The trial wavefunction of eqs. (4,5) is a superposition of hadron clusters at relative distances R. In the interaction region it is advantageous to choose $\chi_{\vec{R}}(r)$ as a Gaussian function peaked around this generator coordinate $\underset{\sim}{R}$ and let the center of mass R_G free to oscillate. With this choice the wavefunction $\psi_{\vec{R}}$ can be reexpressed in terms of quark coordinates $x_1,\ldots,x_5$ (see table 1). ψ_R contains two shell model wavefunctions peaking around $-(1 + K)/(4 + K)\ \vec{R}$ and $+\ 3\ \vec{R}/(4 + K)$, where $K = m_s/m$ is the mass asymmetry.

$$\psi_{\underset{\sim}{R}}(x_1,x_2,\ldots x_5) = A\left\{\prod_{i=1}^{3} \left(\frac{2\Omega}{\pi}\right)^{3/4} e^{-\Omega(\vec{x}_i + \frac{1+K}{4+K}\vec{R})^2} \left(\frac{2\Omega}{\pi}\right)^{3/4} e^{-\Omega(\vec{x}_4 - 3\vec{R}/(4+K))^2} \left(\frac{2K\Omega}{\pi}\right)^{3/4} e^{-\Omega K(\vec{x}_5 - \frac{3\vec{R}}{4+K})^2}\right\} \tag{6}$$

Note the heavier strange quark is more localized at the K-cluster because of $K \approx 1.7$ it will not participate as much in peripheral partial waves.

The key (ref. [1]) to the solution of the time-independent Schrödinger equation

$$(H - E)\ \psi = 0 \tag{7}$$

is the projection of this equation onto a variation of the many quark wavefunction (4) with respect to $c(\vec{R})$:

$$\int \delta\psi\ (H - E)\ \psi\ d^3r d\tau^{12} = 0 \tag{8}$$

$$d\tau^{12} = \prod_{i=1}^{3} d^3\xi_i\ d^3R_G$$

For large distances r the relative motion of the two clusters is described by a free wavefunction. Hence by applying Green's formula to the integration over d^3r we obtain

$$\int_G d^3r \quad d\tau\ [\delta\psi(H - E)\ \psi - \psi(H - E)\delta\psi\ ] =$$

$$- \frac{1}{2\mu} \int_{\partial G} d^2s \quad d\tau\ (\delta\psi \frac{\partial\psi}{\partial n} - \psi \frac{\partial\delta\psi}{\partial n}) = - \frac{1}{2\mu k}\ \delta S \tag{9}$$

where $\mu = \frac{3(1+K)}{4+K}$ m is the reduced mass of the KN-system, G is a bounded region in R^3, ∂G its surface, and $\partial/\partial n$ the normal derivative. Using a superposition of incoming and outgoing spherical waves the r.h.s. of eq. (7) is proportional to the variation of the S-matrix δS. The Schrödinger equation can be reformulated as a variational equation $\delta I = 0$ with

$$I = \frac{1}{2\mu k}\ S - \int_G d^3r \quad d\tau\ \psi(H - E)\psi \tag{10}$$

which leads to a linear system of equations for a discretized set of coefficient $c(R_i) = c_i$ in eq. (4). The whole physics is contained in the expectation values of $(H - E)$ between cluster states located at distances $\underset{\sim}{R}$ and $\underset{\sim}{R}'$.

$$H(\underset{\sim}{R}',\underset{\sim}{R}) - N(\underset{\sim}{R}',\underset{\sim}{R})\ E = \int \psi_{\vec{R}'}\ (H - E)\ A\ \psi_{\vec{R}}\ d\tau \tag{11}$$

In general there are direct and exchange parts for each kernel from the two parts of the antisymmetrizer $A = 1 - 3\ P_{34}$. These are represented in fig. 1. A cross symbolizes a kinetic energy term $p_i^2/2m$, the dotted lines stand for a quark-quark interaction.

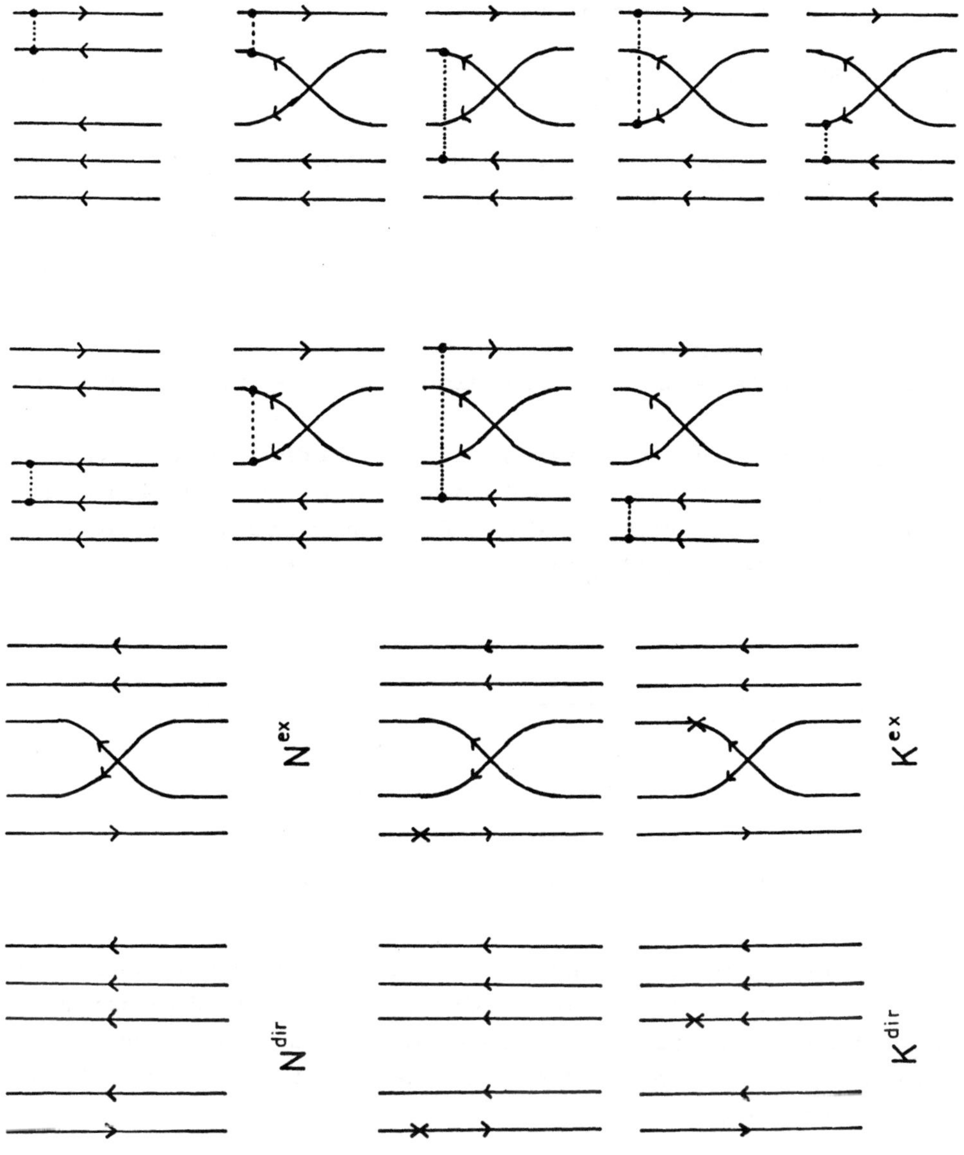

Fig. 1 Contributions to the normalization kernel $N(\underset{\sim}{R},R')$, kinetic energy kernels $K(\underset{\sim}{R},\underset{\sim}{R}')$ and potential kernel $V(R,R')$, where $H(R,R') = K(R,R') + V(R,R')$.

The spin color overlap factors are such that

$$< \chi^{SIC} \mid P_{34}^{SIC} \mid \chi^{SIC} > = 1/9 \; \delta_{I1} , \tag{12}$$

i.e., the exchange operator vanishes for the KN-cluster function in the I = 0 state. Therefore quark exchange alone from N^{ex} and K^{ex} (fig. 1) does not contribute to the I = 0 state. The effect of quark exchange is very strong, however, in the I = 1 state, as one can see from a solution of the problem without QCD-residual interaction, i.e. $\alpha_s = 0$.
In fig. 2 we show the phase shift obtained from the calculation in the I = 1 channel with $\alpha_s = 0$. The theoretical phase shift already reproduces the data (crosses).

The origin of this repulsion can be visualized by regarding the spin-isospin symmetry of the wavefunction. Because of the color symmetry [2, 1, 1] of the 4 light quarks only the SU(4) $[3, 1]_{IS}$ states are allowed for a spatially symmetric wavefunction [1 1 1 1]. The SU(4) state $[4]_{IS}$ has to be associated with a spatially excited 4-quark state:

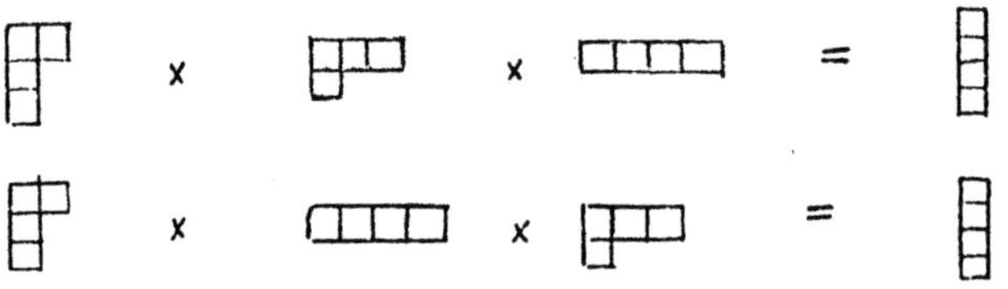

(13)

Color- $\qquad$ $SU(4)_{IS}$- $\qquad$ spatial $\qquad$ total wavefunction

In line with this symmetry argument the quark exchange due to antisymmetrization generates different potentials for the two SU(4) states: the sign of the potential is positive for the $[4]_{IS}$ and negative for the $[3, 1]_{IS}$ spin-isospin wavefunction. Denoting by P_{34}^{SI} the exchange operator of particle 3 and 4 acting on the spin-isospin wavefunction one obtains

$$\begin{aligned} &< [4]_{IS} \mid P_{34}^{SI} \mid [4]_{IS} > = 1 \\ &< [3, 1]_{IS} \mid P_{34}^{SI} \mid [3, 1]_{IS} > = -1/3 \end{aligned} \tag{14}$$

The I = 0 K-N wavefunctions can be expressed as a superposition of SU(4) supermultiplet states coupled to a strange antiquark

$$\begin{aligned} &\mid KN; I = 0, l = 0 > = -1/2([4]_{00}\, \bar{s}) + \frac{\sqrt{3}}{2} ([3, 1]_{01}\, \bar{s}) \\ &\mid KN; I = 1, l = 0 > = \frac{1}{\sqrt{2}} ([4]_{11}\, \bar{s}) - 1/2([3, 1]_{10}\, \bar{s}) - 1/2([3, 1]_{11}\, \bar{s}) \; . \end{aligned} \tag{15}$$

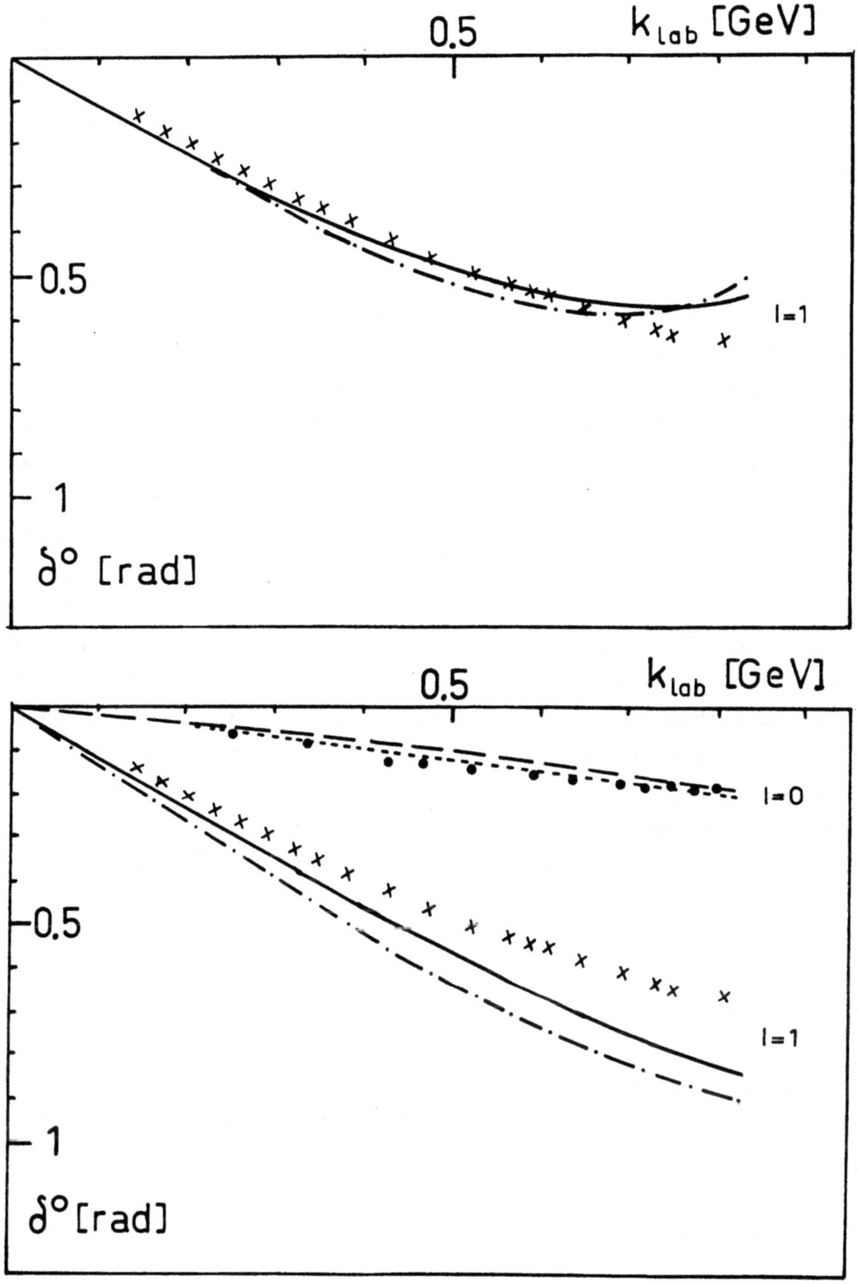

Fig. 2b Theoretical evaluation of the KN-S-wave phase shifts for I = 0 and I = 1 and experimental data from ref. [8]. The upper curves for each isospin correspond to the smaller cluster size $\langle r^2\rangle^{1/2} \approx 0.43$ fm compared to $\langle r^2\rangle^{1/2} \approx 0.47$ fm.

Combining eqs. (14) and (15) we see that for the I = 0 state the attractive $[3, 1]_{IS}$ and repulsive $[4]_{IS}$ contributions of the exchange potential cancel, whereas for the I = 1 state the repulsive $[4]_{IS}$ contribution dominates. This explains the strong repulsion in the I = 1 channel which is produced by the antisymmetrization of the 4 light quarks.

In ref. 1 we have calculated a local approximation to the Hill-Wheeler equation; it gives an r-dependent effective mass which acts like a repulsive shell prohibiting the two hadrons to approach. Including the QCD-residual interaction in addition (fig. 2b)we obtain weakly repulsive I = 0 phase shifts in agreement with the data, but somewhat too large phase shifts for the I = 1 state. We hope that a coupled channel calculation with the K^* and Δ-cluster states can improve these results. In general the agreement between theory and experiment is rather good. Comparing to NN-phase shifts one wonders what is the major difference between the two cases? We think that the attractive 2π-exchange which is sizable in NN-scattering but weak in KN-scattering is causing the different behavior. It remains a challenging task to incorporate such non-perturbative QCD effects in the NN-problem. On the other hand the KN-system has the good feature of weak long-range forces. Thereby it presents a much better candidate to study the properties of the perturbative QCD-interaction for multiquark states. We have also calculated the P-waves and compared to an average of the experimental amplitudes taken in such a way that a weak spin-orbit force cancels. There is no correlation between theory and experiment, however. It is therefore interesting to see whether agreement between theory and experiment can be obtained by including QCD-residual spin-orbit and tensor terms, or whether exotic $(qqqq\bar{q})$ configurations play an important role in the P-waves.

III. Spin-Orbit Interactions

Previous investigations have shown the P-wave KN-interaction can be fitted by a pure spin-orbit interaction obtained from quark-additivity in the limit of strong SU3-breaking.

$$\mathcal{V}^{KN}_{LS} = 30 \text{ MeV fm}^2 \frac{1}{r}\frac{\partial}{\partial r}\frac{\tilde{\rho}(r)}{\rho_o} \left(\sum_{Q\epsilon N} \vec{s}_Q (1 + \vec{\tau}_K \vec{\tau}_Q) \right) (\Sigma \vec{l}_Q) \qquad (16)$$

In the expression of eq. (16) we find the familiar Thomas-form of the spin-orbit potential with a strength adjusted to the baryon-nucleus spin-orbit interaction. The exchange of the quarks leads to the isospin exchange operator $1/2(1 + \tau_K \tau_Q)$, since the isospin of the K is equal to the isospin of the exchanged light quark. Taking the expectation value of eq. (16) with respect to the nucleon we find

$$\mathcal{V}^{KN}_{LS} = 30 \text{ MeV fm}^2 \frac{1}{r}\frac{\partial}{\partial r}\frac{\tilde{\rho}(r)}{\tilde{\rho}_o} \underset{\sim}{s}\cdot\underset{\sim}{l}(1 + 1/3\ \underset{\sim}{\tau}_K\underset{\sim}{\tau}_N) \tag{17}$$

The 5/3 appearing in eq. (17) is the same as in the naive quark model prediction for the axial coupling constant g_A. The normalized quark distribution has been assumed to be $\Theta(R_o - r)$ with $R_o \simeq 0.5$ fm. This model fits the data quite well (ref. [2]). It is built in analogy to the baryon-nucleus spin-orbit interaction which is obtained in ref. [3] by adding quark-nucleon interactions in the nucleus and averaging over the quark-wavefunction of the valence baryon.

$$\mathcal{V}^{B}_{LS} = 90 \text{ MeV fm}^2 \frac{1}{r}\frac{\partial}{\partial r}\frac{\rho(r)}{\rho_o}\ 1/n \sum_{u,d}^{n} \underset{\sim}{s}_Q\cdot\underset{\sim}{l}_Q \tag{18}$$

giving the spin-orbit potentials for the N, Δ, Λ and Σ-nucleus interaction of table 2.

Two questions have to be discussed:

(1) Are the recently measured Σ-hypernuclear data in agreement with this phenomenological model based on quark-additivity?

(2) Can quark-additivity be derived in the constituent quark model from the ΛN and ΣN spin-orbit interactions?

Concerning the first question we have come to the following conclusion. If the peak in the spectrum on $^{12}C(K^-,\pi^+)^{12}C_\Sigma$ coincides with the lower energy ($m_{Hy} - m_A$ smaller) peak in the $^{16}O(K^-,\pi^+)^{16}O_\Sigma$-spectrum, then a residual interaction of the form

$$\mathcal{V}^{\Sigma N} = C_o\ \delta(\underset{\sim}{r}_\Sigma - \underset{\sim}{r}_N)\ [f_{\Sigma N} + f'_{\Sigma N}\ \underset{\sim}{\tau}\underset{\sim}{T} + g_{\Sigma N}\ \underset{\sim}{\sigma}_\Sigma\underset{\sim}{\sigma}_N + g'_{\Sigma N}\ \underset{\sim}{\sigma}_\Sigma\underset{\sim}{\sigma}_N\ \underset{\sim}{\tau}\underset{\sim}{T}] \tag{19}$$

has to give very small mixing of $|3/2\ 3/2^{-1}>$ and $|1/2\ 1/2^->$ states, i.e.

$$<\Sigma_{3/2}\ p^{-1}_{3/2}\ \ |\mathcal{V}^{\Sigma N}|\ \Sigma_{1/2}\ p^{-1}_{1/2}> \quad \approx 0 \tag{20}$$

which puts strong constraints on the coefficients f,f',g,g'. More importantly the peak corresponding to the lower mass in O^{16} has to be the $\Sigma_{3/2}\ p^{-1}_{3/2}$ state. The necessary spin-orbit strength of the Σ then is $\approx$ 5/3 as big as the nucleon spin-orbit strength. The Coulomb energy differences do not change the above statement. The argument is very simple, since we know the undisturbed mass $m^o_{3/2}$ in C^{12} and the sum of the masses $(m_{3/2} + m_{1/2})$ in O^{16} we can deduce the undisturbed level $m^o_{1/2}$

$$m^o_{1/2} = m(\text{lower}) + m(\text{upper}) - m^o_{3/2} \approx m(\text{upper}) \tag{21}$$

No residual interaction can invert the assignment m(lower) = $m_{3/2}$ and m(upper) = $m_{1/2}$, having taken into account the information from C^{12}. For the residual interaction above

M. Sommermann [4] gets the relation of the coefficients

$$f_\Sigma - f'_\Sigma + 3g_\Sigma - 3g'_\Sigma = 0 \tag{22}$$

which is compatible with an extra-polation of the NN-Migdal-interaction using quark-additivity

$$f_{\Sigma N} = 2/3\ f_{NN} \quad , f'_{\Sigma N} = 2\ f_{NN} \quad , g_{\Sigma N} = 4/3\ g_{NN} \quad , g'_{\Sigma N} = 4/5\ g'_{NN} \tag{23}$$

and modifying the I = 1/2 channel because of the presence of the $\Sigma N \to \Lambda N$ channel by a renormalization factor Z. The new Migdal constants $\tilde{f},\tilde{f}'$ are given by

$$\begin{aligned} \tilde{f}_{\Sigma N} &= 2/3(f_{\Sigma N} + f'_{\Sigma N}) + 1/3\ Z(f_{\Sigma N} - 2f'_{\Sigma N}) \\ \tilde{f}'_{\Sigma N} &= 1/3(f_{\Sigma N} + 2f'_{\Sigma N}) + 1/3\ Z(-f_{\Sigma N} + f'_{\Sigma N}) \end{aligned} \tag{24}$$

and $\tilde{g},\tilde{g}'$ obey corresponding equations. The Σ-well depth of half the well depth of normal nuclei corresponds to $Z \approx 1.5$.
It must be admitted, however, that the isospin splitting of "Σ^+p^{-1} and Σ^on^{-1}" states observed in the $^{12}C(K^-,\pi^-)$ reaction would come out too big with the above coefficients.

Taken the above interpretation with two peaks in the ^{16}O-spectrum there is no way other than to conclude that the Σ-hypernuclei have a large spin-orbit splitting. The meson exchange parametrizations have to take into account stronger SU3-breaking and/or include the damping of two-pion exchange also in the real part, since it does not lead to the correct width of the Σ-hypernuclear states.

The second question about a microscopic derivation of the additive quark model for the spin-orbit interaction in the constituent quark model is rather interesting, too. Recently two papers [5, 6] have appeared which have calculated the spin-orbit BN-forces in the quark model. In ref. [5] the spin-symmetric part of the spin-orbit one-gluon exchange is used only. It gives a rather good spin-orbit NN-interaction compared to phenomenological interactions. The most general spin-orbit baryon-nucleon potential has two contributions: a symmetric $L(\underset{\sim}{s}_B + \underset{\sim}{s}_N)$ and antisymmetric $L(\underset{\sim}{s}_B - \underset{\sim}{s}_N)$ part. For the ΛN-interaction both terms are big but their sum almost cancels. Also for the ΣN-potential the antisymmetric part is considerable. The final ratios are given in table 2.

The work of ref. [6] treats both the symmetric and antisymmetric spin-orbit interactions from one-gluon exchange. In addition it includes possible confinement terms which help to solve spin-orbit splittings of mesons but actually have a very bad effect on baryon spin-orbit splittings. The calculations are done in the flavor SU3-limit and yield an almost as strong ΛN and ΣN-potential. Both references correctly criticize the relativistic calculation of ref. [3] which only takes into account one possible diagram. However, only the outcome of a calculation with SU3-breaking will answer the question of whether the nuclear spin-orbit splitting is a remnant of the quark-quark interaction or not. We are working on this question for the P-wave KN-interaction which is so similar to the baryon-nucleus case. Hopefully then, one can also settle the question of the baryon-nucleus spin-orbit potential.

References:

[1] I. Bender, H.G. Dosch, H.J. Pirner and H.G. Kruse, to be published in Nucl. Phys.
[2] H.J. Pirner and B. Povh, Phys. Lett. 114B 308 (1982)
[3] H.J. Pirner, Phys. Lett. 85B (1979) 190
[4] M. Sommermann, to be published
[5] O. Morimatsu, S. Ohta, K. Shimizu and K. Yazaki, University of Tokyo, preprint
[6] Y. Suzuki and K.T. Hecht, University of Michigan, preprint
[7] Heidelberg-Saclay Collaboration in Proceedings of the Int. Conf. on Hypernuclear and Kaon Physics, Heidelberg (1982)
[8] B.R. Martin and G.C. Oades, Contribution to the IV Conf. on Baryon Resonances (1980)

Tables:

table 1 $K = m_s/m$

quark coordinates	masses	cluster coordinates
$\underset{\sim}{x}_1$	m	$\underset{\sim}{\xi}_1 = \underset{\sim}{x}_1 - \underset{\sim}{x}_2$
$\underset{\sim}{x}_2$	m	$\underset{\sim}{\xi}_2 = x_3 - \frac{x_1+x_2}{2}$
$\underset{\sim}{x}_3$	m	$\underset{\sim}{\xi}_3 = x_4 - x_5$
$\underset{\sim}{x}_4$	m	$\underset{\sim}{r} = \frac{1}{3}(x_1+x_2+x_3) - \frac{1}{1+K}(x_4+Kx_5)$
$\underset{\sim}{x}_5$	Km	$\underset{\sim}{R}_G = \frac{1}{4+K}(x_1+x_2+x_3+x_4+Kx_5)$

table 2

Spin-orbit potential for the N,Δ,Λ and Σ nucleus and the KN(I = 1) and KN(I = 0) systems as $V_{LS} = \omega \times 30\ \mathrm{MeV\ fm}^2\ (1/r)(\partial/\partial r)[\rho(r)/\rho_o]\underset{\sim}{S}\cdot\underset{\sim}{L}$.

System						
Spin-orbit: potential (ω)	N	Δ	Λ	Σ	KN (I=1)	KN (I=0)
Additive quark model	1	1	0	4/3	8/3	-4
Constituent quark model (ref.5)	1	?	0.21	0.55	?	?

THE NUCLEON-NUCLEON INTERACTION IN THE QUARK MODEL*

Amand Faessler
Institut für Theoretische Physik
Universität Tübingen
D-7400 Tübingen, West-Germany

Abstract

The nucleon-nucleon interaction is calculated starting from quarks and gluons and the interaction determined by QCD. The short range part is determined by quark and gluon exchange, while the long range part is calculated by allowing π and σ meson exchange between the quarks. The parameters of the model are adjusted to the nucleon mass, the Δ mass and the charge root mean square radius of the proton (including the pion cloud). The π and σ quark-quark coupling constants are adjusted to the known meson proton coupling. Using the resonating group method the ^3S and ^1S phase shifts are calculated in surprisingly good agreement with the data.

1. Introduction

In the last fifteen years the nucleon-nucleon interaction has been described by the exchange of mesons between two nucleons[1,2,3]. These meson exchange models for the nucleon-nucleon interaction contain of the order of 8 parameters being mainly cut-off masses for the form factors and for some of the exotic mesons also the meson-nucleon coupling constant which is experimentally not known. The diagrams calculated include also intermediate Δ-excitations and cross and stretch box diagrams[4] (see figure 1). In this talk we want to ask the question: Will the eighties be the time where we start to understand the nucleon-nucleon interaction using quantum chromodynamics? This means by calculating diagrams of the exchange of gluons and of quark-antiquark pairs between the nucleons.

Every particle physicist will tell you that this program will not be easy, since we nuclear physicists live in infrared slavery. This means that at the low momentum transfers which we have usually in nuclear physics the coupling constants between quarks and gluons is so large that we can not use perturbation theory. At very small distances between the two nucleons one expects already from the uncertainty principle

*Invited talk at the Workshop on Quarks in Nuclei in Bad Honnef, June 13-16, 1983

that one exchanges large momenta and thus one can get there at least a qualitative idea of what is happening using the perturbation approach.

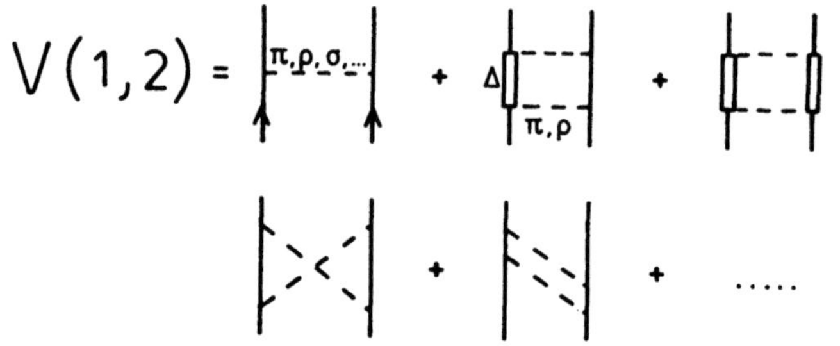

Figure 1: Meson exchange diagrams for the description of the nucleon-nucleon interaction as they have been used in the seventies. The first diagram describes the exchange of one meson (π, ρ, ε, δ, ω). The second diagram contains the intermediate excitation of one nucleon into a Δ-resonance. The last two diagrams are called cross and stretch-box diagrams.

In this talk we shall first see how we can describe the quark-quark interaction at short distances and then we will follow Liberman[5)] and De Tar[6)] using the Born-Oppenheimer approach in calculating the short range part of the nucleon-nucleon interaction. Following this we will study the difficulties of these approaches. This will show that they yield at short distances incorrect results. These difficulties can be cured by using the Resonating Group approach as in the Cluster model[7)]. An important ingredient will be the correct treatment of the orbital symmetry. We shall show that the [42] symmetry (which is mainly described with four quarks in the s- and two quarks in the p-state at distance r=0 between the two nucleons) is essential for the understanding of the short range repulsion. The inclusion of the [42] orbital symmetry yields a phase shift which is equivalent to a hard core (or a strongly repulsive soft core) potential.

The behaviour at large distances can definitely not be described by perturbation theory using gluon and quark exchange. In the past there have been many speculations that the long range part of the nucleon-nucleon interaction in quantum chromodynamics has the wrong r-dependence. This wrong r-dependence comes from the exchange of two gluons between the two nucleons. Such a perturbative approach is not allowed at large distances. There between the two nucleons the two gluons will start to interact with each other by additional gluons and they will form a gluball, an object which has a mass of the order of 1 GeV. Thus it is not contributing to the long range part of the nucleon-nucleon interaction. This part is described in our approach by meson exchange since we describe the short range part by gluon and quark exchanges we use only the lightest mesons (π and σ) since the short range part of the interaction which is described in meson exchange potentials by heavier mesons is treated in the quark-gluon picture directly. To ensure asymptotic freedom the mesons are not allowed to couple to the quarks inside the potential. Thus we use a meson quark coupling constant $g(r) = Cr^3$ which depends on the distance r from the centre of the bag and supresses the meson quark coupling inside the bag. Contributions from π and σ exchanges are added to the Hamiltonian overlap in the Resonating Group equation and the ^3S and ^1S nucleon-nucleon phase shifts are calculated.

The model uses the following experimental data to fit some parameters: nucleon mass, Δ mass, root mean square radius of the nucleon including the π-cloud, the π-nucleon and the σ-nucleon coupling constant at zero momentum transfer. Thus we adjust only five parameters. In addition we use for the π and σ the experimental mass m_π=139 MeV and the value adjusted in one-meson-exchange potentials m_σ=520 MeV[8], respectively. The calculated phase shifts agree extremely well with the experimental data in view of the simple model and the fact that none of the parameters is adjusted to the phase shifts directly.

2. The Born-Oppenheimer Approach

In this chapter we want to discuss the conventional approaches for calculating the nucleon-nucleon interaction in the quark model using the Born-Oppenheimer (adiabatic) approximation. This approach has been pioneered by Liberman[5] in the potential and by De Tar[6] in the MIT bag model.

To calculate the nucleon-nucleon interaction in the quark model we need first the interaction between two quarks. For short distances (and high momentum transfer) we describe this interaction by the one-gluon exchange. Everyone working in this field is aware that this assumption can be only good for high momentum transfers which we normally do not have in nuclear structure physics. We have here the same situation as with the meson-exchange potentials which have been used extensively in the seventies. The coupling constants are larger than unity and one does not expect that a perturbative expansion is converging. But one hopes that the lowest order diagram describe correctly the functional dependence on exchanged momentum and energy. One hopes that the higher order terms do not give a pronounced dependence but yield due to their more complicated structure a more smooth behaviour. This can then be described by adjusting a few parameters (totally five in our approach) to experimental data. To find the quark-quark interaction in the one-gluon exchange approximation we need a quark-gluon vertex. This vertex is modelled according to the photon-electron vertex in quantum electrodynamics.

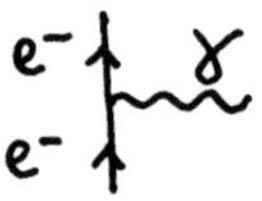

$$= e\,\bar{\psi}_{e'}\,\gamma^{\mu}\psi_{e}\,A_{\mu} \tag{1}$$

Here ψ_e is a Dirac spinor which describes the annihilation of an electron e at the vertex. $\bar{\psi}_{e'}$ is the creation operator of the outgoing electron e' which is described as a Dirac spinor. γ^{μ} are the Dirac matrices and A_{μ} is the vector potential for the electromagnetic field which describes the creation or the annihilation of a photon γ. The quark-gluon vertex is modelled in the same way. The interaction is only slightly more complicated due to the fact that quarks can have colour. Thus one needs to introduce the 3x3 colour matrices.

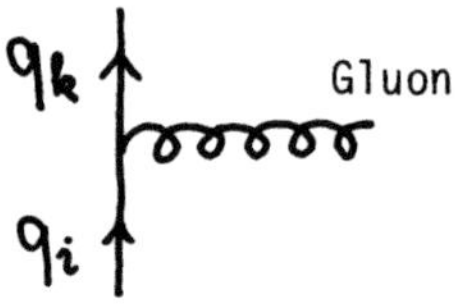

$$= \frac{1}{4}\,g\,\bar{q}_k\,\gamma^{\mu}\,\lambda_{ki}^{(a)} q_i\,G_{\mu}^{(a)} \tag{2}$$

Here q_i are Dirac spinors which describe the annihilation of the incoming quark with colour i=red, blue, yellow. $\lambda_{ki}^{(a)}$ are the matrix elements of the eight (a=1,...8) 3x3 colour matrices. The creation or annihilation of a gluon is described by the vector potential $G_{\mu}^{(a)}$. The coupling constant g is momentum dependent. At small momentum transfer which we have usually in nuclear physics it is larger than unity. Thus we are not expecting that the perturbative approach will be able to describe the data. As already stressed above we have a similar situation as in the mescn-exchange models where one has also coupling constants larger than unity. The lower diagrams are expected to give the more rapidly varying dependence on energy and momentum transfer while the higher order diagrams which can not be neglected, are included by adjusting a few parameters to experimental data.

From vertex (2) one sees immediately that the gluon must carry the colour i and the anticolour k. The outer product of the Young tableaus in colour space yields:

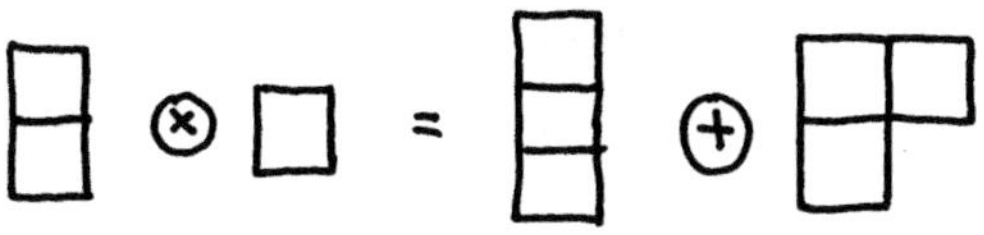 (3)

The product yields a colour singulet and a colour octet. Experiments show that the gluon exists only as an octet. The index a runs over to the different states in this octet.

The quark-quark interaction using one-gluon exchange can now be written down in mo-momentum space using the Feynman rules. In the constituent quark model, where quarks have a mass of the order of 350 MeV, we expand this expression nonrelativistically and Fourier-transform it into r-space. This yields the coloured Fermi-Breit interaction[9].

$V_{qq}(1,2) = \lambda_1 g$ 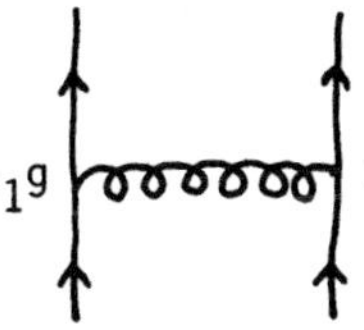 $\lambda_2 g$ (4)

$$\approx \frac{g^2}{4\pi} (\lambda_1 \cdot \lambda_2) \frac{1}{4} [\frac{1}{r_{12}} - \frac{\pi}{m_q^2 c^2} \{1 + \frac{2}{3}\} (\sigma_1 \cdot \sigma_2) \delta(r_{12}) + \ldots]$$

The "scalar product" of the colour matrices λ_1 and λ_2 has to be performed in the octet space. The tensor and two-body spin orbit contributions to this qualitative approach to the quark-quark interaction are not given explicitly since we do not need them to describe the S-wave interaction between two nucleons. The expression $g^2/(4\pi) = \alpha_S$ is often called the strong hyperfine structure constant in analogy to $e^2/(4\pi)$.

The above nonrelativistic expansion can not be justified in general. But at small values of r the small component of the single quark wave function is indeed small compared to the large one. The situation is totally different at the surface of a nucleon. There the large g(r) and the small f(r) components of the Dirac wave function of a quark are comparable size. Thus for the long range part of the NN interaction we use fully relativistic wave functions for the quarks. For small distances the error from a nonrelativistic treatment is small but not zero. On the other side this approximation allows to remove the spurious center of mass motion exactly. Thus at short distances our treatment might be of a similar quality as a fully relativistic treatment with only an approximate treatment of the center of mass motion. One could think that the lowest order contribution of this force (4) to the nucleon-nucleon interaction is given by the left hand side of figure 2. But if one reads the diagram from the left to the right one finds immediately that it has to be identically zero: The nucleon N_1 is a colour singulet if one adds a gluon being a colour octet one automatically should get a colour octet.

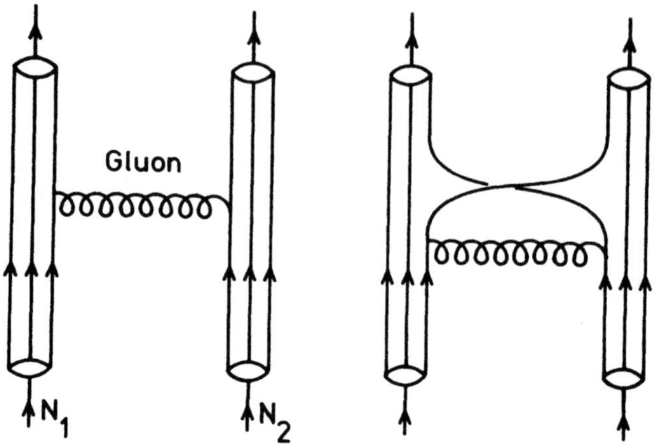

Figure 2: Diagrams which could contribute to the nucleon-nucleon interaction in the quark model. The left diagram is forbidden due to colour selection rules since the exchanged gluon is a colour octet. The right hand side describes the exchange of a quark pair and a gluon. It is the lowest order diagram which is different from zero.

But the nucleon N_2 is again a colour singulet and thus the diagram is forbidden in the same way as there exists no electric dipole transition between two angular momentum zero states. The lowest order contribution must therefore include also the exchange of two quarks as given on the right hand side of Fig. 2. This exchange of two quarks is automatically included due to the antisymmetrization of the six quarks in the two nucleons.

The quark-quark interaction (4) does not contain the colour confinement. Since we are convinced that colour confinement is a nonperturbative effect we have to add it ad hoc to our interaction. The detailed form of colour confinement turns out to be not important for the description of the nucleon-nucleon interaction. Calculations with a quadratic, a linear or a logarithmic confining potential yield the same result. We will lateron give results for calculations with quadratic and linear confining potentials. With a linear confinement the quark-quark interaction we need for S-waves is:

$$V_{qq}(1,2) = \frac{1}{4} (\lambda_1 \cdot \lambda_2) [-ar_{12} + \frac{\alpha_s}{r_{12}} - K \sigma_1 \cdot \sigma_2 (r_{13})] \quad (5)$$

The factor $\lambda_1 \cdot \lambda_2$ guarantees confinement only between coloured objects. Its value is between two colour antisymmetric quarks in a nucleon -8/3. Thus inspite of the negative sign in front of the parameter "a" the confining potential is attractive as it has to be.

Liberman[5] was the first one to use a quark-quark interaction similar to (5) to calculate in an adiabatic approximation a nucleon-nucleon interaction.

$$V_{NN}(r) = \langle A\{\phi_3(r/2)\phi_3(-r/2)\}| T_q + V_{qq} |A\{\phi_3(r/2)\phi_3(-\frac{r}{2})\}\rangle - 2m_N c^2 \quad (6)$$

$\phi_3(r/2)$ is the three quark wave function of a nucleon with the center of mass at the position r/2 and $\phi_3(-r/2)$ the wave function of the second nucleon at position -r/2. T_q is the kinetic energy of the six quarks for which one usually uses the relativistic free energy expression for a quark. V_{qq} represents the interaction (5) between all the six quarks of the two nucleons. The expression (6) yields the energy expectation value of two nucleons at a given distance r. The result of such a calculation by Liberman[5] for even and odd partial waves with total spin S=0 and S=1 is given in Fig. 3. The exciting result in 1977 was that this simple Born-Oppenheimer calculation was able to reproduce a short range repulsion between the two nucleons. Naturally one sees immediately that this repulsion is by far too longranged but nevertheless one was convinced that one has understood the nature of the repulsive soft core.

A similar result was obtained a year later by De Tar in the two center MIT bag model[6]. This result is shown in Fig. 4. Again it seems that one can understand the nature of the short range repulsion of the nucleon-nucleon interaction.

But a closer inspection of the approach used by Liberman[5] and De Tar[6] shows serious difficulties:

(i) The use of the Born-Oppenheimer approximation is an excellent approach in molecular physics where one calculates the intramolecular potential (for example between carbon and oxygen) in this way. The carbon and the oxygen nuclei are very heavy compared to the electrons and thus it is a good approximation to fix the distance between the two nuclei and assume that the electrons move so fast that for each distance between nuclei one can calculate the energy and use this energy as a function of this distance for the intromolecular potential. The situation in the nucleon-nucleon interaction is quite different. In the constituent quark model, the mass of the nucleon is built up essentially by the mass of the three quarks.

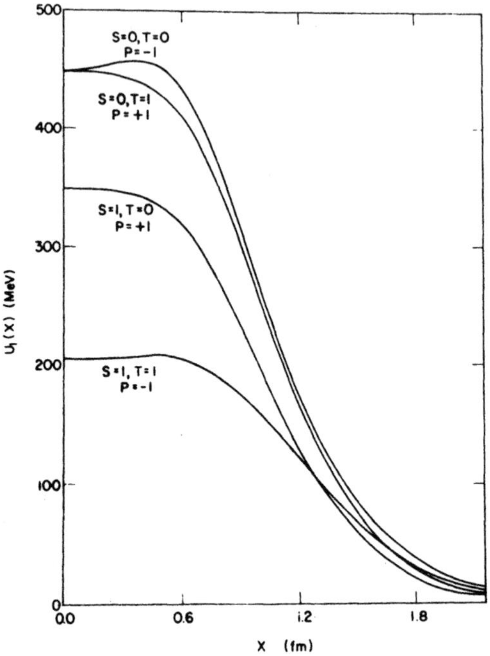

Figure 3: Short range part of the nucleon-nucleon interaction according to Liberman[5]. The spin S=1 and spin S=0 potential for even P=+1 and odd P=-1 partial waves are calculated in the Born-Oppenheimer approximation.

In the current quark model the bare quark mass is zero but the energy eigenvalue of a current quark is again of the order of 300 MeV. Thus in both models we can not assume that the nucleon is heavy and the quark is light, so that the nucleon motion is slow and the quark is fast. Therefore one has no justification for using the Born-Oppenheimer approximation for this problem.

(ii) In the adiabatic expression (6) for the nucleon-nucleon interaction one assumes the same distance of the two nucleons on the left and the right hand side of the matrix element. In reality one still gets large contributions if the distance between the two nucleons on the left r and on the right hand side r' are different. As expected one gets a large nonlocality for the nucleon-nucleon interaction which has not been taken into account previously.

(iii) After one has calculated the nucleon-nucleon interaction $V_{NN}(r)$ of eq. (6) one wants to use it for calculating for example nucleon-nucleon scattering. One introduces it into a Schrödinger equation.

$$[\frac{p^2_{rel}}{2\mu} + V_{NN}(r)]\psi_{NN} = E\ \psi_{NN} \tag{7}$$

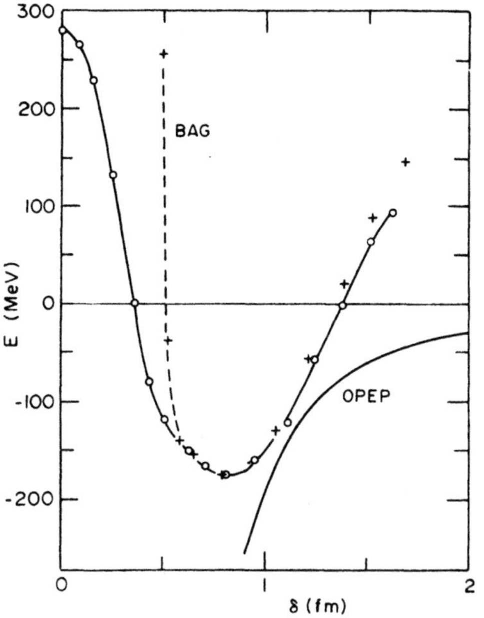

Figure 4: Nucleon-nucleon interaction according to De Tar[6]. The solid line with the circles constraints the variable r as the distance between the two center of masses of the two nucleons. On the right hand side the solid line indicates the one pion exchange potential. The interaction is calculated for S=1, M_s=1 and T=0.

In this equation one then uses the reduced mass μ of the two nucleons. But everyone who has experience with the description of heavy ion collisions knows that at short distances the reduced mass is an incorrect description for the mass parameter for the interaction of two many-body systems. We expect a mass which is r-dependent and even nonlocal $\mu(r,r')$.

(iv) For the interaction between two nucleons we want to describe the relative kinetic energy of the two nucleons separately as for example indicated in eq. (7). This means that in the interaction (6) we should only take into account the kinetic energy of the six quarks minus the kinetic energy of the relative motion of the two nucleons. This point has been treated correctly by Liberman using oscillator wave functions. De Tar used in the two center MIT bag model the total kinetic energy of all the six quarks for T_q. Thus if his potential is used for nucleon-nucleon scattering one counts the relative kinetic energy between the two nucleons twice.

(v) In the adiabatic approach of Liberman and De Tar (and a large number of authors who followed their line) the six quarks are put into the s-state for distance r=0 between the two nucleons. That this is an incorrect approach has first been pointed out by Neudatschin and coworkers. The outer product of two orbital symmetries [3]·[3] is for the S-wave interaction between two nucleons illustrated in Fig. 5. One sees that one obtains not only the [6] symmetry but also the [42] symmetry with 4 quarks in the 2s- and 2 quarks in the 1p-state.

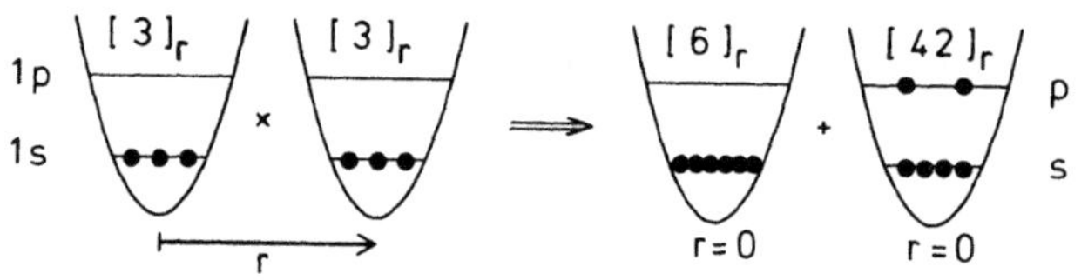

Figure 5: The left hand side shows the product of quark wave functions of two nucleons at distance r. The orbital symmetry in each nucleon is [3]. All quarks are in the 1s-state. The product [3]·[3] = [6] + [42]+[51]+[33] can be characterized according to irreducible representations of the permutation group of six objects S_6. For the S-wave interaction only representations which are symmetric under the exchange of two nucleons are allowed. These are the [6] and [42] orbital symmetry. For the distance r=0 of the two nucleons these symmetries correspond to all six quarks in the 1s-state for the [6] and to 4 quarks in the 1s- and 2 quarks in the 1p-state for the [42] symmetry as indicated on the right hand side.

(The orbital symmetries [51] and [33] which are allowed by the permutation group are forbidden by the condition that in the S-wave the total orbital wave function has to be symmetric under the exchange of the two nucleons). One could argue that the [42] orbital symmetry lies $2\hbar\omega \approx 1.2$ GeV higher than the state with the [6] symmetry. But this cannot be correct if one includes the residual interaction between the quarks: Table 1 shows that the two nucleon configuration which is the lowest state of 6 quarks consists by 89% out of the [42] symmetry.

Table 1 Probabilities for the decomposition of two nucleons, two Δ's and two hidden colour states CC into the orbital symmetries [6] and [42] according to the permutation group S_6.

	$[6]_r$	$[42]_r$
NN	1/9	8/9
ΔΔ	4/45	41/45
CC	4/5	1/5

This is a purely geometrical result given by S_6-Clebsch-Gordan coefficients. It is of the same quality as the statement that the product of two spin 1/2 particles is composed out of a triplet and a singulet where the probabilities are given by the Clebsch-Gordans squared. Since we know that two nucleons are the lowest six quark state it means that the residual interaction between the quarks has to lower the [42] orbital symmetry below the [6] symmetry. Table 1 demonstrates clearly that any calculation using only the [6] orbital symmetry can not be able to describe the nucleon-nucleon interaction correctly.

In addition to this purely geometrical preference the [42] symmetry is also favoured dynamically. Thus one has a similar situation as in spin space when one considers the scattering of two nucleons. The weights of the triplet and singulet contribution is given by two Clebsch-Gordan coefficients squared. But in addition to this pure angular momentum geometry we know that the tensor force acts only in ^3S and favours the contribution of the triplet state to the nucleon-nucleon cross section.

The colour magnetic interaction is mainly responsible for the short range part of the quark-quark interaction (5). The total wave function in orbital space r, spin-colour space SC and isospin-(flavour)-space T has to be antisymmetric. If we assume for the moment (just to get a reasonable first guess for the action of the colourmagnetic interaction), that the radial matrix elements of the δ-function are the same, we can write the colour and spin part of this interaction as Casimir operators in the combined colour spin space and in spin and colour spaces separately.

$$\sum_{i<j} (\sigma_i \cdot \sigma_j)(\lambda_i \cdot \lambda_j) = 4\, P_{CS} - \frac{4}{3} P_S - 2\, P_C \qquad (8)$$

The operators P count the symmetric minus the antisymmetric pairs of quarks in the corresponding spaces. The colour symmetry of two nucleons is fixed (P_C=-3). If we consider either triplet or singulet spin interaction between the two nucleons the colourmagnetic matrix elements are determined by the number of even and odd pairs in colour-spin space (8). The Pauli principle determines now the symmetry in colour spin space. (There exist several solutions. We consider here the one which is lowest in energy). The symmetry in isospin (flavour space) are the same for the orbital symmetries [6] or [42]. Since the total wave function has to be antisymmetric we must have in SC-space a more antisymmetric wave function for the orbital symmetry [6] than for the orbital symmetry [42]. With the minus sign of the colourmagnetic interaction in eq. (5) and eq. (9) this means that the energy of the orbital symmetry [6] is pushed up while the energy for the orbital symmetry [42] is lowered. This additionally favours the orbital [42] symmetry.

Harvey[12)] was the first one to include in a more refined calculation the orbials symmetry [42]. But he still used the Born-Oppenheimer approximation to calculate a local nucleon-nucleon potential. The result is shown in Fig. 6. The dotted line gives the adiabatic approach without the orbital [42] symmetry at r=0.

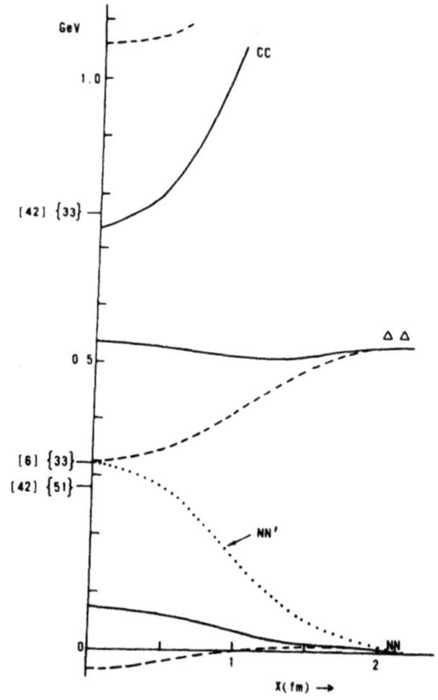

Figure 6: Potential energy as a function of the distance (r=X) fo the two nucleons according to Harvey[12)]. The dotted line gives the nucleon-nucleon interaction energy without normalizing the [42] symmetry separately. This means that this symmetry is not contained in the wave function at r=0. The dashed line gives the energy for the interaction of NN, ΔΔ and the hidden colour state CC as a function of the distance r. The solid line includes also the coupling between the three channels.

The solid line includes the orbital [42] symmetry also for r=0 while the dashed line takes also the coupling to the ΔΔ and to the hidden colour channel CC into account. One sees that the nucleon-nucleon interaction is not any more repulsive at short distances. It is even slightly attractive. Thus improving the calculations of Liberman[5] by including the orbital [42] symmetry, the short range repulsion disappears. (A closer analysis of the results of Harvey shwos that this is partially due to the adiabatic treatment of the nucleon-nucleon interaction and partially due to a quark-quark interaction which does not yield the nucleon at the right radius in the energy minimum).

3. The Short Range Nucleon-Nucleon Repulsion

Harvey's calculation[12] shows that the inclusion of the orbital [42] removes essentially the short range repulsion which was praised before as a major success of the quark model for the description of the nucleon-nucleon interaction. So we are back at the beginning and have to ask again: What is the nature of the short range nucleon-nucleon repulsion? But in the meantime we have learned that the orbital [42] symmetry is very essential. Let's for the qualitative discussion thus consider for a moment only the [42] symmetry of the six quarks (see Fig. 7).

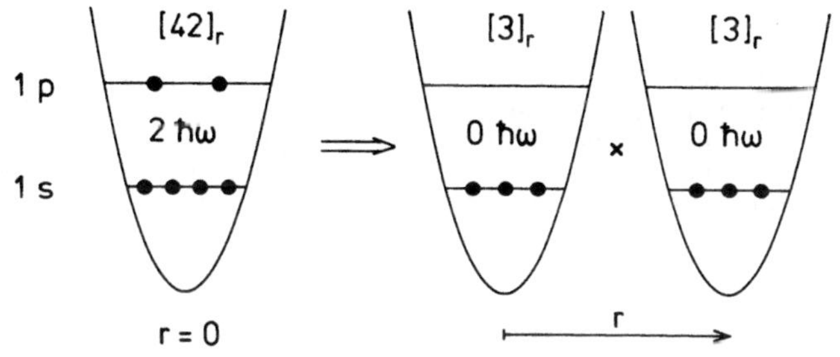

Figure 7: The left hand side shows the configuration of two nucleons at distance r=0 in the [42] orbital symmetry. It contains the excitation of two oscillator quanta. The right hand side shows the situation of two nucleons at distance r. Each nucleon has no internal excitation and therefore no excitation of oscillator quanta which have to be conserved. These quanta can only be contained in the relative motion between the two nucleons. Thus in a S-state the relative wave function must have a node.

At distance r=0 of the two nucleons the [42] symmetry includes an excitation of two oscillator quanta. If we take the nucleons apart (right hand side of Fig. 7) at distance r, we have no oscillator quanta excited internally in the nucleon. Since the number of the oscillator quanta has to be conserved in an oscillator basis the relative wave function between the two nucleons $u_N(r)$ has to contain these two oscillator quanta. Thus if we describe the S-wave interaction between the two nucleons, the relative wave function has to have a 2s character in the interaction region (Fig. 8).

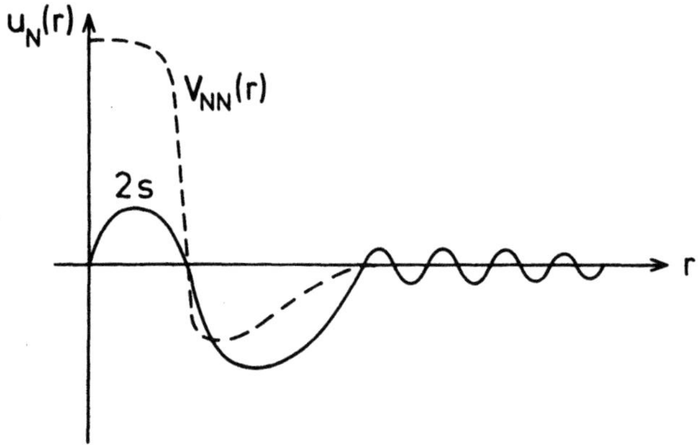

Figure 8: The relative wave function between two nucleons in a state with the orbital symmetry [42]. The wave function must have a node in the interaction region. Thus it has to be in this region a 2s-wave function. The node yields a hard core phase shift which can only be measured asymptotically. The same phase shift would be enforced by a soft core potential $V_{NN}(r)$ (dashed line).

A node yields an asymptotic phase shift of the relative nucleon-nucleon wave function which is the same as the one of a hard or a soft core in the interesting range of the relative energy. Fig. 8 shows how such a zero of the relative wave function could be enforced by a soft core nucleon-nucleon interaction (dashed line). In reality the zero is not due to a hard core but to the [42] orbital symmetry which enforces a node in the relative wave function in the range of the nucleon-nucleon interaction. Thus we want to suggest that the short range repulsion is not due to a repulsive potential but due to a group theoretical symmetry property in the orbital space of the six quarks. Naturally the [6] orbital symmetry is not forbidden. This symmetry allows a relative wave function without a node. But as explained in the previous chapter the [42] symmetry is by about 89% contained in the nucleon-nucleon configuration which is the lowest six quark state.

Now let us try to describe quantitatively the nucleon-nucleon interaction circumventing the difficulties which we have pointed out above: (i) The relative coordinate between the two nucleons and the quark coordinates have to be described on the same footing. (ii) The mass parameter has to be allowed to be nonlocal. (iii) In the kinetic energy and the interaction between the two nucleons we have to allow for nonlocalities. (iv) The relative kinetic energy between the two nucleons should not be counted twice. (v) The orbital [42] symmetry has also to be included. All these points can be taken into account using the Resonating Group method which is so successful in describing reactions between light nuclei in the cluster model[7]. The ansatz for the wave function in the Resonating Group approach

$$\psi_{6q} = A\{|NN,r\rangle\, u_N(r) + |\Delta\Delta,r\rangle\, u_\Delta(r) + |CC,r\rangle\, u_C(r)\} \tag{9}$$

allows also for polarization of the two nucleons in the interaction region. This polarization is described by admixing the ΔΔ and the hidden colour CC configurations. The quark wave function of these three configurations with the two three quark clusters at distance r are well known. The only unknown functions are the three relative wave functions $u_N(r)$, $u_\Delta(r)$ and $u_C(r)$ in the three different channels. If we use the ansatz (9) in a Kohn-Hulthen variational principle we get a coupled system of integral equations for the determination of the three relative wave functions. In this calculation we restrict ourselves to ^3S and ^1S waves. First results of this approach have already been published[13,14,15,16]. A similar appraoch for describing the short range part of the nucleon-nucleon interaction has also been used by Oka and Yazaki[24] and similar work using the generator coordinate method is in preparation by Harvey and coworkers[17]. Oka and Yazaki[24] included only the NN and ΔΔ channels. They used furhter a quark-quark force which does not yield the nucleon in the energy minimum at the right mass with the correct radius (see below Fig. 11).

The hidden colour channel is in principle a closed channel. Below 600 MeV in the centre of mass system of the two nucleons also the ΔΔ channel is closed. This simplifies greatly the solution of the three coupled integral equations for which we use a method given by Kamimura[18]. Before we look to the phase shift obtained in this way we want to see what we would expect if we would have only a hard core potential with the hard core radius r_0.

$$\delta_{l=0}(E_{rel}) = -r_0 \sqrt{m\, E_{rel}}\, /\hbar \tag{10}$$

The S-wave phase shift is then negative and proportional to the hard core radius r_0. It is described by the negative branch of a parabola rotated by 90 degrees if one plots $\delta_{l=0}(E_{rel})$ as a function of the relative energy E_{rel} of the two nucleons in the centre of mass system.

The ^3S and ^1S phase shifts obtained in this way are shown in the figure 9 and 10. Naturally the result depends on the detailed parameters in the quark-quark interaction (5) and on the oscillator length b for the nonrelativistic oscillator wave function for the constituent quarks. The constituent quark mass used is m=355 MeV, the coupling constant α_S=0.97 is adjusted to the Δ-nucleon mass difference while the oscillator length b = 0.475 fm fixes the root mean square radius of the mass distribution of the quark content of the nucleon and is a free parameter. The linear a=61.6 MeV fm^{-1} and the quadratic confinement parameter a'=34.5 MeV fm^{-2} are adjusted to obtain the minimum for the mass of the nucleon at the above root mean square radius $<r^2>^{1/2}$ = b = 0.475 fm for the quark content of nucleon. The results given in Fig. 9 and 10 are identical for the linear and the quadratic confinement potential.

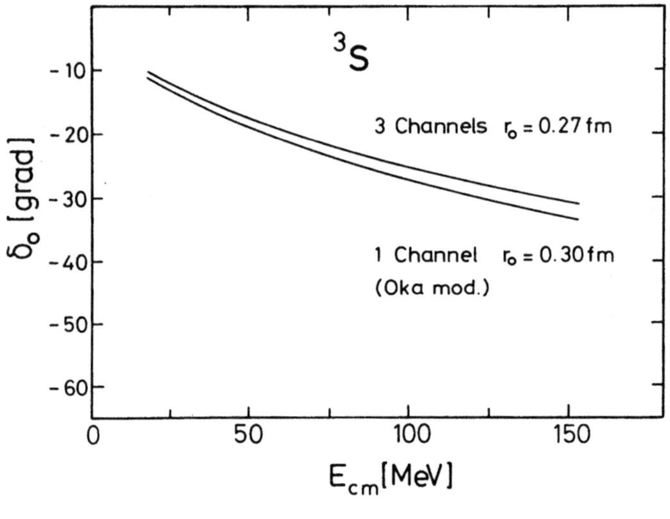

Figure 9: Triplet S phase shifts in degrees as a function of the center of mass energy of the two nucleons for the NN channel only and for the coupling of the three channels NN, ΔΔ and CC (hidden colour). The parameters of the quark-quark force are as in the work of Oka and Yazaki[24] with the exemption that the root mean square radius b $<r^2>^{1/2}$ = 0.475 fm is varied to minimize the nucleon energy. The hard core phase shift (11) can be adjusted to these results to yield an equivalent core radius r_0 which is given in the figure.

For the moment it is not essential what type of parameters are used to calculate these phase shifts since they are anyway not realistic due to the fact that the medium and long range attraction of the nucleon-nucleon interaction is not included in this calculation.

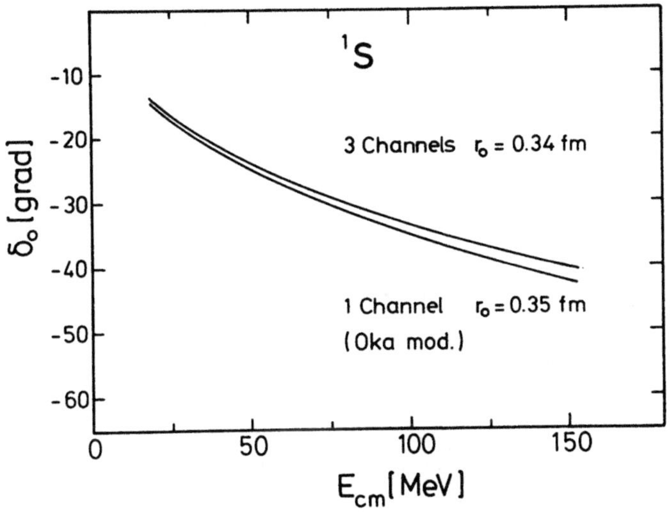

Figure 10: Singlet S phase shift for the nucleon-nucleon interaction calculated in the Resonating Group method. For details see Fig. 9.

But for all the forces which we used the results look qualitatively like in Fig. 9 and 10 for one and three channel calculations. One verifies easily that within the line width the curves in Figs. 9 and 10 can be fitted by the hard core phase shift (10) adjusting the hard core radius r_0. Thus indeed the Resonating Group calculation including the orbital [42] symmetry yields a phase shift which is identical to the one of a hard core. In Fig. 11 we used a large number of forces to solve the resonating group equation and to adjust to the so calculated phase shifts an equivalent hard core radius r_0 according to eq. (10). One sees in Fig. 11 that this equivalent hard core radius r_0 depends only on the root mean square radius of the quark content of the nucleon and not on the detailed parameters of the quark-quark interaction.

A few of the results are not lying on the straight lines for the one and three channel cases. This can be traced back to the fact that these forces (see for example Oka, Yazaki and Ellwanger) do not give the minimum of the energy of the nucleon as a function of the radius at the right mass. Thus the admixture of ΔΔ and CC channels tries first to modify the radius to come closer to the minimum. But for all the "good" forces the equivalent hard core radius obtained from the ^3S phase shifts lie on a straight line (one line for the 1 channel and one for the 3 channel case) plotted as a function of the root mean square radius of the quark content of the nucleon.

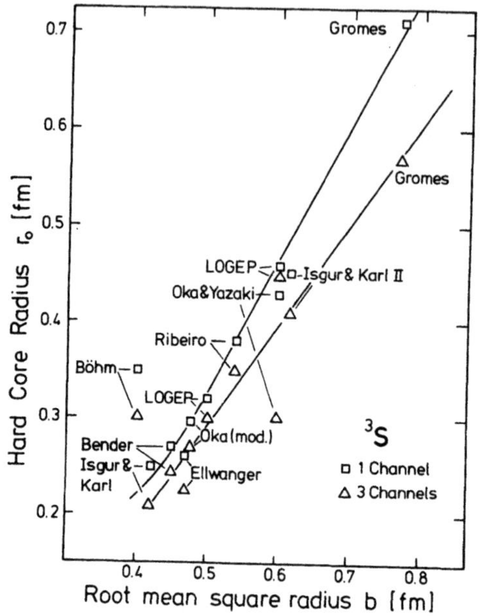

Figure 11: Hard core radius r_0 as a function of the root mean square radius $<r^2>^{1/2} = b$ for different quark-quark forces with linear and quadratic potentials which are referenced in the work of Faessler and coworkers[13]). The adjusted hard core radius r_0 is given for the one and the three channel calculations. The forces of Oka, Yazaki and of Ellwanger are not chosen to minimize the nucleon mass as a function of the root mean square radius. Thus they drip out of the systematics.

At the end of this chapter I want to come back to the question about the nature of the repulsive core of the nucleon-nucleon interaction. We would say that there exists no repulsive potential in the naive sense. The six quark configurations of two nucleons consist by about 90% out of the orbital [42] symmetry. This symmetry requests a node in the relative s-wave function between the two nucleons in the interaction region. This node is phenomenologically described by a hard or a soft core potential.

4. The Full S-Wave Interaction

In the previous chapter we described the short range part of the nucleon-nucleon interaction by quark and gluon exchange and a phenomenological confinement potential. The nucleon-nucleon interaction at larger distances can not be described perturbatively. In the region where the two "bags" of the nucleons are not overlapping only colourless objects should be exchanged. In this region we describe the nucleon-nucleon interaction by meson exchange. To guarantee asymptotic freedom, the mesons should interact with the quarks in the nucleons only at the surface. Thus we assume for the meson-nucleon coupling a r-dependent coupling constant g(r) which suppresses the coupling of the meson to the quarks inside the nucleon. For the moment we are only taking into account the two light mesons: the π and σ-meson. The heavier mesons may also be important but they describe mainly the behaviour of the nucleon-nucleon interaction at shorter distances where their coupling is reduced by the r-dependent coupling constant g(r). Especially we neglect the exchange of the ω-meson which describes in the meson exchange model the short range repulsion[3]. To get the right strength for the short range repulsion in the meson exchange model on needed to use a $g_{\omega NN}$ which is more than a factor 2 larger than the flavour SU_3 symmetry would allow. This already indicates that the short range repulsion due to ω-meson exchange is artificially blown up to describe the data. But in reality another mechanism should be responsible for this short range repulsion. As we discussed in the previous chapter this mechanism is the [42] orbital symmetry. In addition to this reduction we get a further reduction by the r-dependence of the coupling constant which suppresses the coupling to the quarks inside the nucleon. Thus it seems reasonable for a first orientation as mentioned already above only to use the π and σ-exchange.

We assume for the meson-nucleon coupling the form[19,20]:

$$g_{\pi qq}(r) = C_\pi\, r^3 \;;\; g_{\sigma qq}(r) = C_\sigma\, r^3 \tag{11}$$

This r-dependence suppresses the meson-quark coupling inside the nucleon. But also outside the nucleon the coupling gets automatically zero since the quark wave function disappears there. Thus the coupling (11) means that the mesons are mainly cop-coupled to the nucleon at the surface. The two parameters C_π and C_σ are adjusted by calculating the meson-nucleon coupling constant for zero momentum transfer and adjusting the C's so that the experimental data are reproduced.

$$\frac{g^2_{\pi NN}(q=0)}{4\pi} = 14.1 \qquad \frac{g^2_{\sigma NN}(q=0)}{4\pi} = 5.65 \tag{12}$$

The σ-nucleon coupling parameter is taken from the one-meson exchange model[8]. (These numbers are in Faessler, Fernandez[14] wrongly quoted as $g_{\pi NN}$ and $g_{\sigma NN}$).

In our model for calculating the nucleon-nucleon interaction in the quark model we adjust parameters to the following data: Nucleon mass, Δ mass, root mean square radius of the charge distribution of the proton including a meson cloud,pion-nucleon coupling constant $g_{\pi NN}$ and $g_{\sigma NN}$. In addition we use the experimental values for the pion mass m_π=139 MeV and for the σ mass m_σ=520 MeV. The last value is taken from the one-meson exchange model[8].

The root mean square radius of the charge distribution of the proton is used to adjust the oscillator constant for the oscillator potential which gives the quark wave functions. If we include π and σ meson exchanges we need to have a correct behaviour of the quark wave functions also at the surface. This means that we have to calculate the quark wave functions fully relativistically since at the surface the small amplitude f(r) is of the same order as the large amplitude g(r) of the relativistic quark wave function. At small distances the small amplitude can be neglected compared to the large one so that a nonrelativistic treatment is there justified. To be able to bring the treatment of the short range and the long range behaviour to a unified model, we request that the relativistic and the nonrelativistic treatment must yield the same root mean square radius of the quark content of the nucleon. To obtain the correct root mean square radius of the quark content of the nucleon and by that also the oscillator constant for the nonrelativistic b_{nr} and the relativistic b_r quark wave functions we use the electric form factor of the proton obtained by electron-proton scattering. The meson cloud is included in the way given by Höhler and co-workers[23]. He allows that the photon exchange between the electron and the nucleon is transformed into a vector meson. Fitting the data he obtains a vector meson with the quantum numbers of the ρ meson and a mass m_V=730 MeV.

$$\tilde{G}_E^p(q^2) = \frac{m_V^2}{m_V^2-q^2} G_E^p(q^2)$$

$$G_E^p(q^2) = \int_0^\infty dr\, j_0(qr)\, [g^2(r) + f^2(r)] \tag{13}$$

Here $G_E^p(q^2)$ is the electric form factor of the quark conent of the proton. $j_0(qr)$ is the spherical Bessel function for orbital angular momentum l=0. The center of mass correction[19] complicates formula (13) a little bit. One transforms the single quark wave function $\psi_\alpha(r)$ into momentum space $\Phi_\alpha(p)$ and constructs then a wave function of the three quarks with a good total momentum P.

$$\Psi_P(\vec{r}_1,\vec{r}_2,\vec{r}_3) = \frac{N(P)}{(2\pi)^6} \int d^3p_1 d^3p_2 d^3p_3$$

$$\delta\left(\sum_{l=1}^{3} \vec{p}_l - \vec{P}\right) e^{i\sum_l \vec{p}_l \vec{r}_l} \phi_1(\vec{p}_1)\phi_\alpha(\vec{p}_2)\phi_\alpha(\vec{p}_3) \tag{14}$$

The electric form factor for the quark content of the nucleon is then with the recoil correction given by[19,20]:

$$G_E^P(q^2) = \int d^3r_1 d^3r_2 d^3r_3$$

$$\Psi_{pf}^+(\vec{r}_1, \vec{r}_2, \vec{r}_3) \sum_{j=1}^{3} e_j\, e^{i\vec{q}\vec{r}_j}\, \Psi_{pi}(\vec{r}_1,\vec{r}_2, \vec{r}_3) \tag{15}$$

The only free parameter in expression (15) is the oscillator length b_r which is adjusted to the form factor in Fig. 12 to be b_r=0.59 fm. This corresponds to a root mean square radius of the relativistic single quark wave function in the 1s-state of $\langle r^2\rangle^{1/2}$ = 0.66 fm.

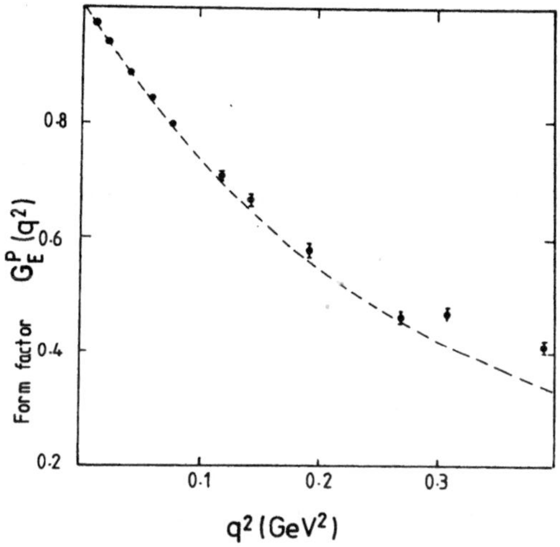

Figure 12: Electric proton form factor calculated from a relativistic quark wave function of the proton as a function of the momentum transfer squared (q^2). Recoil corrections and the contributions of a pion cloud are included[19]. The rms radius of the quark content is adjusted with $\langle r^2\rangle^{1/2}$= 0.66 fm to give a good fit to the electric from factor of the proton. This corresponds to an oscillator length b_{nr}=0.54 fm and b_r=0.59 fm for the nonrelativistic quark and relativistic wave functions, respectively.

If we adjust now the oscillator parameter b_{nr} in the nonrelativistic case in such a way that we have the same root mean square radius of the single quark wave function we obtain b_{nr}=0.54 fm. In this way we connect the nonrelativistic treatment for the short range part with the fully relativistic calculation of the long range part described by meson exchange between the quarks.

Now we have all the parameters of our model and can calculate the π and σ exchange potentials in momentum and r space. The pion-nucleon form factor is given by:

$$F_\pi^\lambda(\vec{q}^2) = \sum_{j=1}^{3} \int d^3r_j \; g_{\pi qq}(r_j) \bar{\Psi}_j(\vec{r}_j)\gamma_5(j)\tau_\lambda(j)\psi_j(\vec{r}_j)e^{-i\vec{q}\vec{r}_j} \tag{16}$$

For the π-meson exchange potential we obtain:

$$V_\pi(q) = -\frac{|F_\pi^\lambda(\vec{q}^2)|^2}{\vec{q}^2+m_\pi^2}$$
$$V_\pi(q) \approx -\frac{g_{\pi NN}^2(q^2)}{4m_N}\frac{(\vec{\sigma}_1\cdot\vec{q})(\vec{\sigma}_2\cdot\vec{q})}{\vec{q}^2+m_\pi^2} \tag{17}$$

The last expression is the nonrelativistic reduction which contains also explicitly the pion-nucleon coupling constant. Fig. 13 and 14 show the Fourier transform into r space of the π and σ exchange. In addition to the result for r=r' corresponding to eq. (17) we give also (dashed line) the result for a point nucleon. Fig. 15 shows the phenomenological form factor (dashed line) which one uses usually

$$F(q) = \frac{\Lambda^2-m_\pi^2}{\Lambda^2+\vec{q}^2-m_\pi^2} \tag{18}$$

if one treats the nucleon as a point particle (Λ=1 GeV). The form factor

$$g_{\pi NN}(r) = \frac{5}{3}\frac{4M}{|\vec{q}|}\int_0^\infty dr \; j_1(qr)f(r)g(r)g_{\pi qq}(r) \tag{19}$$

of our microscopic model is also given in this figure (dotted line). It cuts off the high momentum transfer much stronger than the phenomenological form factor. The mass parameter which fits the microscopic form factor using the phenomenological formula (18) would be Λ=475 MeV. But this behaviour is expected: The high momentum transfer in our model is not described by the meson exchanges but by quark and gluon exchange. The coupling of the mesons to the quarks inside the nucleon is suppressed by (11) and thus the mesons cannot exchange a too large momentum.

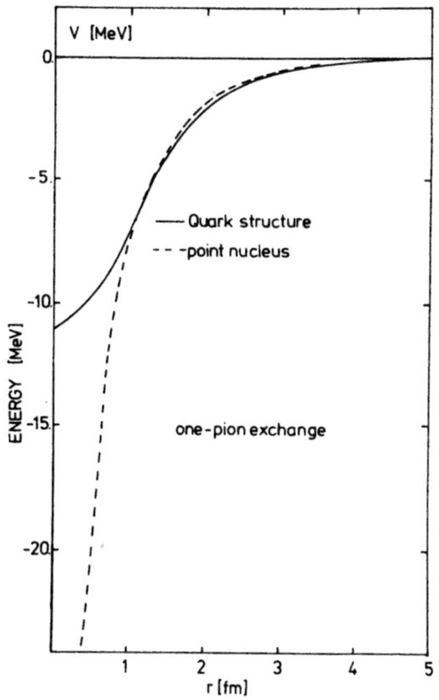

Fig 13

Figure 13: One-pion exchange potential as a function of the distance between the two nucleons for the quark model (solid line). The dashed line gives the equivalent potential for a point nucleon. In both cases the δ-function is removed. Since the quarks are point particles there is no need to remove the δ-function. In the calculation it is included. The π exchange potential gets then repulsive below around r=1.3 fm. σ and π exchange together are always attractive.

We now want to describe the short range part and the long range part of the nucleon-nucleon interaction on the same footing using the Resonating Group Method. We start from the trial wave function (9) and use the Kohn-Hulthen variational principle

$$\langle \delta\Psi_{6}q | \; \hat{H}_6 - E | \Psi_{6q} \rangle = 0 \tag{20}$$

to derive a system of three coupled integral equations.

$$\sum_{j=1} \int dr' \; [K_{ij}(r,r') - E \, N_{ij}(r,r')] \; u_j(r') = 0 \tag{21}$$

$$\text{for: } i,j = NN, \; \Delta\Delta, \; CC$$

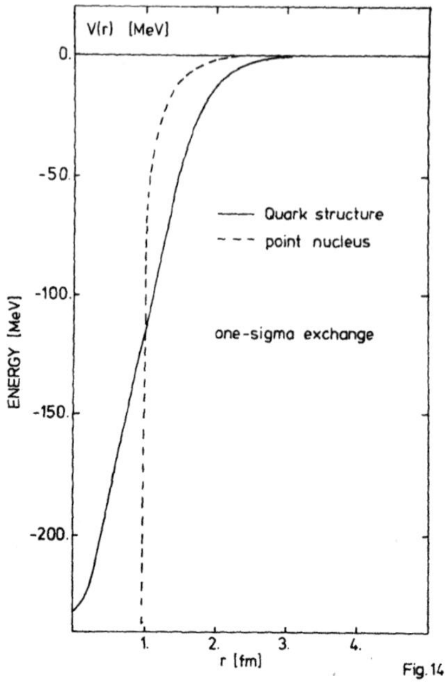

Figure 14: σ-meson exchange potential in the quark model (solid line) and for a point nucleus (dashed line).

The Hamiltonian overlap

$$K_{ij}(r,r') = (i,r|\hat{H}_6|j,r') + (i,r|\hat{V}_\pi|j,r') + (i,r|\hat{V}_\sigma|j,r') \tag{22}$$

and the norm overlap

$$N_{ij}(r,r') = (i,r|j,r') \tag{23}$$

are both nonlocal and are 3·3 matrices in the three channels: NN, ΔΔ and CC. The round brackets of the matrix elements in (22) and (23) indicate that one integrates over all variables apart of r and r'. We again use the method of Kamimura[18] to solve the coupled system of Resonating Group equations (21). The ^3S and ^1S phase shifts are given in Fig. 16 and 17. The experimental phase shifts are from Arndt and coworkers[21,22].

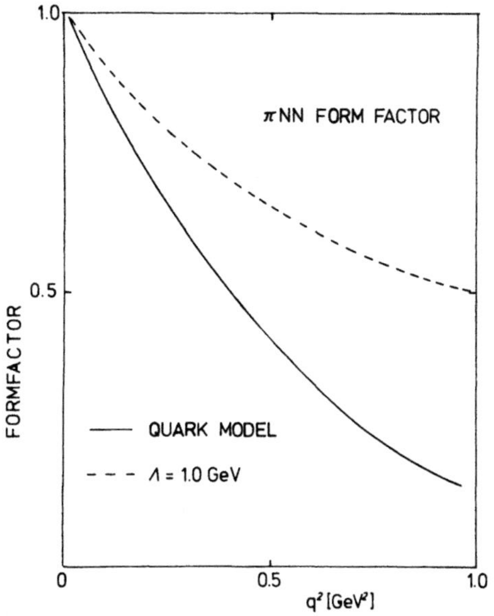

Fig. 15

Figure 15: Phenomenological (18) dashed line) form factor with Λ=1 GeV and microscopic form factor $g_{\pi NN}(q)$ (19) as a function of the momentum transfer q^2. A microscopic form factor could be described by a phenomenological one with the mass parameter Λ=475 MeV.

The agreement between theory and experiment is surprisingly good in view of the simple minded model and the use of perturbation theory for the short range part of the nucleon-nucleon interaction. This good agreement is especially surprising since we have not adjusted the parameters of the model to the ^3S and ^1S phase shifts. In this view the good agreement is even more surprising. It might perhaps indicate that the doubtful perturbation approach for the short range part of the interaction might be justified in the same way as one uses the perturbation approach in the meson exchange model for the nucleon-nucleon interaction. The lowest order diagrams reproduce the fast varying dependence on the momentum transfer quite well. The higher order diagrams yield a more smeared-out dependence which can be simulated by adjusting a few free parameters.

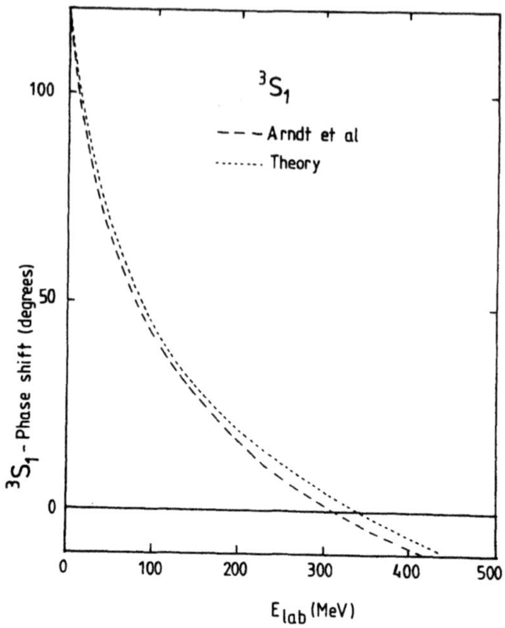

Figure 16: ^3S nucleon-nucleon phase shift as a function of the laboratory energy of an incoming nucleon. The dotted curve is the present calculation and the dashed curve represents the experimental phase shifts[21,22]. The parameters are: α_s=1.3; a=41.0 MeV fm^{-1}; b_r=0.59 fm; b_{nr}=0.54 fm; m_q=336 MeV and eq. (12).

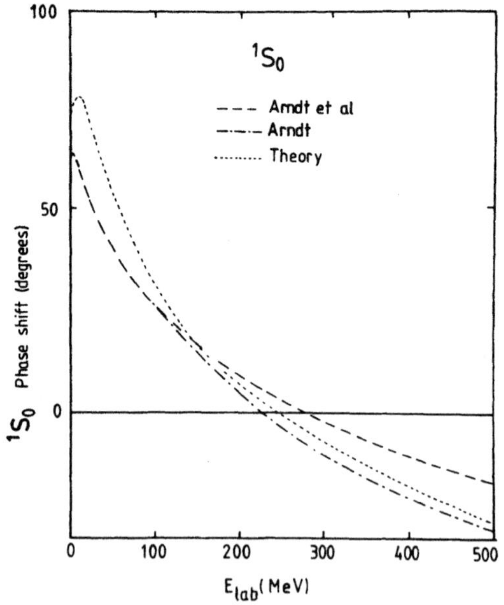

Figure 17: ^1S nucleon-nucleon phase shift as a function of the laboratory energy. The dotted curve gives the results of the theory presented. The dashed and the dashed-dotted line are two different sets of experimental phase shifts[21,22].

5. Conclusions

In these lectures we discussed the short and the long range part of the nucleon-nucleon interaction using for the description of the nucleons the quark model. The short range part is described by gluon and quark exchanges. We understood the nature of the short range repulsion as a consequence of the orbital [42] symmetry. The residual interaction between the quarks prefers this symmetry and lowers it in energy so that it plays a major role. Purely group theoretical arguments from the permutation group of 6 objects show, that in the configuration of two nucleons the [42] symmetry is contained by 89%. At distance zero of the two nucleons the [42] orbital symmetry is described by four quarks in the 1s and two quarks in the 2p-state. If one moves the two nucleons apart to a finite distance r, each nucleon contains the excitation of zero oscillator quanta. Since the number of oscillator quanta has to be conserved if one uses an oscillator basis, the missing two oscillator quanta have to reside in the relative motion. For the S-wave interaction this means that the relative wave function has to be in the interaction region a 2s function with a node. This node yields asymptotically a hard or a soft core phase shift in the same way as a hard or a soft core potential would do it.

For the long range part we use the π and σ meson-exchange. To guarantee asymptotic freedom we suppressed the coupling of the π and σ-meson with a r^3-dependence inside the nucleons.

The nucleon-nucleon scattering is then treated as the scattering of a cluster with three quarks on a cluster with three quarks by using the Resonating Group method. The parameters of the model are fitted to the nucleon mass m_N=938 MeV, to the Δ mass m_Δ=1236 MeV, to the root mean square radius of the charge distribution of the proton $\langle r^2\rangle_p^{1/2}$ = 0.86 fm, to the pion-nucleon $g_\pi^2NN/(4\pi)$=14.1 and the σ-nucleon coupling constant $g_\sigma^2NN/(4\pi)$=5.65. Although none of the parameters of the model is fitted directly to the phase shifts we find a surprising good agreement for the ^1S and ^3S phase shifts as a function of the energy.

If now someone asks to tell in a simple way what is the nature of the short range repulsive core of the nucleon-nucleon interaction one could refer him to the following story: Let us assume we have here a bird's cage covered by a black fabric. Someone wants to know if there is a bird in the cage or not. To answer this question we do the following experiment: We stick a finger under the black fabric into the cage and if we get picked, then there is a bird in the cage and if not there is no bird in the cage. We stick the finger in and get picked (we measure hard core phase shifts). We conclude there is a bird (hard core) in the cage. Now we remove the black fabric and see there is no bird. We got picked by a nail sticking out in the cage.

Thus there is not a bird (hard core) in the cage but we got picked (measured a hard core phase shift) anyway.

References

1) G.E. Brown, A.D. Jackson, "The Nucleon-Nucleon Interaction (Horth-Holland), Amsterdam (1976).
2) K. Erkelenz, Phys. Rep. 13 (1974) 191.
3) K. Holinde, Phys. Rep. 68C (1981) 121.
4) K. Holinde, R. Machleidt, M.R. Anastasio, A. Faessler, H. Müther, Phys. Rev. C19 (1979) 948.
5) R. Liberman, Phys. Rev. D16 (1977) 1542.
6) De Tar, Phys. Rev. D17 (1978) 326.
7) K. Wildermuth, Y.C. Tang "A Unified Theory of the Nucleus", Vieweg, Braunschweig (1977).
8) K. Holinde, R. Machleidt, Nucl. Phys. A256 (1976) 479.
9) A. De Rújula, H. Georgi, S.L. Glashow, Phys. Rev. D12, (1975) 147.
10) V.G. Neudatschin, I.T. Obukhovsky, V.I. Kukulin, N.F. Golovanova, Phys. Rev. C11, (1975) 128.
11) V.G. Neudatschin, Yu.F. Smirnov, R. Tamagaki, Prog. Theor. Phys. 58 (1977) 1072. I.T. Obukhovsky, V.G. Neudatschin, Yu.F. Smirnov, Yu.M. Tschuvil'sky, Phys. Lett. 88B (1979) 231.
12) M. Harvey, Nucl. Phys. A352 (1981) 301 and 326.
13) A. Faessler, F. Fernandez, G. Lübeck, K. Shimizu, Phys. Lett. 112B (1982) 201.
14) A. Faessler, F. Fernandez, Phys. Lett. 124B (1983) 145.
15) A. Faessler, F. Fernandez, J. Phys. G9, (1983) L39 and 471.
16) S. Furui, A. Faessler, Nucl. Phys. A397 (1982) 413.
17) M. Harvey, Proc. Int. Symposium on Clustering Phenomena in Nuclei. Page 179. Eds. P. Kramer, R. Schultheis, Attempto Verlag, Tübingen (1982).
18) K. Kamimura, Prog. Theor. Phys. 62 (1977) 236.
19) R. Tegen, R. Brockmann, W. Weise, Z. Phys. A307 (1982) 339.
20) W. Weise, Proceedings of the 5th Topical School, Motril (Granada), Spain, Sept. 6-10, 1982, to be published.
21) R. Arndt and coworkers, Phys. Rev. C15 (1977) 1002.
22) R. Arndt, Proceedings of the Conference on Nucleon-Nucleon Interaction, Vancouver, p. 117 (1977).
23) G. Höhler and coworkers, Nucl. Phys. B114 (1975) 505.
24) M. Oka, K. Yazaki, Phys. Lett. 90B (1980) 41 and Prog. Theor. Phys. 66 (1981) 556 and 572.

A STUDY OF NN PHASE-SHIFT CALCULATIONS IN A QUARK POTENTIAL MODEL

M. Harvey

Chalk River Nuclear Laboratories, Atomic Energy of Canada Limited, Chalk River, Ontario, Canada KOJ 1JO

Great advances have been made in recent years, through the gauge symmetry models, in our understanding of three (electromagnetic, weak and strong) of the four (with gravity) fundamental interactions recognized in nature[1]. Since nuclear physicists can only ignore these advances at great peril to their subject, it is particularly encouraging for me to attend this workshop designed specifically to address the new facets of our field.

Clearly one of the questions that must be asked is what new insights into the structure of nuclei, and into reactions involving nuclei, do these new concepts yield. The question involves discerning how large a role the internal structure of nucleons (and hence higher resonances) plays in determining properties of nuclei. At Chalk River, together with my colleagues J. LeTourneux and B. Lorazo from the Université de Montréal, we have been considering the question with regard to the interaction between nucleons.

Of course, quantum chromodynamics (QCD) being the accepted fundamental form of the strong interaction (at least for the moment), should be responsible for the whole of the strong interaction between nucleons. In practice, the complexities of QCD will surely prevent us from describing the whole of the NN-interaction starting from the fundamental Lagrangian. Fortunately, there appears to be a possible "symbiosis" between the new ideas introduced by QCD and the old meson exchange picture of the NN-interaction given by Yukawa. QCD has the property of being "asymptotically" free at high energies i.e. the short-distance interaction between quarks can be treated perturbatively. It is, therefore, in the short-distance behaviour of the nucleon-nucleon force that we should hope to recognize QCD effects first. This is the very part of the force that has traditionally[2] been the most difficult to describe and which is often simulated by high momentum regularisation[3,4] or a phenomenological hard core potential[5,6]. At high energies QCD is confining and even today, with the vast amount of computer time spent in the lattice gauge calculations, we do not have a successful description of this mode. The confinement mode does not mean that quarks cannot leave the baryon, only that colour is confined. This seemingly illogical statement (since quarks carry colour) is possible since the quark, on leaving the environs of a baryon, can polarise the vacuum to form an additional quark-antiquark pair (Fig. 1): together with the original quarks the physical system can now resemble an observable baryon-meson pair. Thus confinement is intimately tied to meson production - and meson exchange at long distances is the most successful part of our traditional description of the nuclear force. The problem, as I see it, is how to successfully marry a perturbative QCD description of the short range part of the NN-interaction, with the long range meson exchange. How short is short and how long is long? It is to be hoped that the two approaches overlap at some range so that we can go from one to the other.

In Canada we have concentrated our efforts up to now on the short range components. More detail beyond that given below appears elsewhere[7-9]. We have asked whether we can correlate the known properties of nucleons (N) and deltas (Δ) with the known properties of the NN-interaction at short distances - and how accurately can we determine this correlation. Our work has been stimulated by the earlier studies of Isgur & Karl[10] who showed that the properties of single hadrons could be correlated by a non-relativistic potential model. Although the non-relativistic aspects of the model are surprising, the success suggests that the short range behaviour of the NN-interaction may be studied in the same model. Our present approach follows that of Oka & Yazaki[11] and is an outgrowth of my earlier studies[12,13]. Although we make extensive use of the fractional parentage techniques[12] the generator coordinate and resonating group techniques now used are superior to, and should be considered to supersede, the old Born-Oppenheimer approach[13].

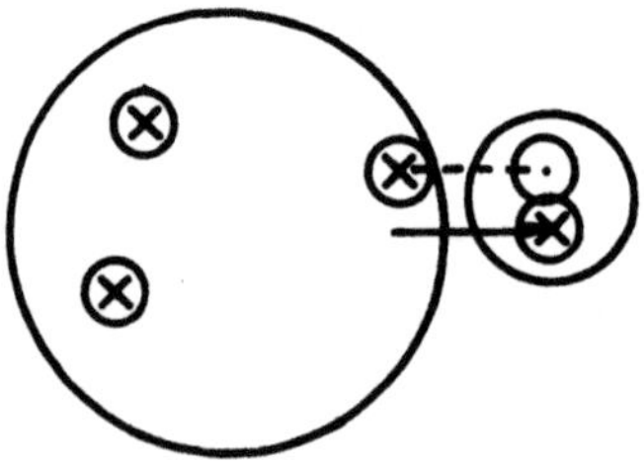

Fig. 1. A schematic diagram depicting the polarisation of the vacuum into a quark-antiquark pair when a quark leaves the baryon thus yielding an observable baryon and meson.

In the generator coordinate (GC) approach[14] we start with the ansatz that the eigensolutions to the Schrodinger equation,

$$H\Psi = E\Psi \,, \tag{1}$$

can be approximated by

$$\Psi = \sum_{\alpha} \int \phi_{\alpha}(\underset{\sim}{x};\underset{\sim}{R})\, f_{\alpha}(\underset{\sim}{R})\, d\underset{\sim}{R} \,. \tag{2}$$

Here $\phi_{\alpha}(\underset{\sim}{x};\underset{\sim}{R})$ is a cluster state representing six quarks in two three-quark clusters in oscillator wells separated by a distance $\underset{\sim}{R}$. All other coordinates are described schematically by $\underset{\sim}{x}$. The function $f_{\alpha}(\underset{\sim}{R})$ represents (crudely) the relative wavefunction of the wells and the parameter α represents the various channels (NN, NΔ, CC etc.) in terms of which the solution is to be represented. We write CC for a "hidden colour" state[12,13] which is a colour singlet six-quark state in which each three-quark cluster is colour octet.

Actually the function Ψ is equivalent to that in the resonating group (RG) approximation,

$$\Psi = \sum_{\alpha} \chi_{\alpha}(\underset{\sim}{x}) \quad g_{\alpha}(\underset{\sim}{\rho}) \quad , \tag{3}$$

where $\underset{\sim}{x}$ are internal coordinates and $\underset{\sim}{\rho}$ is the distance between the clusters. The function $g_{\alpha}(\underset{\sim}{\rho})$ is the relative wave function of the clusters which, for the α = NN channel, will be written in terms of the phase shifts for scattering solutions. We have used both the GC and RG techniques and have got identical results. In general we have found the RG approach to be numerically more stable. The problem is to solve the Schrödinger equation with the ansatz in eq. 2 or eq. 3 for the unknown $f_{\alpha}(\underset{\sim}{R})$ or $g_{\alpha}(\underset{\sim}{\rho})$ subject to given boundary conditions for the various channels. Thanks to a proof by deTakacsy[15], we know that the functions $f_{\alpha}(\underset{\sim}{R})$ and $g_{\alpha}(\underset{\sim}{\rho})$ have similar forms for large values of their arguments.

The Hamiltonian[8] is chosen to have a central confining term $v_c(r_{ij})$ and a spin-spin interaction introduced to describe the splitting between the N and Δ states. Thus

$$V(r_{ij}) = \sum_{i<j} \lambda_i \cdot \lambda_j \; [v_c(r_{ij}) + v_{\sigma}(r_{ij}) \; \sigma_i \cdot \sigma_j] \tag{4}$$

where λ_j^{α} are the ($\alpha = 1,2,\cdots 8$) generating operators of the SU_3 colour group and $\lambda_i \cdot \lambda_j$ denotes scalar product. For simplicity we take

$$v_c(r_{ij}) = A \exp(-r_{ij}^2/\beta^2) + Br_{ij}^2 + C \tag{5a}$$

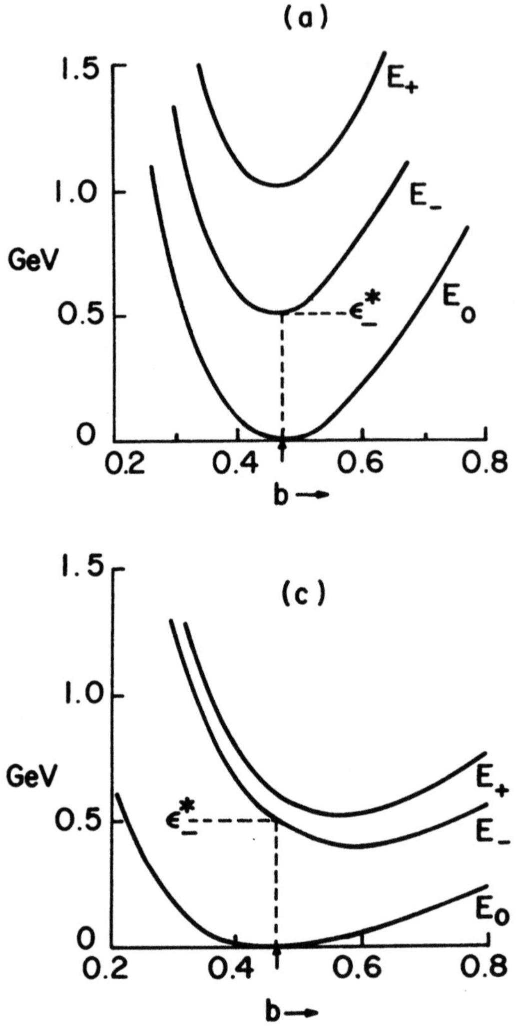

Fig. 2. Variational curves for the mean ground state (E_0), negative parity state (E_-) and excited positive parity state of [3] symmetry with force parameters chosen such that ($E_- - E_0$) at $b = b_0$ fits experiment. Shown for various values of E_+ (see text).

and

$$v_\sigma(r_{ij}) = D\ \delta(r_{ij}) \tag{5b}$$

where, in the confining term, the Gaussian is introduced so that we can vary the mean energy (in a variational calculation) of the excited positive parity states of the N and Δ relative to the mean energy of the

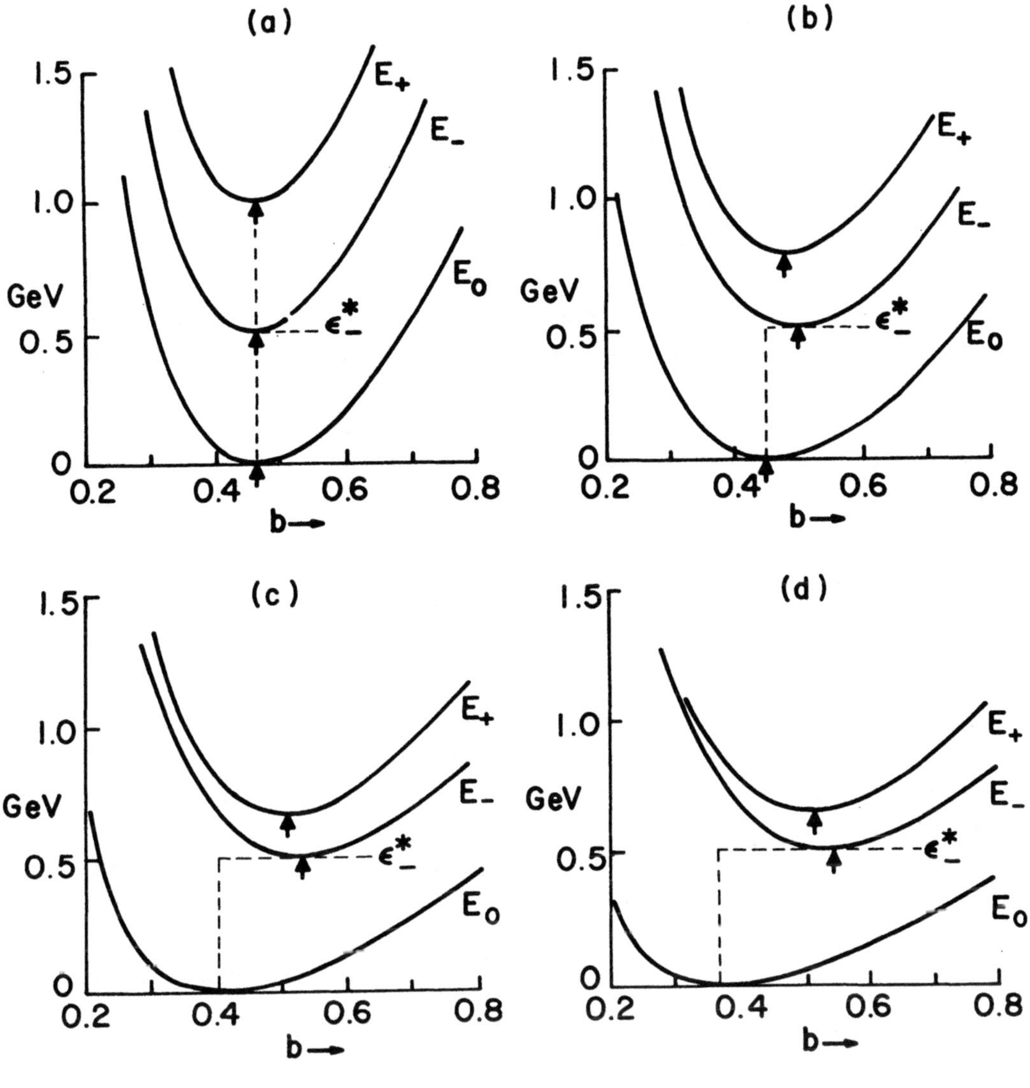

Fig. 3. As in Fig. 2 but where the difference between the minima of the curves for E_ and E_0 fit experiment.

excited negative parity states. By mean energy, we mean that in the absence of the spin-spin interaction.

It is crucial that harmonic oscillator states be used in the GC- or RG-approach. It is natural therefore to choose parameters of the Hamiltonian such that three-quark oscillator states represent the eigensolutions of the baryon spectrum in a variational calculation treating the oscillator parameter (b) as the variational parameter.

This requirement does not determine the parameters of the potential in a rigorous way. Thus the parameters could be chosen such that only the mean energy (E_0) of the ground state (1086 MeV) is variationally stable ($b = b_0$) and the mean energy (E_-) of the set of lowest negative parity states fitted to the experimentally observed value ($\sim$1600 MeV) at the same value of $b = b_0$ (c.f. Fig. 2). We refer to this type of fit by the Roman numeral II. Alternatively, we could demand fits to the experimental spectrum at the variational minima of both the ground and negative parity states (c.f. Fig. 3): then the variational parameters at the minima need not be the same. We refer to this type of fit by the Roman numeral I. The stability of the ground state - and whether the negative parity states and ground states minimise at the same or different points - depends on the mean energy (E_+) of the excited positive parity states of [3] orbital symmetry. If these latter states happen to have twice the excitation energy of the negative parity states, the spectrum is "oscillator like", the parameter A in eq. 5 is zero and all states minimise with the same value of b (see Fig. 2 and 3 for case (a)). As the value for the excited positive parity states decreases (cases b, c and d in Fig. 2 and 3), minima for the various states occurat different values of b and, moreover, the minima become "softer" to variations in b (i.e. smaller curvature). Experimentally the mean energy of these excited positive parity states (close to fit d) is almost degenerate with the negative parity states and this is not only the hardest to fit, but leads to a very "soft" ground state. A most undesirable state of affairs.

We have investigated whether the phase shifts (δ) are at all dependent on the above features. In Figs 4 and 5 we show the S-wave phase shifts for extreme values of the parameters i.e. (a) (an oscillator-like baryon spectrum) and Id and IId (which yield the observed baryon spectrum in a variational calculation). The calculation was for the three channel (NN,$\Delta\Delta$,CC) approximation where the channels involve the orbital symmetries found in the NN-channel at large nucleon separation[12]. Similar approximations are taken by Faessler et al.[16-18]. As is seen in Figs. 4 and 5, substantial changes ($\sim$50%) can result, warning us that the model is not to be trusted for quantitative results in its present form. Nevertheless, the phase shifts are negative for all energies (similar features are found for the higher partial waves) in accord with what has been deduced about the nature of the short range interaction[4]. The phase shifts are of comparable magnitude (to within a factor of 2) to those calculated from the $\omega+\rho$-exchange used in the Bonn potential[3,4]. We conclude that the quark contribution at short ranges is not negligible and cannot be ignored.

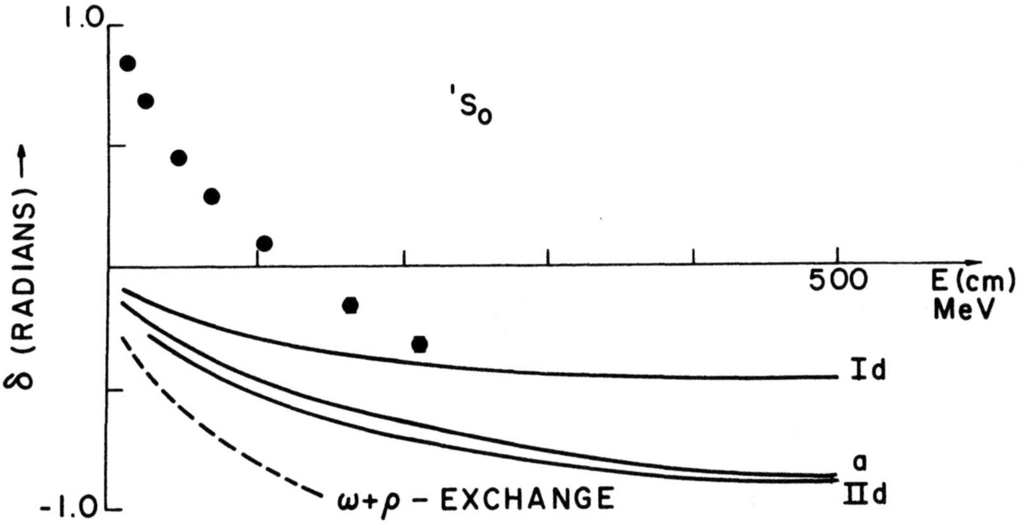

Fig. 4. $^1S_0$ phase shifts from the quark potential calculation with three different potentials. Phase shifts from the experimental scattering data are shown by error bars[19].

The negative phase shifts have been interpreted as those resulting from a phenomenological, local, repulsive potential at short range. The character of this phenomenological potential can be deduced by looking at the wave shift $(-\delta/k)$ where $(\hbar k)^2/M_N = E_{CM}$ (c.f. Figs. 6 and 7). For an infinite hard core repulsion the wave shift is independent of k and equal to the hard core radius. For a soft-core repulsion the wave shift decreases with k as shown in Fig. 6 where the $^1S_0$ wave shift with Force IId could be fitted for $k < 1.5$ MeV by a soft core with height 513 MeV and range ~.63 fm. With some of the forces used the wave shift increases with k for $k \gtrsim 0$ MeV and this indicates some effective attractive potential or a momentum dependent repulsion. It would be nice if the local potential could be deduced directly from the quark calculation, but we argue (c.f. ref. 7) that this cannot be done unambiguously, and therefore, attempts that do so, do not seem to have physical relevance.

We conclude by considering whether the quark potential model is capable of yielding positive phase shifts at low energy in greater accord with experiment. Recent calculations by Maltman & Isgur[20] suggest that this is indeed possible. In a variational calculation for

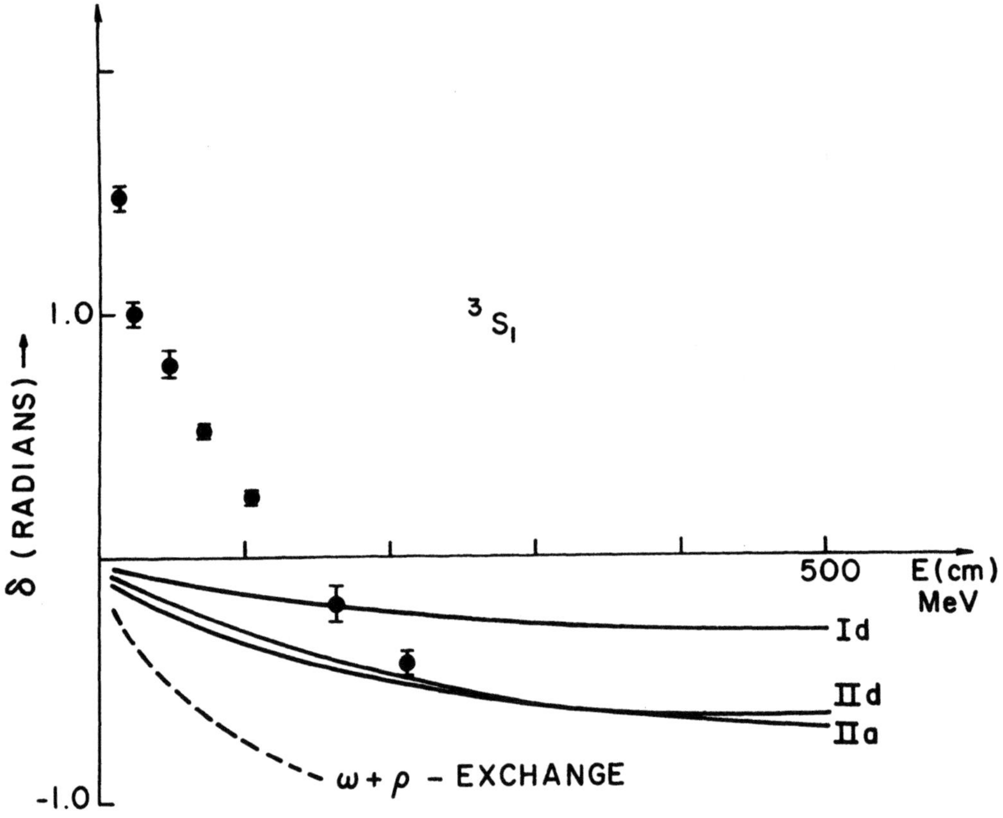

Fig. 5. As in Fig. 4 but for $^3S_1$.

the deuteron ground state, Maltman & Isgur found significant attraction, equivalent to that of a central attractive potential of intermediate range, from certain (additional) six-quark, hidden-colour (CC) states. These are the configurations that also lead to the long range attractive van der Waals force. Although this latter effect must be a spurious product of the potential model (as discussed by Greenberg & Lipkin[21]), the intermediate range attraction is of the magnitude of that of the sigma exchange in meson exchange models[3,4]. We suggest that this is no accident and that the two models are, at this intermediate range, describing the same physical feature. We argue that the separation of a CC state must, at some intermediate range, emit (colour octet) gluons which can then decay into mesons (c.f. Fig. 8). The gluons carry away the colour-octet nature of the hidden colour states

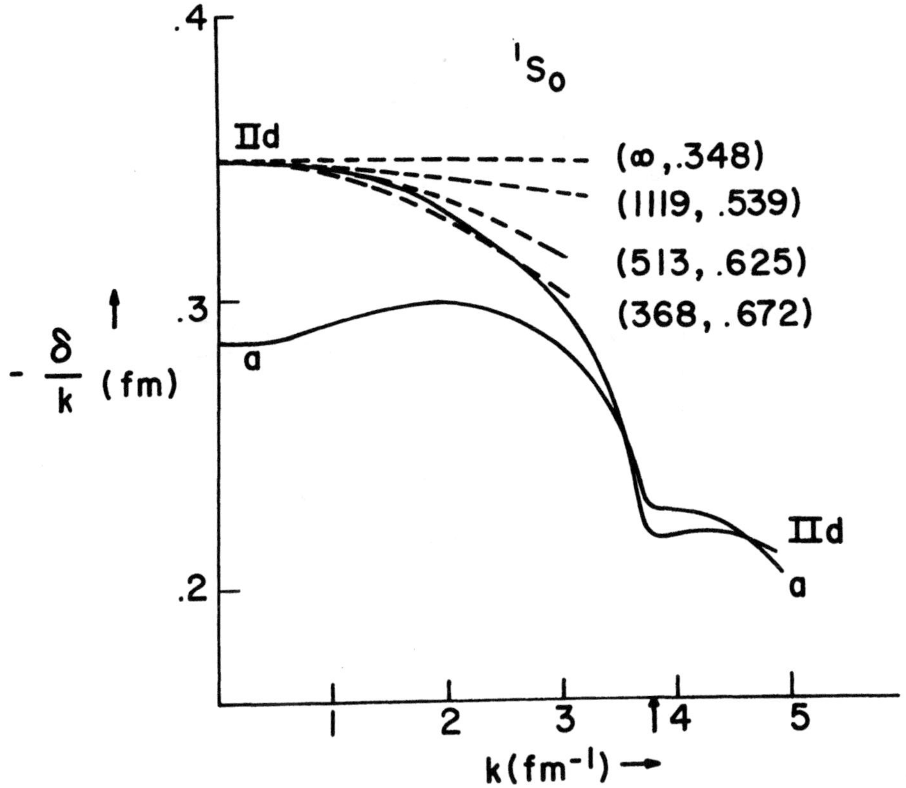

Fig. 6. $^1S_0$ wave shifts from the quark potential calculation for various assumed potentials. The dashed curves are the wave shifts from various local repulsive core potentials with height and range (V_0 (MeV), a (fm)).

leaving colour singlet baryons. Thus the hidden colour degrees of freedom may just be the short range manifestation of some of the mesonic degrees of freedom. The non-relativistic potential model, not being a field theory, cannot describe this equivalence (hence the spurious long range van der Waals force) and it is not clear at the moment how the model can be modified to accommodate it. To leave these hidden colour configurations out of the phase shift calculation (as we have done up to now) may be throwing away the significant medium range attraction in an effort not to have the spurious long range van der Waals interaction. We are now attempting the RG-calculation including these configurations but we cannot tell at this stage whether a relevant calculation is feasible.

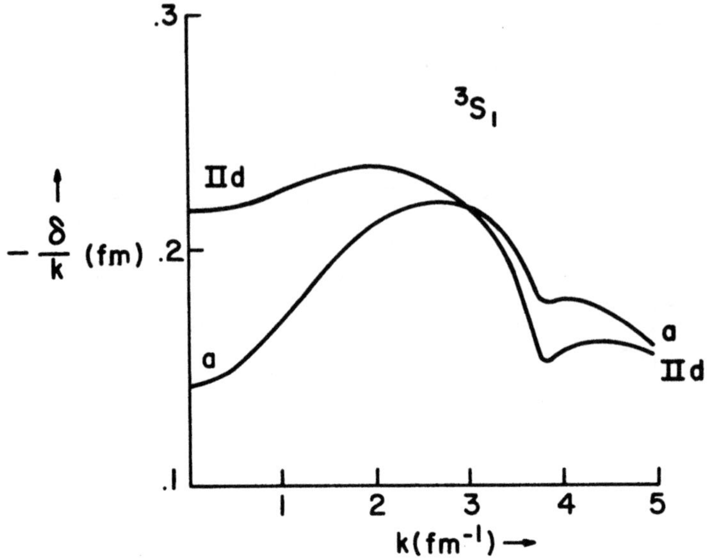

Fig. 7. $^3S_0$ waveshifts.

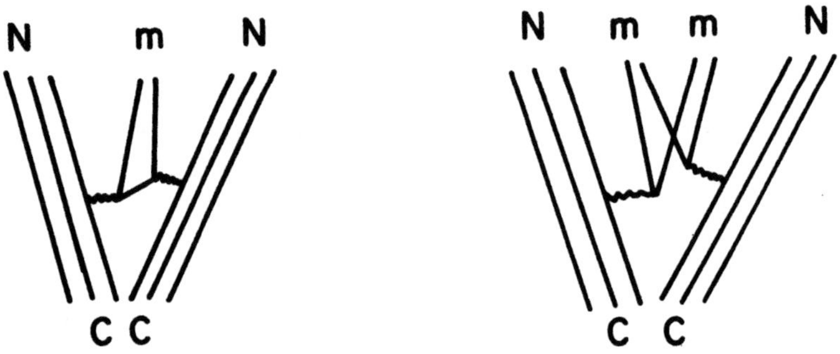

Fig. 8. A schematic diagram illustrating how the "fissioning" of a hidden-colour state leads to colour singlet baryons and mesons.

References

1. F.E. Close, "An Introduction to Quarks and Partons", Academic Press, 1979.
2. P. Signell, Adv. in Nucl. Phys., Vol. 2, 223 (Ed. M. Baranger and E.W. Vogt), Plenum Press, 1969.
3. K. Holinde and R. Machleidt, Nucl. Phys. A247(1975)495; A256(1976)479.
4. K. Holinde, Phys. Reports 68(1981)121.
5. M. Lacombe, B. Loiseau, J.M. Richard, R. Vinh Mau, J. Coté, P. Pires and R. de Torreil, Phys. Rev. C21(1980)861.
6. R. Vinh Mah, Nucl. Phys. A335(1980)211.
7. M. Harvey, J. LeTourneux and B. Lorazo, to be published.
8. M. Harvey and J. LeTourneux, to be published.
9. M. Harvey, "The NN-Interaction in the Quark Potential Model", Lecture Notes at the Summer School on "Nucleon-Nucleon Interaction and Nuclear Many Body Problems", Chang Chun, China, August 1983; also at the Canadian Assoc. of Physicists Summer School, Pearson College, Vancouver, 1982, Plenum Press.
10. N. Isgur and G. Karl, Phys. Rev. D18(1978)4187; D19(1979)2653.
11. M. Oka and K. Yazaki, Prog. Theor. Phys. 66(1981)556.
12. M. Harvey, Nucl. Phys. A352(1981)301.
13. M. Harvey, Nucl. Phys. A352(1981)326.
14. M. Onsi and J. LeTourneux, Can. J. Phys. 58(1980)612.
15. N.B. deTakacsy, Phys. Rev. C5(1972)1883.
16. A. Faessler, F. Fernandez and G. Lübeck, Phys. Lett. 112B(1982)201.
17. A. Faessler, F. Fernandez, G. Lübeck and K. Shimizu, Nucl. Phys. A402(1983)555.
18. A. Faessler and F. Fernandez, Phys. Lett. 124B(1983)145.
19. M. McGregor, R. Arndt and R. Wright, Phys. Rev. 182(1969)1714.
20. K. Maltman and N. Isgur, Phys. Rev. Lett. 50(1983)1827; Phys. Rev. D (to be published).
21. O.W. Greenberg and H.J. Lipkin, Nucl. Phys. A370(1981)349.

MANY-NUCLEON FORCES AND CURRENTS DERIVED FROM QUARK DEGREES OF FREEDOM⁺

P.U. Sauer and U.-J. Wiese
Theoretical Physics, University Hannover
3000 Hannover, Germany

1. Introduction

Nuclear physics is still on the search for a workable and quantitatively successful microscopic theory of nuclear phenomena. This paper wants to contribute to this on-going search.

The traditional concept of nuclear theory views the nucleus as a system of elementary nucleons interacting through instantaneous two-nucleon forces according to the rules of nonrelativistic quantum mechanics. This many-body problem of the nucleus can be solved for three nucleons with satisfying numerical accuracy. Its solution demonstrates: The classic theory of nuclear phenomena is qualitatively successful for bound and scattering states, but it does not yield[1)] quantitative agreement with experimental data. Its successes warn to give up the traditional concept as wrong altogether, but the persisting discrepancies with experiment necessitate corrections.

The prominent problem of traditional microscopic nuclear structure is the theoretical description of binding energies and electromagnetic (e.m.) properties. The standard corrections take into account the virtual excitation of a nucleon to an isobar and mesonic exchange currents. These special correction mechanisms yield[2)] encouraging improvements of the theoretical results, but - in the three-nucleon system, the test case for the more complicated heavier nuclei - they are unable to fully cure the existing discrepancies. However, the nucleon is itself a composite system consisting of quarks, antiquarks and gluons. This substructure of the nucleon only plays an insignificant role for the correction mechanisms considered so far. Because of the remaining discrepancies it therefore appears reasonable to investigate corrections of traditional microscopic nuclear structure, which only arise in many-nucleon systems and which are due to the internal nucleonic degrees of freedom. This is the motivation for the work reported on in this paper.

In a quantum-chromodynamical (QCD) description most nuclear problems are phenomena of strong quark-gluon coupling. The absence of an explanation, why nature chooses colorless quark clusters only, i.e., why quarks are confined, and the absence of a technically simple and micro-

---

⁺Talk presented by P.U. Sauer

scopically convincing phenomenology of confinement appear to be the stumbling blocks for a successful treatment of the internal nucleonic degrees of freedom in nuclear physics. The employed methods for generating confinement by "fiat" are bag models with relativistic, practically massless quarks and potential models of nonrelativistic quarks with an effective constituent mass. In the latter models gluon degrees of freedom are frozen into quark-quark potentials and only the quark degrees of freedom are kept.

The internal nucleonic quark degrees of freedom are usually envoked to understand[3)] known nucleonic properties, e.g., e.m. form factors, meson-nucleon couplings and the nucleon-nucleon interaction, on a more fundamental level. However, these nucleonic properties are only the input parameters for the nuclear many-body problem, which then remains to be solved in a Hilbert space of nucleons only as in traditional microscopic nuclear structure. In this way qualitatively new results for the many-nucleon systems and a cure of existing discrepancies between theory and experiment are not to be expected: The dependence of nuclear-structure results on a variation of these input parameters has already been tested phenomenologically, it is the dependence on the off-shell properties[4)] of the two-nucleon potential.

The present work is complementary to the procedure just mentioned. It aims at those many-nucleon phenomena which can only arise from the quark degrees of freedom in nucleons. The phenomena to be investigated are based on nonnucleonic hidden-color six-quark states: When the three-quark clusters of two nucleons in the nucleus have a nonvanishing overlap, six-quark configurations arise through gluon exchange and the indistinguishability of identical quarks which can no longer be interpreted as states of two physically separable baryons. For these nonnucleonic six-quark states the quark property color is essential. Whereas all physically observed quark systems are colorless, the colorless six-quark states may also contain spatially localized substructures with nonvanishing color[5)], hidden-color six-quark states. Such quark states have been employed in the two-nucleon system for the description of elastic nucleon-nucleon scattering[6)] at low energies and in the analysis of the e.m. deuteron properties[7)]. Such quark states yield the basis for the attempted quark interpretation of the structures seen in nucleon-nucleon scattering at intermediate energies as dibaryon resonances[8)]. The present work extends their use to many-nucleon systems. In the nuclear medium these six-quark states yield new effects: (i) Many-nucleon forces arise when a hidden-color six-quark state interacts with the colorless three-quark clusters of other nucleons. (ii) The hidden-color six-quark states have a spin-isospin structure, which differs from that of two interacting nucleons, and therefore also a different coupling to the photon field. Thus, a new type of many-nucleon currents arises. This paper formulates these new quark phenomena in nuclear physics, i.e., many-nucleon forces and currents, as corrections for traditional microscopic nuclear structure and tests their consequences for understanding nuclear properties by the example of the bound three-nucleon system.

2. Two-nucleon interaction model with hidden-color six-quark states

The aim of this section is to develop a model for the two-nucleon interaction with novel characteristics. The description of the interaction should remain traditional and macroscopic, i.e., it should use the coordinate $\vec{r}$, i.e., the separation between two three-quark clusters, the total spin S and isospin I as dynamic variables. However, the description should also contain physically relevant features of the composite nature of nucleons, i.e., of subbaryonic degrees of freedom. The latter goal will be acchieved by coupling channels with nonbaryonic quark content to two-baryon channels traditionally employed.

We view two baryons in interaction as a system of six quarks interacting nonrelativistically through instantaneous quark-quark potentials. The interaction between two quarks is parametrised in the usual form $\vec{\lambda}_i \cdot \vec{\lambda}_j\ V(i,j)$. The $\vec{\lambda}_i$ are the eight component vectors of SU(3) generators in color space and the subscripts i and j denote the particle labels of the quarks. The potential V(i,j) contains the Fermi-Breit one-gluon exchange[9], nonrelativistically approximated and reduced to its local central pieces and a harmonic confinement potential, i.e.,

$$V(i,j) = \frac{1}{4}\hbar c\alpha_s \left[\frac{1}{|\vec{r}_i-\vec{r}_j|} - \pi\left(\frac{\hbar}{mc}\right)^2 \delta(\vec{r}_i-\vec{r}_j)\left(1+\frac{2}{3}\vec{\sigma}_i\cdot\vec{\sigma}_j\right)\right] - \frac{3}{16}k(\vec{r}_i-\vec{r}_j)^2. \quad (2.1)$$

The emerging quark hamiltonian of internal motion has the form

$$H = \sum_i \left(\frac{p_i^2}{2m} + mc^2\right) - \frac{(\sum_i \vec{p}_i)^2}{2\sum_i m} + \sum_{i<j} \vec{\lambda}_i\cdot\vec{\lambda}_j\ V(i,j). \quad (2.2)$$

It only makes sense in a restricted space of states with vanishing total color.

The hamiltonian (2.2) is first applied to three-quark systems in order to determine its parameters. The three-quark systems are assumed to consist of up and down quarks only and to cluster around a center in space located, say, at $\vec{R}_a$. The three-quark wave function of the baryon A

$$<\vec{r}_1\vec{r}_2\vec{r}_3|A_q\vec{R}_a> = <\vec{r}_1|\vec{R}_a><r_2|\vec{R}_a><\vec{r}_3|\vec{R}_q>|A_q> \qquad (2.3)$$

has a spatially symmetric part, the normalized product of harmonic oscillator wave functions $<\vec{r}_1|\vec{R}_a><\vec{r}_2|\vec{R}_a><\vec{r}_3|\vec{R}_a>$ in the lowest orbit with the oscillator parameter α, and a totally antisymmetric spin-flavor-color part $|A_q>$. Given the spatial wave function there are six different types of three-quark clusters. Two of them are color singlets, they correspond to the nucleon (N) and the delta-isobar (Δ). Three of them are color octets (0,0' and 0"), which differ by isospin and spin, i.e., $(IS)=(\frac{1}{2}\,\frac{1}{2}),(\frac{1}{2}\,\frac{3}{2}),(\frac{3}{2}\,\frac{1}{2})$ respectively. The sixth is a color decuplet. The three-quark cluster wave functions are given in ref. (10). The parameters of the quark hamiltonian (2.2), i.e., the quark mass m, the strong coupling constant α_s, the confinement parameter k and the oscillator parameter α for the spatial wave functions are fixed by properties of the physical baryons nucleon and delta using the nucleonic magnetic moments and the nucleon-delta mass difference $M_\Delta c^2 - M_N c^2$. The masses of the nucleon and the Δ-isobar are required to be stationary under variation of the oscillator size parameter α. The colored three-quark clusters do not have a realization as isolated physical objects. However, the three color octets can serve as building blocks for obtaining all colorless six-quark configurations. This is their function in the present context. The color-decuplet cluster does not form a colorless six-quark state with any other three-quark cluster due to SU(3) coupling rules and therefore will not be discussed any further.

Next the quark hamiltonian (2.2) is applied to six-quark systems. We are very well aware of the fact that this hamiltonian cannot describe the intermediate- and long-range attraction of the two-nucleon potential. In fact it yields a wrong long-range van-der-Waals force[11)], which is contradicted by the low-energy nucleon-nucleon scattering data. The hamiltonian has proven to be successful for short-range phenomena, for which confinement does not play a crucial role. We therefore use it for a description of the dynamic evolution of the six-quark system at small cluster separations and shall phenomenologically correct its faulty aspects at large distances. For the description we assume that three quarks always remain clustered around their individual centers. Admittedly, this is an asumption hard to justify.

We describe the six-quark system with the tools of the generator coordinate method (GCM)[12)]. The physically relevant generator coordinate is the distance $\vec{r}$ between two centers $\vec{R}_a$ and $\vec{R}_b$, i.e.,$\vec{r} = \vec{R}_a - \vec{R}_b$, around which the two three-quark clusters A_q and B_q are built. The used basis states are[10)]

$$|\beta\vec{r}> = |A_q B_q ISC \vec{r}> \tag{2.4a}$$

$$|A_q B_q ISC\vec{r}> =$$

$$\frac{1}{\sqrt{2}} \bar{\mathcal{A}} \sum_{\Theta} \mathcal{N}(\Theta) |\Theta\vec{r}> \frac{1}{\sqrt{2}} \left[|A_q B_q ISC> + (-1)^{I-I_A-I_B+S-S_A-S_B+\Theta} |B_q A_q ISC> \right] \tag{2.4b}$$

The basis states are colorless, i.e., C = 1, and totally antisymmetrised. They are constructed from orbital states $|\Theta\vec{r}>$ of definite permutation symmetry Θ, Θ being 3, 2, 1 and 0 for the Young tableaux [6], [51], [42] and [33] respectively, and from spin-flavor-color states $|A_q B_q ISC>$ in which the individual cluster states $|A_q>$ and $|B_q>$ are coupled to total isospin I, total spin S and total color C. Antisymmetrization within the clusters and cluster exchange is carried out beforehand, thus $\bar{\mathcal{A}} = \frac{1}{10}\left[I - P_{14} - P_{15} - P_{16} - P_{24} - P_{25} - P_{26} - P_{34} - P_{35} - P_{36}\right]$ denotes the remaining antisymmetrization between clusters only. The channel label β stands for the cluster content A_q and B_q and the total quantum numbers; isospin and spin projections are suppressed. There are two-baryon channels β_B, channels $\{N_q N_q IS\}$ with purely nucleonic content and channels $\{N_q \Delta_q IS\}$, $\{\Delta_q N_q IS\}$ and $\{\Delta_q \Delta_q IS\}$, well-known from usual corrections of the classic picture of microscopic nuclear structure, in which one or both nucleons are turned into a Δ-isobar. However, there are also channels β_C with color-octet content, i.e., $\{O_q O_q IS\}$, $\{O_q O_q' IS\}$, $\{O_q O_q'' IS\}$, $\{O_q' O_q IS\}$, $\{O_q' O_q' IS\}$, $\{O_q' O_q'' IS\}$, $\{O_q'' O_q IS\}$, $\{O_q'' O_q' IS\}$, $\{O_q'' O_q'' IS\}$, which appear as novel channels of hidden color. They arise from the quark description and are no realization of simple two-baryon states. The still unspecified weights $\mathcal{N}(\Theta)$ and phases $(-1)^{\Theta}$ are given in ref. (10). The basis $\{|\beta\vec{r}>\}$ is nonorthogonal and noncomplete in the six-quark Hilbert space.

The GCM ansatz for the dynamic many-quark wave function $|\phi_E>$ of the collective motion in the degree of freedom associated with the generator coordinate $\vec{r}$ is a linear superposition of the noncomplete basis states $\{|\beta\vec{r}>\}$, i.e.,

$$|\bar{\Phi}_E> = \sum_{\beta} \int d^3r |\beta\vec{r}><\beta\vec{r}|\phi_E> \tag{2.5}$$

Diagonalizing the hamiltonian H of equ. (2.2) in the subspace spanned by the basis states leads to the Hill-Wheeler integral equations for the weight function $<\beta\vec{r}|\phi_E>$ which contains the information on the dynamics of collective motion

$$\sum_{\beta'} \int d^3r' [<\beta\vec{r}|H|\beta'\vec{r}'> - E <\beta\vec{r}|\beta'\vec{r}'>] <\beta'\vec{r}'|\phi_E> = 0 \qquad (2.6a)$$

$$[\overline{H} - E\overline{N}] \; |\overline{\phi}_E> = 0 \qquad (2.6b)$$

Equ. (2.6b) is the operator abbreviation of the Hill-Wheeler equation in the subspace $\{\beta\vec{r}>\}$ of the Hilbert space. Let $\overline{X}$ be a - not necessarily hermitian and certainly non-unique - square-root of the overlap operator $\overline{N}$ in the considered subspace, i.e., $\overline{X}^+\overline{X} = \overline{N}$, then the Hill-Wheeler equation can be formally rewritten as

$$\sum_{\beta'} \int d^3r' <\beta\vec{r}|\mathcal{H}|\beta'\vec{r}'><\beta'\vec{r}'|\Psi_E> = E <\beta'\vec{r}'|\Psi_E> \qquad (2.7)$$

with $\mathcal{H} = (\overline{X}^{-1})^+ \overline{H}\, \overline{X}^{-1}$ and $|\Psi_E> = \overline{X}|\overline{\phi}_E>$. This is a conventional Schrödinger equation in the collective variable $\vec{r}$ which couples channels of different cluster content β. The states satisfy the usual orthonormality condition $<\Psi_{E'}|\Psi_E> = \delta(E'-E)$. They can be regarded as a traditonal two-baryon wave function with a direct probability interpretation on which boundary conditions for the scattering and bound-state situations can be imposed as usual. Physicswise, the Schrödinger equation (2.7) describes an interaction model which couples two-baryon channels β_B of separable colorless three-quark clusters with hidden-color channels β_C. The latter channels are spatially confined. Thus, the coupling between two-baryon channels and channels of hidden-color and the propagation in the latter channels occur at small relative distances $\vec{r}$.

The nonrelativistic one-gluon exchange hamiltonian of equ. (2.2) has been successful in describing small-distance quark dynamics. Thus the coupling between baryonic and nonbaryonic six-quark channels $<\beta_B\vec{r}|\mathcal{H}|\beta_C\vec{r}'>$ and the propagation in the nonbaryonic channels $<\beta_C\vec{r}|\mathcal{H}|\beta_C'\vec{r}'>$, derived from the hamiltonian (2.2) should be physically reliable. In contrast, the hamiltonian (2.2) cannot account for the intermediate and long range of the two-baryon interaction. We therefore have to repair this defect and replace the part $<\beta_B\vec{r}|\mathcal{H}|\beta_B'\vec{r}'>$ of the derived effective hamiltonian by a traditional one-boson exchange description, i.e.,

$$<\beta_B\vec{r}|\mathcal{H}|\beta_{B'}\vec{r}\,'> = -\frac{\hbar^2}{2\mu_B}\Delta_{\vec{r}}\,\delta(\vec{r}-\vec{r}\,')\,\delta_{BB'} + <\beta_B\vec{r}|V_{OBE}|\beta_{B'}\vec{r}\,'> \tag{2.8}$$

In equ. (2.8) μ_B is the reduced mass of the channel and V_{OBE} the one-boson exchange interaction between nucleons and isobars.

Thus, a two-nucleon force model is obtained in equs. (2.7) and (2.8), which remains in the traditional framework by using baryon separation, spin and isospin as dynamic variables, as well as an ordinary Schrödinger equation for the description of its time evolution. The force model yields an interaction between nucleons and Δ-isobars, but it is amended through coupling to nonbaryonic channels of hidden color for subnucleonic degrees of freedom. By the phenomenological replacement (2.8) the force model becomes a truely realistic one.

3. The interaction with hidden color channels

In this section two nucleons whose dynamics are governed by the force model of equs. (2.7) and (2.8) are emersed in a many-nucleon system and are subjected to external fields. The interaction with other nucleons and with the external sources is discussed. Novel many-nucleon forces and currents arise. The many-nucleon system is described in a Hilbert space of A baryons with a sector in which two baryons are turned into a nonbaryonic bound-state channel of hidden-color characteristics β_C. The baryons may be either nucleons or Δ-isobars. The considered Hilbert space is diagramatically described in fig. 1.

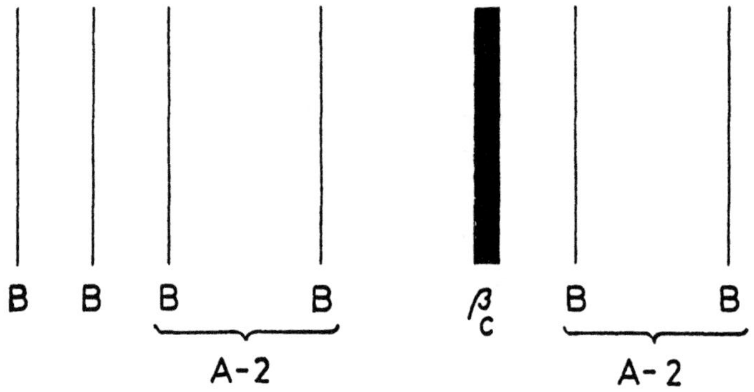

Fig. 1: The Hilbert space of a nucleus with baryon number A. It has two sectors. The first one contains A ordinary baryons B, in the second sector two baryons are excited to a hidden-color state of characteristics β_C.

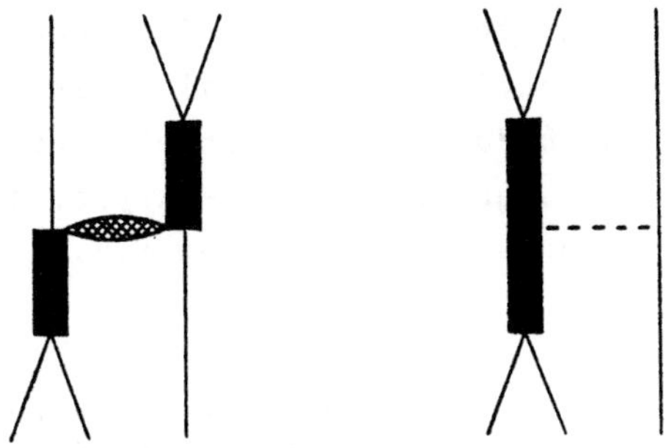

Fig. 2: Three-baryon forces. In the first type (a) the three-baryon force is due to color transport from one pair to another. It is of short range. The second type (b) is of long range arising from meson exchange between a hidden-color pair and an ordinary baryon.

Firstly the quark hamiltonian (2.2) can switch the hidden-color structure β_C from one pair of interacting baryons to another. This process yields a three-nucleon force, in general a many-baryon force, as diagramatically shown in fig. 2a. It can occur only when all three nucleons are close to each other.

When the quark system is subjected to external sources, the quark hamiltonian H of (2.2) has to be extended for additional couplings, i.e.,

$$H_{\phi A} = H + \sum_i g\, \vec{\sigma}_i \cdot \vec{\nabla}_x\; \vec{\tau}_i \cdot \vec{\phi}(\vec{x}) + \sum_i \frac{1}{c}\, j_{i\mu}(\vec{x})\, A^{\mu}(\vec{x}) \tag{3.1}$$

Clearly, quarks are the carriers of the e.m. current of baryons and the third term is the appropriate interaction with an external e.m. source of four-vector potential $A^{\mu}(\vec{x})$. Term two describes the coupling of quarks with a mesonic field. In equ. (3.1) the mesonic field is treated as elementary. This approximation neglects the internal quark-antiquark structure of mesons. As an example the isovector pion field $\vec{\phi}(\vec{x})$ is used in (3.1). It is assumed to originate from those nucleons of the many-nucleon system not described by their quark degrees of freedom through the hamiltonian (2.2) and (3.1). Due to term two of $H_{\phi A}$ the six-quark bound-state structures of hidden-color characteristics β_C, when emersed in a many-nucleon system, can also interact with the colorless three-quark clusters of other baryons in the traditional way. The interaction is of one-boson exchange type. Compared to the first process of fig. 2a the meson coupling in $H_{\phi A}$ yields a second three-nucleon force diagramatically described in fig. 2b which is of longer range and therefore might be physically more important.

Thirdly, the six quark bound-state channels β_C interact with photons. Compared to two nucleons in interaction the spatial location and the spin - isospin structures of the hidden-color channels are different. Thus, they yield e.m. two-nucleon, generally many-baryon, currents of novel characteristics abbreviated by the diagram of fig. 3.

Fig. 3: Two-baryon exchange current arising from the coupling of the photon to a hidden-color pair.

The basic result of this section is: When the two-nucleon system is described by its quark degrees of freedom, corrections for the traditional picture of the nucleus arise. The interaction of the two nucleons with other nucleons in the many-nucleon system yields novel many-nucleon forces, its interaction with an external e.m. probe yields novel many-nucleon exchange currents. The remainder of this paper discusses the practical applicability of these many-nucleon forces and currents in nuclear structure and estimates their effect for the three-nucleon bound states.

4. Simulation of hidden-color six-quark channels by two confined color-excited baryons

The theoretical frame work of sections 2 and 3 is practicable and could be employed, e.g., in the three-nucleon bound state, but, admittedly, it is technically demanding. For first estimates of the importance of the resulting many-nucleon forces and currents we used the following simplification for the hidden-color six-quark states β_C. They are assumed to be approximated by two point baryons C which carry the new quantum number color besides the quantum numbers isospin and spin, i.e., C being O, O' or O" of section 2. The color-octet baryon spans the eight-dimensional representation of color SU(3). Each color-octet baryon requires a color-octet partner in the many-nucleon system to whom it is confined and with whom together it forms a colorless bound two-baryon state. Thus the Hilbert space which describes a physical A-nucleon system consists - besides the ordinary part of colorless baryons - of a sector with two color-octet baryons coupled to a color-singlet. In this approximation the A-nucleon system remains in its microscopic description an A-baryon system. The quark degrees of freedom are entirely eliminated, but they have given rise to baryon states with new properties. The Hilbert space is described in fig. 4. It is to be interpreted as an approximative version of the Hilbert space of fig. 1.

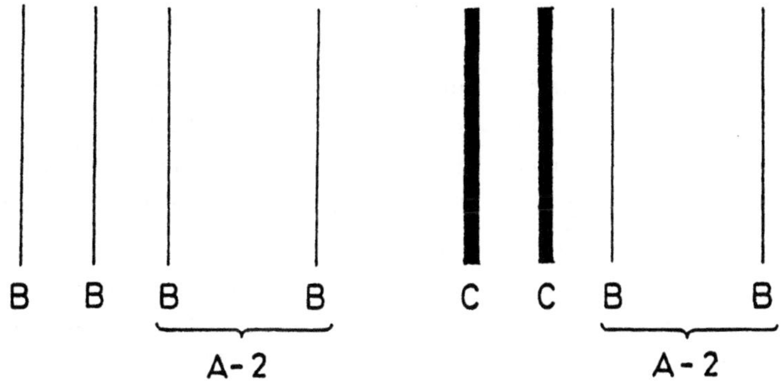

Fig. 4: The Hilbert space with the hidden-color pair simulated by two color-octet baryons C.

In the Hilbert space with color-octet baryons the dynamics of the two-baryon system is approximately described by the following coupled-channel Schrödinger equation.

$$\left[(M_B(1)+M_B(2)-2M_N)c^2 - \frac{\hbar^2}{2}\left(\frac{1}{M_B(1)} + \frac{1}{M_B(2)}\right)\Delta_{\vec{r}}\right]<\beta\vec{r}\,|\Psi_E> +$$

$$+ \sum_{\beta'} \int d^3r' <\beta\vec{r}\,|V|\,\beta'\vec{r}\,'><\beta'\vec{r}\,'|\Psi_E> = E<\beta\vec{r}\,|\Psi_E>. \qquad (4.1)$$

In the channels with nucleons and isobars the interaction used is the one of equ. (2.8). The coupling to color-excited baryonic configurations and the potentials in color-excited channels are identified with the generator-coordinate matrix elements of equ. (2.7). These potentials are only effective at small relative distances between baryons. They are used in a local and - neglecting the necessary transformation $\overline{X}$ to a unit overlap kernel - in an adiabatic approximation.

$$\langle BB\ IS\ \vec{r}|V|CC\ ISF'\rangle = V^{IS}_{BB\to CC}(\vec{r})\ \delta(\vec{r}-\vec{r}') \tag{4.2}$$

$$V^{IS}_{BB\to CC}(\vec{r}) = \langle\{B_q\ B_q IS\}\vec{r}|H|\{C_q C_q IS\}\vec{r}\rangle \tag{4.2a}$$

$$\langle CC\ IS\ \vec{r}|V|CC\ IS\ \vec{r}'\rangle = V^{IS}_{CC\to CC}(\vec{r})\ \delta(\vec{r}-\vec{r}') \tag{4.3}$$

$$V^{IS}_{CC\to CC}(\vec{r}) = \langle\{C_q C_q IS\}\vec{r}|H|\{C_q C_q IS\}\vec{r}\rangle - 2\langle C_q|H|C_q\rangle \tag{4.3a}$$

The mass of the color-octet baryon needed for the formulation of the coupled-channel Schrödinger equation (4.1) is taken to be $M_C c^2 = \langle C_q|H|C_q\rangle$ The questionable notion of a mass of a color-octet baryon is not needed for a single baryon, but only for the propagation of a pair of color-octet baryons in the form $-\frac{\hbar^2}{2}\left(\frac{1}{M_C(1)} + \frac{1}{M_C(2)}\right)\Delta_{\vec{r}}$. This mass M_C drops out everywhere else. The identifications (4.2) and (4.3) specify the content of equ. (4.1) which describes the dynamics of the two-baryon system. Since the transition potentials (4.2) and (4.3) are localized to act at small relative distances, no long-range van-der-Waals force can arise. Admittedly, equ. (4.1) with these identifications is only a primitive approximation of the Born-Oppenheimer type for the full generator-coordinate formulation of equ. (2.7). Without solving (2.7) the validity of its approximative form (4.1) cannot be proven. We simply hope it to be sufficient for a first estimate of many-nucleon forces and currents arising from hidden-color configurations in nuclei.

The coupling potentials between the baryonic channels are abbreviated by the diagrams of figs. 5a to 5c. The Schrödinger equation (4.1) can be generalized in the usual way to many-nucleon systems. There the one-gluon exchange can carry color from one baryon to another. This process is indicated in fig. 5d. If mesons could be interpreted as elementary fields, color-octet baryons have a normal coupling to mesons and therefore a normal one-boson exchange interaction with color-singlet baryons as in fig. 5e. There is an ordinary coupling to the photon given in fig. 6. Quantitatively, the coupling of color-octet baryons with mesons and photons is determined by quark considerations in the same way as the couplings of nucleons and Δ-isobars with mesons and photons are related to each other. Through the couplings of figs. 5 and 6 many-nucleon

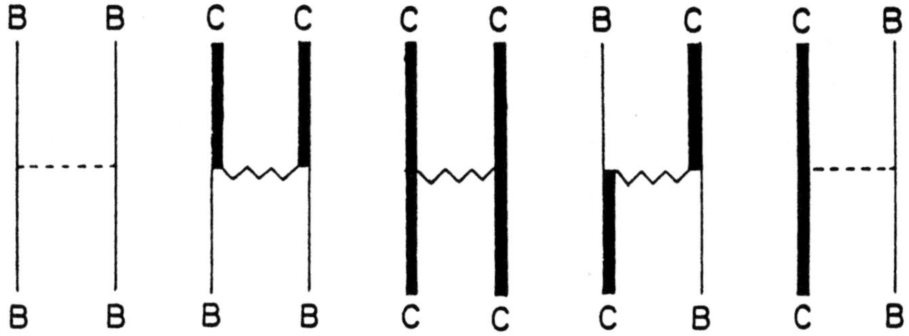

Fig. 5: Force model consisting of meson- and color-exchange between ordinary and color-octet baryons.

Fig. 6: Photon coupling to a color-octet baryon.

forces arise in the many-nucleon system as well as e.m. many-nucleon currents in the interaction with an e.m. probe. The three-nucleon forces are illustrated in fig. 7, they approximate the corresponding three-nucleon forces of fig. 2. The two-nucleon current is illustrated in fig. 8, it approximates the corresponding current of fig. 3.

5. Application in the three-nucleon bound state

The excitation of two nucleons to a hidden-color six-quark state at small relative distances yields novel many-baryon forces and currents. In this section their effect on the ^{3}H binding energy and the ^{3}He charge form factor is studied[10]. The hidden-color state is simulated as described in the previous section.

The hidden-color component in the wave function is calculated in perturbation theory. It is obtained in first order in the potential

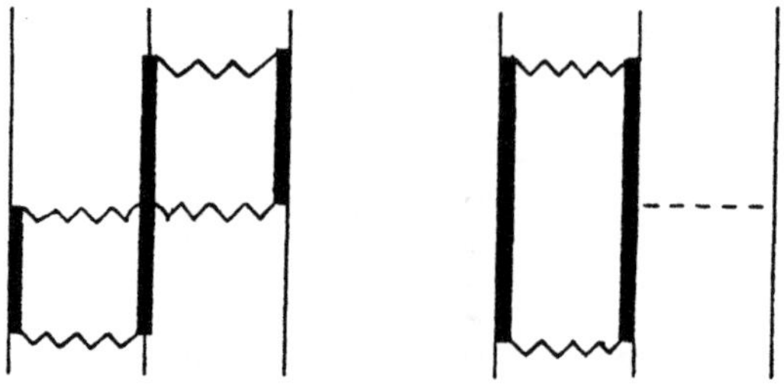

Fig. 7: The three-baryon forces of fig. 2 simulated with color-octet baryons.

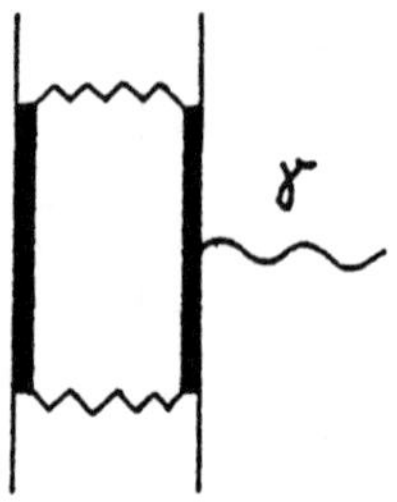

Fig. 8: The two-baryon exchange current of fig. 3 simulated with color-octet baryons.

(4.2) from a purely nucleonic bound-state wave function. In the practical calculation the wave function of the Paris potential is used and together with the potential (4.2) a 0.5% probability for hidden-color components in the three-nucleon bound state arises.

Consistent with this rather small probability the effect of hidden-color configurations on binding energy and e.m. properties of the three-nucleon bound state is modest. In lowest nonvanishing order of perturbation theory the ^3H binding energy is affected by the three-baryon forces of fig. 7 and by the two-baryon dispersion diagramatically shown in fig. 9. The process of fig. 7a is of short range, since it is derived from short-ranged color exchange in each pair. The repulsion in the two-nucleon potential effectively screens this contribution to the three-nucleon force. Its effect is believed to be negligible and therefore has not been calculated. On the other hand the three-baryon force of fig. 7b contains a long-range meson exchange in one pair and should be much more important. Using π-, σ- and ρ-exchange as meson exchange the

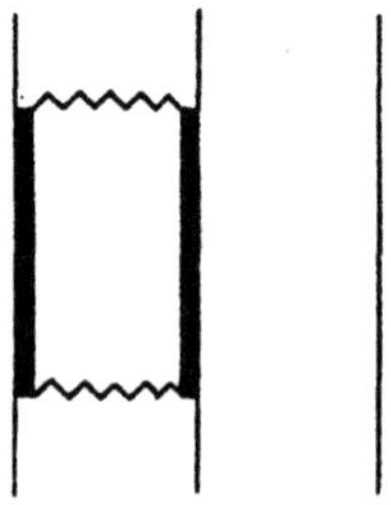

Fig. 9: Two-baryon dispersion weakening the three-baryon force of fig. 7b.

process of fig. 7b in fact yields an additional attraction of -0.8 MeV to the ^{3}H binding energy. This attraction is partly balanced by the two-baryon dispersion of fig. 9 resulting in repulsion of 0.1 MeV. According to this first primitive estimate the theoretical value of -7.4 MeV derived from the pure nucleonic Paris potential is increased by -0.7 MeV through hidden-color effects towards the experimental value of -8.5 MeV. In the same estimate the e.m. exchange current of fig. 8 yields a correction to the ^{3}He charge form factor. In fig. 10 the result including hidden-color contributions (solid curve) is compared to the pure nucleonic one of the Paris potential (dashed curve). The first minimum of the charge form factor is slightly shifted towards higher momentum transfers, whereas the slope around the 2. minimum is improved compared to the result for the pure nucleonic case.

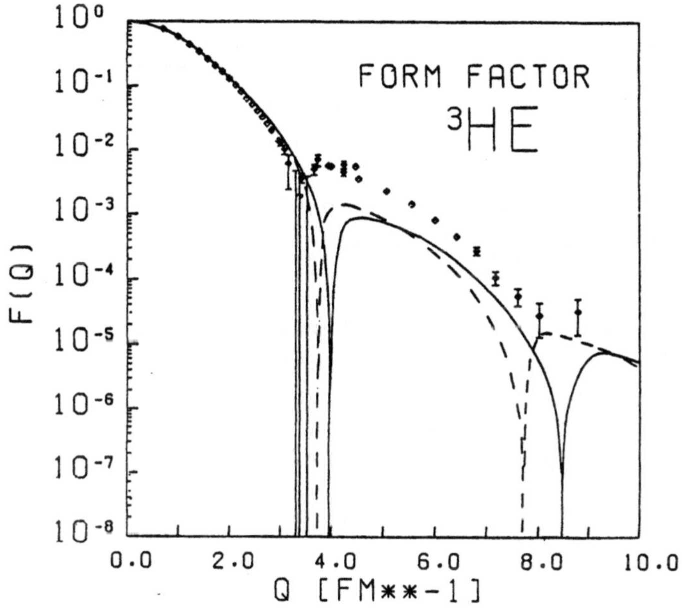

Fig. 10: Charge form factor of ^{3}He. The inclusion of hidden-color corrections (solid curve) is compared to the pure nucleonic result of the Paris potential (dashed curve).

6. Conclusion

The presented model is able to describe many-nucleon forces and currents derived from the quark substructure of nucleons more consistently and more quantitatively than it has been done before. The first results are clearly estimates only, all technical difficulties concerning the correct generator-coordinate treatment of the problem have been avoided in the present description. Many-baryon forces and currents appear small. We take this as a first indication that the traditional picture of the nucleus as an interacting system of nucleons, simple isobars and mesons is basically valid and requires only small corrections from subnuclear degrees of freedom.

References

1. "Mesons in Nuclei", ed. by M. Rho and D. Wilkinson (North-Holland, Amsterdam 1979);
P.U. Sauer, Nuovo Cim. 76A (1983) 309
2. Ch. Hajduk, P.U. Sauer and W. Strueve, Nucl. Phys. A405 (1983) 581;
W. Strueve, Ch. Hajduk and P.U. Sauer, Nucl. Phys. A405 (1983) 620
3. K. Johnson, Acta Physica Polonica B6 (1975) 2060;
N. Isgur, "Soft QCD: Low Energy Hadron Physics with Chromodynamics" in "The New Aspects of Subnuclear Physics", ed. by A. Zichichi, Plenum Press 1980, p. 107
4. P.U. Sauer, "Phase-Equivalent Nonlocal Potentials" in "Nucleon-Nucleon Interactions", AIP Conference Proceedings 41, p. 195
5. M. Harvey, Nucl. Phys. A352 (1981) 301 and 326
6. A. Faessler, F. Fernandez, G. Lübeck and K. Shimizu, Phys. Lett. 112B (1982) 201
7. V.A. Matveev and P. Sorba, Lett. Nuovo Cim. 20 (1977) 435
8. A. Yokosawa, Phys. Rep. C64 (1980) 47;
A. Th. M. Aerts, P.J.G. Mulders and J.J. de Swart, Phys. Rev. D17 (1978) 260
9. A. de Rujula, H. Georgi and S.L. Glashow, Phys. Rev. D12 (1975) 147
10. P.U. Sauer and U.-J. Wiese, to be published
11. M.B. Gavela, A. Le Yaouanc, L. Oliver, O. Pène, J.C. Raynal and S. Sood, Phys. Lett. 82B (1979) 431
12. H. Friedrich, Phys. Rep. C74 (1981) 209

QUANTITATIVE SPECTROSCOPY WITH QUARKS

A. Martin
Theory Division, CERN
Geneva, Switzerland

We discuss heavy quarkonium spectroscopy in potential models, as well as the extension of these potential models to hadrons containing ordinary quarks.

In this talk, I shall not speak at all of nuclear structure in terms of quarks, but only of the description of "elementary" particles in terms of quarks. Initially, the quark model was largely a mnemonic trick to remember which states are allowed and which are forbidden. In the eyes of Murray Gell-Mann, it was a procedure to write currents with the right commutation relations. It is only in 1974, when S. Ting and B. Richter discovered the J/ψ and when people understood that this very narrow particle (60 keV for a mass of 3 GeV !) could not be something else than a heavy quark-antiquark bound state, that things changed. At that moment it was understood that quarks, or, at least heavy quarks like the c quark predicted by S. Glashow-J.Iliopoulos-L. Maiani in 1970, behaved very much like real particles, except for the fact that maybe they cannot be isolated. This was in fact from the very beginning the point of view of George Zweig.

It was natural to describe the $c\bar{c}$ system, of which two levels were known from the very beginning, the J/ψ (mass 3095 MeV) and the ψ' (mass 3685 MeV) as a system of two heavy objects interacting by a potential and obeying the Schrödinger equation. However, unlike what happens in the hydrogen atom, the quark-antiquark potential was not known. However, the theoretical prejudice is that one gluon exchange dominates at short distances while at large distances a linear rising confining potential seems to be the most natural thing to prevent quarks from escaping. So the initial potential used was

$$V = -\frac{4}{3}\frac{\alpha_s}{r} + Cr .$$

The potential model description was tremendously successful in a qualitative and semi-quantitative way:

- the existence of P states between the ψ and ψ' was predicted. A triplet of P states was discovered. The non-observation of the singlet P state is due to the difficulty of producing it.

- A D state, with the same quantum number as the photon, was predicted with an energy slightly higher than the ψ' and was observed in e^+e^- collisions.
- Radial excitations of the S states were predicted and seen.
- Pseudoscalar states η_c and η_c' respectively associated with the vector states $J\psi$ and ψ' were predicted and, after a complicated history, observed experimentally.

In 1977 another heavy quark system was found, the Υ, made of a $b\bar{b}$ pair, as indicated by an indirect measurement of the absolute value of the charge of the constituent quarks. The Υ, with a mass 9459.9 ± ± 0.1 MeV (notice the fantastic precision!) is much heavier than the $J\psi$ and logically the potential description should work still better. Now, in fitting the energy levels of the Υ, we have the extra constraint that the potential should be approximately the same. This is "flavour independence": because QCD is a non-Abelian gauge theory, the quark-quark-gluon coupling is related to the three gluon coupling and, therefore, cannot depend on the particular flavour of the quarks. It is not completely obvious that the resulting potential (if such a notion makes sense) will be exactly flavour independent, but the deviations from flavour independence should go to zero when the mass of the quarks goes to infinity. However, we can try to assume "precocious flavour independence" and try to impose that the $b\bar{b}$ and $c\bar{c}$ potentials be exactly the same.

It is at this point that we enter into the domain of quantitative spectroscopy. The question is what is the quark-antiquark potential. In principle, the quark-antiquark potential could be obtained from lattice gauge theory calculations, and this is on the way. However, lattice gauge theory would have a hard time getting the short distance behaviour, at distances less than the lattice spacing. However, the short distance behaviour can be obtained from asymptotic freedom. This was done first by H. Krasemann and S. Ono[1], J. Richardson[2] and W. Buchmüller, G. Grunberg and S. Tye[3]. One has at short distances:

$$\left.\begin{aligned} V &= -\frac{4}{3}\frac{\alpha_s(r)}{r} \\ \alpha_s(r) &= \frac{12\pi}{25\log\left(\frac{1}{r^2\Lambda_{MS}^2}\right)} \end{aligned}\right\} \qquad (1)$$

Notice that this attenuation of the Coulomb singularity is already not negligible for the T spectrum but becomes crucial for the $t\bar{t}$ system, since we know that the mass of the t quark is superior to 20 GeV.

At larger distances, while H. Krasemann and S. Ono just connect the potential (1) to the linear confining potential with a logarithmic part, as was done before by G. Bhanot and S. Rudaz[4)], J. Richardson and W. Buchmüller (whose more refined short range potential contains two loop effects) have a nice analytic interpolating formula in momentum space.

In contrast with these QCD inspired approaches, one can also take a purely phenomenological point of view and try to find the simplest potential fitting, the $c\bar{c}$ and $b\bar{b}$ spectrum. Historically, the first attempts in this direction were made by Y. Tomozawa, C. Quigg and J.L. Rosner. The latter pointed out that if the spacings between all levels of the $c\bar{c}$ system were the same as the spacings between the levels of the $b\bar{b}$ system, this would be a proof that the potential is exactly $V = C \log(r/r_0^*)$. It is a simple exercise to prove, rescaling T, that if $V = \log r$, the distances between the levels do not depend on the mass of the quarks. The converse is more difficult to prove. In any case, as one can see from Table 1 (which is much more complete now!) this property is not very far from being satisfied by the $c\bar{c}$ and $b\bar{b}$ spectrum. We notice, however, a tendency of the $b\bar{b}$ levels to have a somewhat smaller spacing. To take this into account, we can consider the simplest generalization of the log r potential, $V = A + Br^{\alpha}$, with α slightly positive. This is what I tried[5)], taking

$$V = -8.064 + 6.870\, r^{0.1} \tag{2}$$

where the units are GeV for energies and GeV^{-1} for distances.

In Table 1, we give the level spacings, from the 1^{--} ground state, to the various $L = 0$, $L = 1$ and $L = 2$ states (for $L = 1$, we give the centre of gravity of the triplet states), and the relative leptonic width of the $L = 0$ state, $\Gamma_{ee}(nS)/\Gamma_{ee}(1S)$ calculated "naively" using the formula $\Gamma_{ee} = C|\psi(0)|^2/M^2$. We give the experimental numbers[6)], the theoretical predictions of the Buchmüller potential and the predictions of potential (2). The absolute masses of the various states are not significant since the ground state is adjusted by fixing the effective quark mass.

	$c\bar{c}$			$b\bar{b}$		
1^{--} ground state	3095 MeV			9460 MeV		
Quark mass	Exp.	Buchmüller 1480 MeV	Martin 1800MeV	Exp.	Buchmüller 4870 MeV	Martin 5174 MeV
2S - 1S	589	600	592	561	560	560
$\Gamma_{ee}(2S)/\Gamma(1S)$	0.46 ±0.06	0.45	0.40	0.43	0.44	0.41
3S - 1S	935	1020	937	890	890	900
$\Gamma_{ee}(3S)/\Gamma(1S)$	0.16?	0.32	0.25	0.32	0.32	0.35
4S - 1S	1319?	1380	1185	1116	1160	1140
$\Gamma_{ee}4S/\Gamma(1S)$				0.21	0.26	0.27
1P - 1S	425	420	407	440	430	400
2P - 1S				796	790	780
1D	677	710	692			

Table 1: Excitation energies and relative leptonic widths for the $c\bar{c}$ and $b\bar{b}$ systems compared with the theoretical predictions of Buchmüller et al. and Martin.

Table 1 does not incorporate the data on the pseudoscalar states η_c and η_c' whose spacing to the corresponding vector states are

$$J\psi - \eta_c = 112 \text{ MeV}, \quad \psi' - \eta_c' = 93 \text{ MeV},$$

for which W. Buchmüller et al.[3], as well as previously D. Beavis et al.[7] have predictions of the right order of magnitude but, however, too low by a factor 0.8.

There are many other fits of the data but I have chosen these two as representatives of "theoretical" and "phenomenological" potentials. In both cases, the agreement with experiment is strikingly good. The flavour independence of the potentials is perfect (in fact too perfect maybe!). In the phenomenological potential the number of parameters is only three. If we look closer at the figures, we see that the phenomenological potential is somewhat better for higher L = 0 excitations. This may, however, not be terribly significant since open and closed coupled channel effects are present (remember that the $2m_D = 2 \times 1865$ MeV = 3730 MeV and $2m_B = 2 \times 5274 = 10548$ MeV). What matters more is that the QCD motivated potential reproduces much better the P states, especially the 1P state, than the phenomenological potential. This is presumably due to the fact that a non-singular potential overestimates the mass of the L = 0 states; this, however, is corrected by adjustment of the quark mass and leads to underestimating the mass of the L = 1 states.

In the most ambitious attempts, one tries also to reproduce the separation between the P states due to spin dependent forces. This is especially interesting now that we have two triplets of P states. Generally, the spin forces appear from a reduction of the Breit equation, but the nature of the interaction in the Breit equation is model dependent. We give in Table 2 some theoretical predictions by D. Beavis et al.[7], W. Buchmüller[8] and E. Eichten and F.L. Feinberg[9].

		Exp.	Buchmüller	Eichten Feinberg	Beavis
$c\bar{c}$	$2^{++}-1^{++}$	44 MeV	57 MeV		69 MeV
	$1^{++}-0^{++}$	93 MeV	83 MeV		71 MeV
$b\bar{b}$	$2^{++}-1^{++}$	20 MeV	19 MeV	26 MeV	23 MeV
χ_b	$1^{++}-0^{++}$	21 MeV	28 MeV	25 MeV	24 MeV
	$2^{++}-1^{++}$	16 MeV	18 MeV	17 MeV	17 MeV
χ_b'	$1^{++}-0^{++}$	21 MeV	21 MeV	17 MeV	24 MeV

We see that orders of magnitude are reasonable but agreement is not perfect in any of the three calculations. For $c\bar{c}$, QCD sum rules also give reasonable numbers but we have no time to develop this approach.

We conclude that the potential model description of $b\bar{b}$ and $c\bar{c}$ works very well, that the potential is very well fixed in the interval 0.1 to 1 Fermi, where the two potential models coincide (after a shift due to quark mass differences), and that the potential which has the correct QCD behaviour at short distances is favoured. Also flavour independence is very well tested.

Now we want to move towards lighter quark masses, because, after all, the non-relativistic description of $c\bar{c}$ works very well as far as energies are concerned in spite of the fact that $\langle v^2/c^2 \rangle \simeq 0.25$ (E1 or M1 transition rates are not as satisfactory but relativistic corrections can improve the situation). So if it works for $c\bar{c}$ why not try to fit $s\bar{s}$ with potential (2)? Naturally, one can reproduce the ϕ mass, 1020 MeV by adjusting the s quark effective mass to be 518 MeV. However, there are checks:

1) One predicts a ϕ', radial excitation of the ϕ, 615 MeV above the ϕ, while an experimental candidate from Orsay is 630 MeV above the ϕ.

2) Once the strange quark mass is fixed, one can predict the F and the F* masses:

$$m_F = 1990 \text{ MeV} \qquad m_{F^*} = 2110 \text{ MeV}.$$

Experiments give masses for the F ranging from 2020 MeV to 1970 MeV[10].

This success has encouraged J.-M. Richard[11] to use potential (2), together with a phenomenological spin-spin force of the Fermi type adjusted to the $J\psi-\eta_c$ mass difference to calculate the mass of the Ω^- which is a sss system. The recipe giving the quark-quark potential inside a baryon which is a colour singlet is

$$V_{QQ} = \frac{1}{2} V_{Q\bar{Q}} \tag{3}$$

This is suggested by what happens for one gluon exchange or more generally colour octet exchange, and has the advantage that it is insensitive to the choice of quark masses: if one neglects changes in kinetic energies, a fit of the $Q\bar{Q}$ system with quark mass m_Q and potential $V_{Q\bar{Q}}$ will be stable in the change

$$\begin{aligned} m_Q &\rightarrow m_Q + \Delta \\ V_{Q\bar{Q}} &\rightarrow V_{Q\bar{Q}} - 2\Delta . \end{aligned} \tag{4}$$

The prescription (3) is such that also the three-quark system mass will be left approximately invariant in the change (4). Anyway, J.-M. Richard has taken seriously this prescription and found

$$m_{\Omega^-} = 1665 \text{ MeV},$$

while experiment gives

$$m_{\Omega^-} = 1672 \text{ MeV}.$$

This kind of success may, of course, be fortuitous but it leads to reconsidering the systems containing even lighter quarks, i.e., up and down quarks, mesons and baryons, and trying to fit them in a non-relativistic way with an effective mass of about 300 MeV for the up and down quarks and a flavour independent potential which may or may not be constrained to satisfy property (3). In fact, in the past, this has been done to a certain extent by A. De Rújula, H. Georgi and S. Glashow[12)] and they have been incredibly successful, justifying to lowest order in $m_s - m_{u,d}$ the Gell-Mann-Okubo mass formulae for the octet and the decuplet, showing that the spin-spin force derived from one gluon exchange has the right qualitative features:

$$m_{J=1} > m_{J=0} \qquad \text{for mesons}$$

$$m_{J=3/2} > m_{J=1/2} \qquad \text{for baryons,}$$

and leads, with $m_s/m_u = 0.6$, to a lot of interesting predictions like

$$\frac{M_{K^*} - M_K}{M_\rho - M_\pi} = 0.6 \qquad (\text{exp. } 0.64)$$

$$\frac{M_{D^*} - M_D}{M_{K^*} - M_K} = 0.29 \qquad (\text{exp. } 0.35) \tag{5}$$

$$\frac{M_{\Xi^*} - M_\Xi}{M_{\Sigma^*} - M_\Sigma} = 1 \qquad (\text{exp } 1.12)$$

$$\frac{1}{2}\,\frac{2M_{\Sigma^*} + M_\Sigma - 3M_\Lambda}{M_\Delta - M_N} = 1 \quad (\text{exp } 1.04) \tag{6}$$

$$\frac{M_\Sigma - M_\Lambda}{M_\Delta - M_N} = \frac{2}{3}\left(1 - \frac{m_u}{m_s}\right) = 0.26 \quad (\text{exp } 0.26)$$

$$\frac{M_{\Sigma_c} - M_{\Lambda_c}}{M_\Delta - M_N} = 0.55 \quad (\text{exp } 0.56),$$

with $m_c = 1.8$, and naturally, the magnetic moment predictions of the quark model:

$$\mu_N/\mu_p = -0.67 \qquad (\exp\ -0.68)$$
$$\mu_\Lambda/\mu_p = -0.2 \qquad (\exp\ -0.22)$$
$$\mu_{\Sigma^+}/\mu_p = 0.95 \qquad (\exp\ 0.85 \pm 0.01)$$
$$\mu_{\Sigma^-}/\mu_p = -0.38 \qquad (\exp\ -0.50 \pm 0.09)$$
$$\mu_{\Xi^-}/\mu_p = -0.16 \qquad (\exp\ -0.25 \pm 0.01)$$
$$\mu_{\Xi^0}/\mu_p = -0.49 \qquad (\exp\ -0.45 \pm 0.05)$$

What is missing in the list is μ_{Ω^-} ! It could be measured and this measurement would be very interesting !

However, A. De Rújula et al. did something very crude: they worked to lowest order in the symmetry breaking due to the u, d and s mass differences which meant that they could use a variational approach and disregard the distortions of the space wave function due to the breaking. This might be a very bad approximation, especially if heavy charmed or beautiful quarks are present. Calculations using a flavour independent potential for mesons and baryons, and the Schrödinger equation or a modified version, in which relativistic kinematics are used, have been made by A. Bhadury et al.[13)] and by D.P. Stanley D. Robson[14)] first. However, it is not very easy to get a feeling from these calculations which end with long lists of numbers predicted but are sometimes criticized because of the many parameters introduced. Also, it is difficult to be certain that calculations are sufficiently accurate.

I proposed to present the results in a way in which the improvements, with respect to the "naîve" A. De Rújula et al. treatment are manifest so as to measure the effects of the wave function distorsion by symmetry breaking by the quark effective masses. For instance, the expression in the Gell-Mann-Okubo mass formula is not exactly zero and one should try to reproduce this. Similarly, one can study the deviations of the right-hand side of Eqs. (5) and (6) from unity. Two groups have done this S. Ono and F. Schöberl using variational methods with superpositions of Gaussians[15)], J.-M. Richard and P. Taxil using an expansion in hyperspherical harmonics and solving a system of coupled differential equations[16)] (a very efficient procedure for quark-quark potentials). Here, I report the J.-M. Richard-P. Taxil results (the others are very similar).

The Gell-Mann-Okubo mass formula for the octet gives:

$$2(M_N + M_\Xi) - M_\Sigma - 3M_\Lambda = -25 \pm 5 \text{ MeV}$$

experimentally, where the uncertainty is due to the spread due to electromagnetic effects.

A naïve estimate disregarding the distortion of the wave functions but only the spin-spin interaction of A. De Rújula et al. gives + 35 MeV ! However, J.-M. Richard and P. Taxil using a reasonable soft potential of the type (2) get - 22 MeV, in perfect agreement with experiment. Similarly, they get

$$\frac{M_{\Xi^*} - M_\Xi}{M_{\Sigma^*} - M_\Sigma} = 1.08 \quad (\text{exp. } 1.12)$$

$$\frac{1}{2}\,\frac{2M_{\Sigma^*} + M_\Sigma - 3M_\Lambda}{M_\Delta - M_N} = 1.07 \quad (\text{exp } 1.04)$$

It is encouraging to see that these results all go into the right direction

Finally, I want to report another calculation of J.-M. Richard and P. Taxil to obtain the mass of the newly discovered hadron (csu) at CERN[17], with a mass of 2.46 GeV. A previous calculation by N. Isgur[18], using a harmonic oscillator potential gave 2.50 MeV for this state, a result already quite nice.

J.-M. Richard and P. Taxil[19] have taken a soft potential of the type (2) allowing for a change of the coefficients and fitting the nucleon, the Δ, the Ω^- and the Λ_c, and choosing $m_{u,d}$ = 300 MeV. They get

$$m_s = 600 \text{ MeV}, \quad m_c = 1895 \text{ MeV}$$

$$V_{qq} = \frac{1}{2}\left[-8.3377 + 6.9923\, r^{0.1}\right]. \tag{7}$$

Notice the constants in (7) are not very different from those in (2). Their results are listed in Table 3.

	theory	exp.		theory	exp.
N	input	0.939 GeV	Ω^-	input	1.672 GeV
Δ	input	1.232 GeV	Λ_c	input	2.283 GeV
Λ_0	1.111 GeV	1.115 GeV	Σ_c	2.443 GeV	2.450 GeV
Σ	1.176 GeV	1.193 GeV	Σ_c*	2.542 GeV	
Ξ	1.304 GeV	1.318 GeV	A(csu)	2.457 GeV	2.460 GeV
Σ*	1.392 GeV	1.383 GeV	S(csu)	2.558 GeV	
Ξ*	1.538 GeV	1.533 GeV	S*(csu)	2.663 GeV	

The states A and S have spin-1/2 but correspond to different spin structures. In the limit $m_s = m_u$, A reduces to Λ_c while S reduces to Σ_c. S* has spin-3/2.

What do we conclude from these results ? A soft potential is better than a harmonic oscillator potential. However, J.-M. Richard and P. Taxil have also tried another exercise. They take three-body forces inspired by the string picture instead of two-body forces, with a potential which is the sum of the distances of the three quarks to a point minimized over the location of the point, plus a constant. They get results almost as good.

We see therefore that this kind of approach has a great predictivity but, on the other hand, it does not allow us to solve all problems concerning the nature of the interaction. We think that it should be pursued in spite of the fact that relativistic effects are disregarded. Of particular interest are, for instance, the corrections to the naîve quark model predictions for magnetic moments because at present we begin to have very precise measurements of the magnetic moments, for instance, in the case of Σ^+ [20].

REFERENCES

1) H. Krasemann and S. Ono, Nucl. Phys. B154, 283 (1979).

2) J.L. Richardson, Phys. Lett. 82B, 272 (1979).

3) W. Buchmüller, G. Greenberg and S.H.H. Tye
W. Buchmüller and S.H.H. Tye, Phys. Rev. D24, 132 (1981).

4) G. Bhanot and S. Rudaz, Phys. Lett. 78B, 119 (1978).

5) A. Martin, Phys. Lett. 100B, 511 (1981).

6) The latest data on the Υ system can be found in the review by P.M. Tuts, Electron Photon Conference, Cornell (1983), to be published.

7) D. Beavis et al., Phys. Rev. D20, 743 and 2345 (1979).

8) W. Buchmüller, Phys. Lett. 112B, 479 (1982).

9) E. Eichten and F.L. Feinberg, Phys. Rev. Lett. 43, 1205 (1979); Phys. Rev. D23, 2724 (1981).

10) S.L. Wu, Nato Advanced Study Institute on Particles and Fields, Cargèse 1983, to be published.

11) J.-M. Richard, Phys. Lett. 100B, 515 (1981).

12) A. De Rújula, H. Georgi and S. Glashow, Phys. Rev. D12, 147 (1975).

13) A. Bhadury et al., Phys. Rev. Lett. 44, 3180 (1980).

14) D.P. Stanley and D. Robson, Phys. Rev. D21, 3180 (1980);
D.P. Stanley and D. Robson, Phys. Rev. Lett. 45, 235 (1980).

15) S. Ono and F. Schöberl, Phys. Lett. 118B, 419 (1982).

16) J.-M. Richard and P. Taxil, Marseille CPT preprint (1983).

17) S.F. Biagi et al., Phys. Lett. 122B, 455 (1983).

18) N. Isgur, New Flavours and Hadron Spectroscopy, Proc. XVIth Rencontre de Moriond, (Les Arcs, France, 1981), eds. J. Tran Tranh Van (Editions Frontières, Dreux, France, 1981).

19) J.-M. Richard and P. Taxil, Phys. Lett. 128B, 453 (1983).

20) C. Ankelbrandt et al., Fermilab preprint 83/52 Exp, to appear in in Phys. Rev. Lett.

NUCLEAR FORCES WITHIN A CONSISTENT MESON-EXCHANGE MODEL

R. Machleidt
Institut für Theoretische Kernphysik der Universität Bonn
Nußallee 14-16, D-5300 Bonn, West-Germany

Abstract

In this contribution we present a meson-exchange model for the NN-interaction which includes irreducible diagrams up to fourth order in a consistent way. These diagrams contain in particular an explicit determination of the 2π-exchange taking into account virtual Δ-excitation and direct $\pi\pi$-interaction. This part of the model agrees quantitatively with results obtained from dispersion theory which in turn are based on the analysis of πN- and $\pi\pi$-scattering data. A detailed interpretation of the lower partial wave phase-shifts of NN-scattering requires, however, (apart from the well-known OBE contributions) the introduction of additional irreducible diagrams containing also heavy boson exchange, in particular the combination of π and ρ. In the framework of this consistently extended meson model (it contains all terms up to an exchanged mass of 1 GeV) an accurate description of the NN--scattering data below 300 MeV laboratory energy as well as the deuteron data is achieved; the numerical results are definitely superior to well-known simplified boson exchange models, e.g. the one--boson exchange. Our enlarged viewpoint has also important consequences within the theory of nuclear matter due to our explicit determination of the 2π-exchange.

1. Introduction

Nowadays it seems to be generally accepted that the fundamental theory of strong interaction is, in fact, quantum chromodynamics (QCD). On that basis, the nucleon-nucleon (NN) interaction has to be considered as being in principle completely determined by quark and gluon exchanges. However, at the moment it appears by no means possible to give a fully satisfactory quantitative expression for the whole nuclear force in this way. So far, there are various promising attempts to describe the short range part of the NN-interaction[1], whereas the medium and long range part which plays the major role within nuclear physics can presently not be understood from this viewpoint.

For this reason, we have - at least for the time being - to step back from the high-brow aim of a complete quantum-chromodynamical treatment of the NN-interaction. However, we can attempt to pursue our aim in the framework of a suitable approximation scheme which is based on QCD. Noticing first of all that the masses of all relevant hadrons are now in general understood within the framework of lattice QCD[2], we may use them as parameters of the approximative concept. In very much the same way, we may further consider boson-nucleon vertices as a natural and effective description of a most complicated n-quark-reaction[3]. Thus, we are lead to a boson-exchange model of nuclear forces which, however, is related to the fundamental theory of strong interaction. It differs from the old boson theory, which dates back to Yukawa[4], in that it does not assume the mesons as fundamental fields. Within the framework of this new interpretation where the quantum fields are just used for a phenomenological description of the various bosons, it appears natural (and even compulsory) to introduce within the explicit expressions for the vertices characteristic form-factors (so far called regularization terms) which by now play the role of relevant physical parameters: They may be visualized as a consequence of the extension (in space) of the nucleons and might be - to some extent - related to the bag-radii (compare, however, the contribution of G. Miller in this volume). The use of cutoffs for the meson-nucleon vertices in our model appears natural from this new viewpoint and has its origin in the bag-structure of hadrons.

A consistent meson model taking into account all relevant exchange-diagrams has, up to date, never been accomplished. Most existing models are restricted to single meson exchanges and the more extended ones may include just the two-pion-exchange. It is, therefore, the aim of our work presented in this report to attempt a meson exchange model for the NN-interaction which is consistent inasmuch as it includes all diagrams up to an exchanged total meson-mass of ≃ 1 GeV. This upper limit is reasonable because the cutoffs, which have to be applied according to our discussion in the last paragraph, are in the order of 1.0-1.5GeV. The necessity of this limitation appears as a direct consequence of the new (phenomenological) interpretation of boson-exchange. On the basis of such a model it now becomes possible to check reliably the extent to which this 'effective meson theory' of nuclear forces is valid and constitutes a realistic concept.

Apart from a really satisfactory description of all relevant phase-shifts of NN-scattering in the interval of 0-300 MeV (the effects on the various phases of the different exchange-terms, which according to our general rule have to be included, will be discussed in detail) this model leads at the same time to the interpretation of characteristic properties of nuclear matter: The 2π-exchange terms lead in a natural way to a characteristic density dependent quenching of the nuclear force within matter which is needed for a consistent and simultaneous description of light and heavy nuclei. Furthermore, the model offers the opportunity of taking into account the effect of virtual nuclear Δ-excitations within matter in a systematic way: This yields a natural explanation for the recent quark structure function measurements in iron[5).]

2. Formalism

As in subsequent sections many diagrams will occur, we have to state once the rules of the game by which we evaluate them. We use the non-covariant or "old-fashioned" perturbation theory. It has been suggested to us by D. Schütte[6)], and allows us to go, in a consistent way, from the two-body to the many-body problem. It further provides the possibility to take effects of mesonic variables in nuclear matter into account. The retardation in the propagator of non-covariant perturbation theory turns out to be essential for a realistic size and range of isobar contributions to the NN-interaction[7] and for the strength of the tensor force provided by pion-exchange. In non-covariant perturbation theory all possible time-orderings have to be taken into account separately. Therefore, the diagrams depicted in this report have to be considered as an abbreviated notation, just standing for all possible time-orderings. Anti-particle contributions in intermediate states are left out ("pair-suppression"[8]).

The Lagrangians of subsequent interactions are listed here in conventional notation (explanations can be found e.g. in ref. 7):

$$L_{ps} = \sqrt{4\pi} g_{ps} i\bar{\psi}\gamma_5\psi\varphi_{ps}$$

$$L_s = \sqrt{4\pi} g_s \bar{\psi}\psi\varphi_s$$

$$L_v = \sqrt{4\pi}\{g_v\bar{\psi}\gamma_\alpha\psi\varphi_v^\alpha + \frac{f_v}{4M}\bar{\psi}\sigma_{\mu\nu}\psi(\partial^\mu\varphi_v^\nu - \partial^\nu\varphi_v^\mu)\}$$

$$L_{N\Delta\pi} = \sqrt{4\pi}\,\frac{f_{N\Delta\pi}}{m_\pi}\,\bar{\psi}\vec{T}\psi_\mu\partial^\mu\vec{\varphi}_\pi + \text{h.c.}$$

$$L_{N\Delta\rho} = \sqrt{4\pi}\, i\,\frac{f_{N\Delta\rho}}{m_\rho}\,\bar{\psi}\gamma_5\gamma_\mu\vec{T}\psi_\nu(\partial^\mu\vec{\varphi}_\rho^\nu - \partial^\nu\vec{\varphi}_\rho^\mu) + \text{h.c.}$$

(In the 3 first lines the well-known isospin dependence is suppressed). The simplest contribution is the one-meson-exchange which is displayed in fig. 1; as those contributions are well-known, they will not be discussed further.

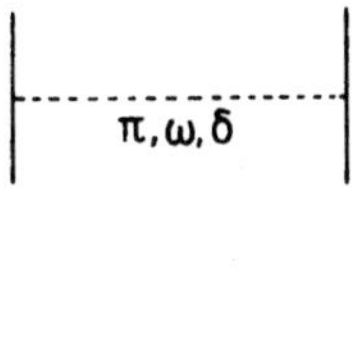

Fig. 1: One-meson-exchange contributions to the NN--interaction taken into account in this work.

3. The 2π-exchange model

Our model for the 2π-exchange contribution to the NN-interaction is shown in fig. 2

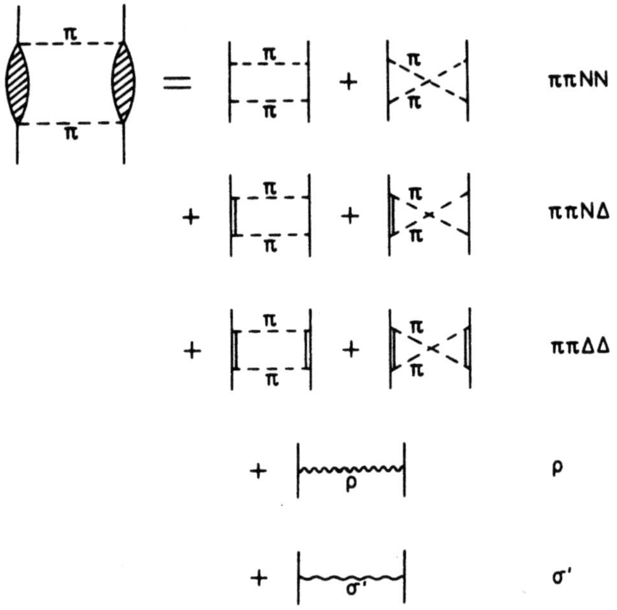

Fig. 2

Our model for the 2π-exchange contribution to the NN-interaction. A full line represents a nucleon and a double line a Δ(1232)-isobar. Further explanations are to be found in the text.

The various crossed box diagrams have to be taken into account as they are non-negligible[9)] and help to provide an iso-scalar character for the 2π-exchange contribution (at least in higher partial waves). An almost iso-scalar character is suggested by results from dispersion theory. The six upper diagrams of fig. 2 represent the uncorrelated 2π-exchange. However, there also exists a strong interaction between two pions: This gives rise to the well-known ρ-meson, if the two interacting pions are in a relative P-wave. On the other hand the relatively strong $\pi\pi$-S-wave interaction does not lead to a resonance. However, Durso et al.[10] have shown that the three diagrams on the left hand side of fig. 3 may well be approximated by the exchange of a scalar-isoscalar boson which we will denote by σ'. In the work quoted here the $\pi\pi$-S-wave interaction had first been determined from the empirical $\pi\pi$-S-wave phase-shifts of $\pi\pi$-scattering. σ' is thus introduced in a well-defined way in contrast to the σ used in former boson--exchange models.

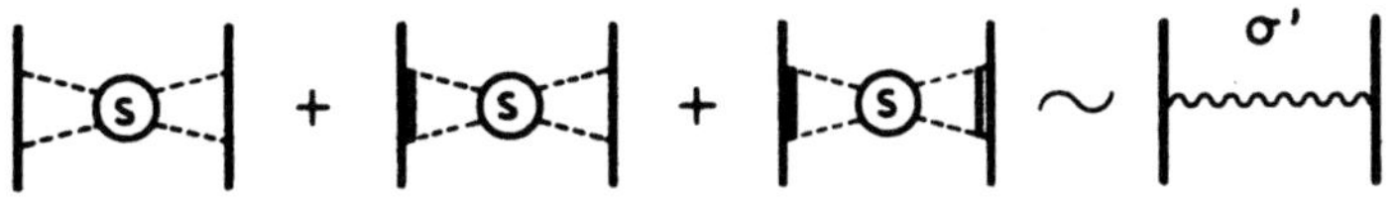

Fig. 3

Correlated 2π-exchange contributions considered in the work of Durso et al.[10]. The circled S symbolizes the $\pi\pi$-S-wave interaction adjusted to empirical $\pi\pi$-S-wave scattering. Further notation as in fig. 2.

The alternative way to derive the 2π-exchange contribution to the NN-interaction consists in the introduction of dispersion theory, in which empirical information of πN- and $\pi\pi$-scattering is used in order to evaluate the amplitude $N\bar{N} \rightarrow \pi\pi$[11]. We perform a quantitative comparison with the results from this latter approach. For this purpose we consider sufficiently high partial wave phase-shifts of NN-scattering, such that there is no cutoff-dependence. The contribution from our model in these higher partial waves is then uniquely determined by the πNN- and πNΔ-coupling constant, the ρNN- and ωNN-coupling as well as the $\pi\pi$-S-wave contribution given by σ'. (All meson parameters used in this work are listed in section 6).

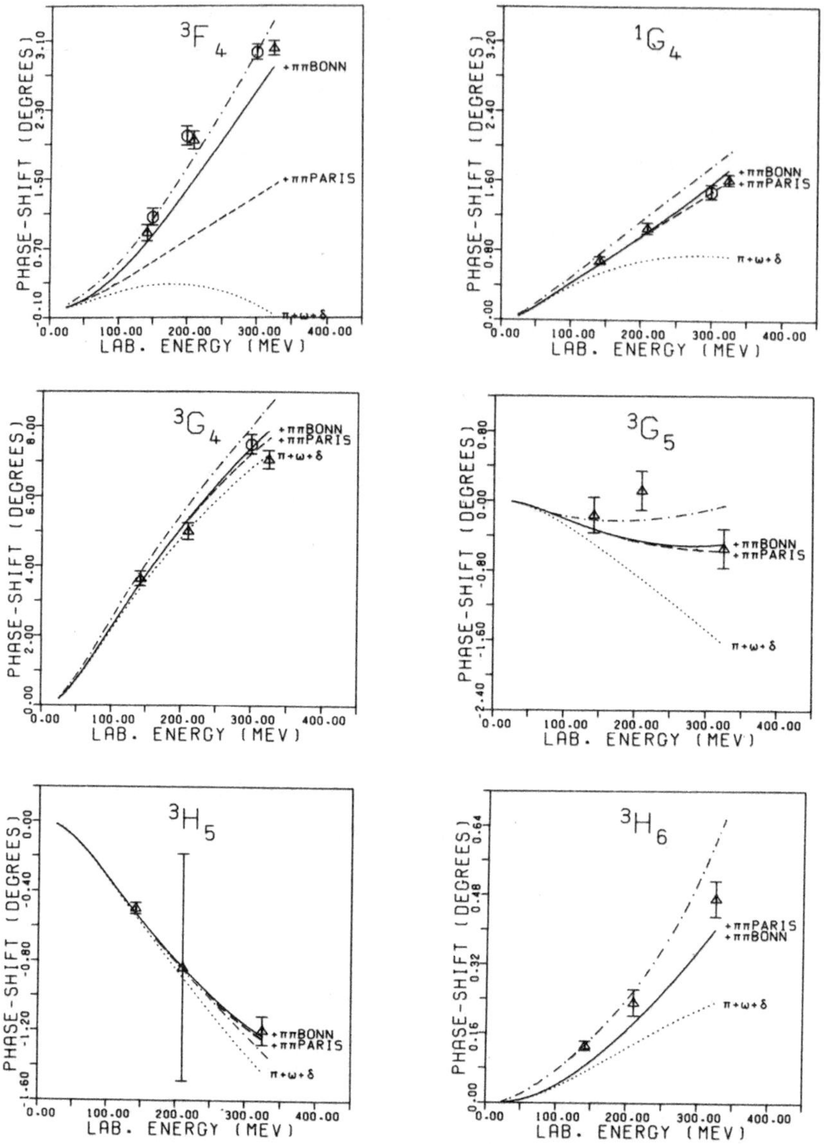

Fig. 4: Some higher partial wave phase-shifts of NN-scattering. The full line contains the contribution from our model for the 2π-exchange shown in fig. 2 (indicated by +ππ Bonn), the dashed line the dispersion--theoretic result taken from ref. 11 (indicated by +ππ Paris). In both curves the same one-π, -ω and -δ exchange, which is also displayed on its own in the dotted curve, is included. (The empirical data plotted are explained in the caption of fig. 11). The dash-dot curve denotes the result for the parametrized Paris-Potential which should be in quantitative agreement with the dashed curve. However, in e.g. $^3F_4$ and $^3H_6$ the original result for the 2π-exchange obtained by that group has obviously been changed by up to 100% to achieve a better agreement with the phase-shift data in some partial waves.

Fig. 4 demonstrates a close agreement of our 2π-exchange contribution with the dispersion-theoretical result as well as the empirical NN--phase-shifts. This double agreement confirms that we have chosen a physically reasonable model. At the same time this proves the consistency of our basic assumptions. In addition, our method has the advantage to permit an explicit determination of the virtual Δ-admixture which plays a decisive role in the theory of nuclear matter. In fact, nuclear binding is to an overwhelming part due to this admixture and the ππ-S-state interaction represented here by σ'. After having checked our model for the 2π-exchange successfully in higher partial waves, we proceed now to states of lower angular momentum. We show first the various contributions of the 2π-exchange diagrams in a lower partial wave phase-shift compared to a higher one in fig. 5.

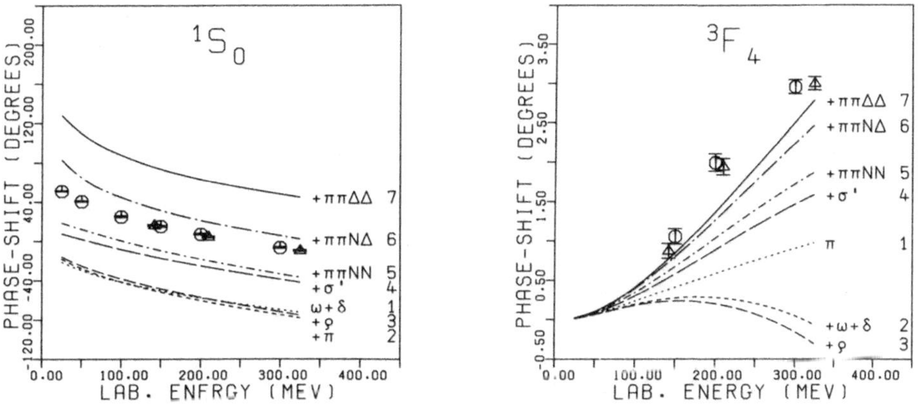

Fig. 5

$^1S_0$ and $^3F_4$ partial wave phase-shifts displaying the individual contributions to the 2π-exchange, which are added up successively in the denoted order (numbers at the left hand side). The short notation for the various contributions is defined in fig. 2.

What can be seen in the $^1S_0$-state is typical for all lower partial waves: The 2π-exchange is too attractive, i.e. the curve Nr. 7 is too high (in contrast to the behaviour in higher phases like $^3F_4$). As the contributions are sensitive to the cutoff regularization of the πNN- and πNΔ-vertex in low partial waves, one may argue that the over-

-attraction might be due to a wrong choice of the cutoff-mass. Therefore, we demonstrate in fig. 6 the cutoff-dependence of the whole 2π--exchange contributions.

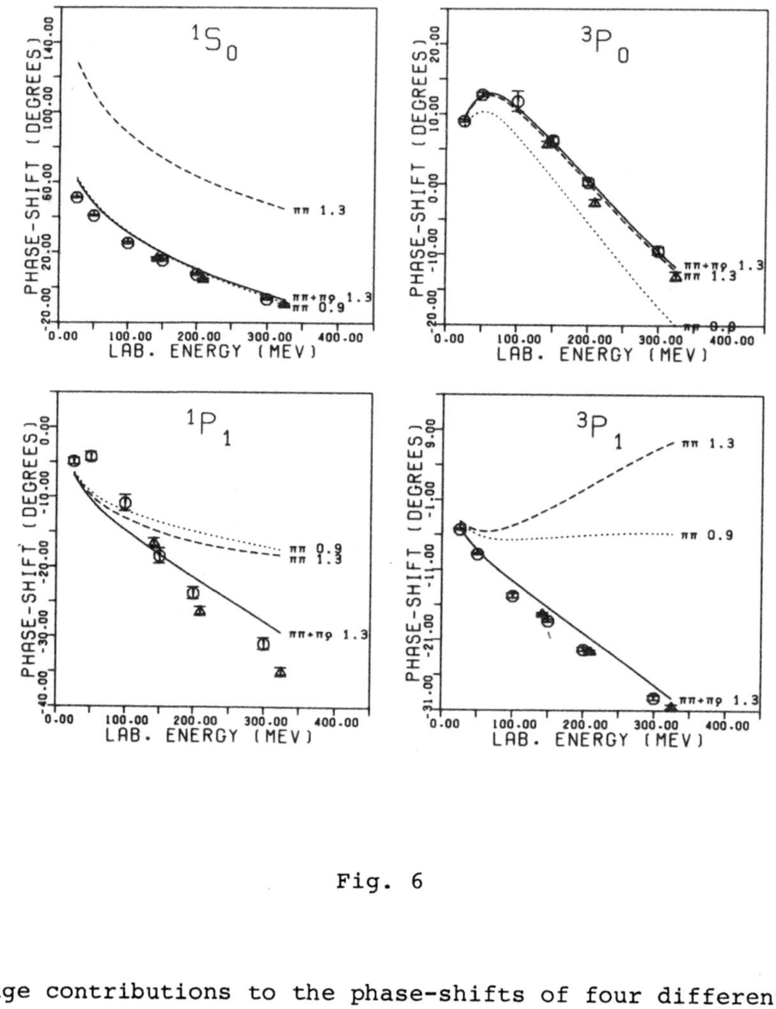

Fig. 6

2π-exchange contributions to the phase-shifts of four different low partial waves with two different choices of the cutoff-mass, Λ_π, regularizing the pion-vertex, namely Λ_π = 1.3 GeV (denoted in fig. 6 by ππ 1.3) and Λ_π = 0.9 GeV (ππ 0.9). For the former case the effect of the additional πρ-contribution (compare section 4) is also demonstrated (ππ+πρ 1.3). In addition all curves contain the one-meson--exchange of π, ω and δ.

It is seen from fig. 6 that e.g. a simultaneous fit of the $^1P_1$ and the $^3P_1$-phase-shift can never be achieved, as a stronger cutoff lowers the $^3P_1$ and raises the $^1P_1$, although both phase-shifts had to be lowered for a closer agreement with the empirical phase-shift data. Also a consistent fit of ^1S_o and $^3P_1$ is obviously impossible: When the ^1S_o-phase-shift is correct (namely for $\Lambda_\pi = 0.9$ GeV), the $^3P_1$ is still considerably too high.

The conclusion states that a model consisting of one-meson- and 2π-exchange only is unable to describe the empirical NN-data if the concept of regularizing the vertices by cutoffs is applied in a consistent way.

4. The $\pi\rho$-contribution

The failure in low partial waves of the model developed so far is no reason to give up. Various meson-exchange contributions, which have to be taken into account according to our consistency rule developed in the introduction, have been neglected so far. As we already realized on the one-meson-exchange level, π and ρ play the role of counterparts. It seems therefore reasonable to include next the two boson-exchange diagrams of π and ρ expecting them to counterbalance $\pi\pi$-contributions.

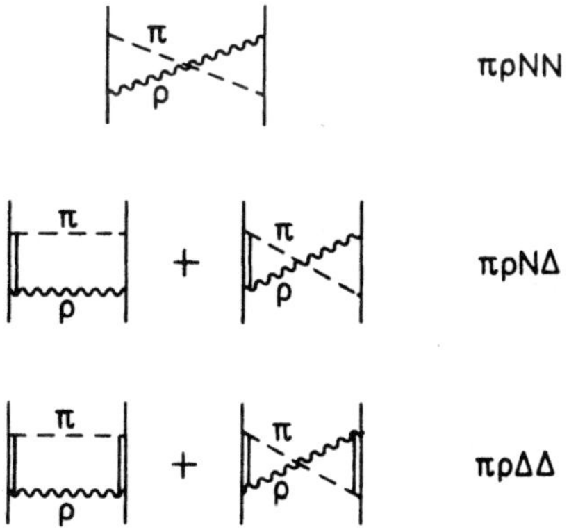

Fig. 7: $\pi\rho$-contributions to the NN-interaction considered in this work.

Fig. 7 displays the contributions to be considered in analogy to the diagrams of uncorrelated 2π-exchange. (The diagrams shown represent the contributions to the kernel in the Lippmann-Schwinger-type equation).

(The correlated 3π-exchange is represented by the ω which we took already into account on the one-boson-exchange level). The typical and large contributions are displayed in detail in fig. 8 for two examples.

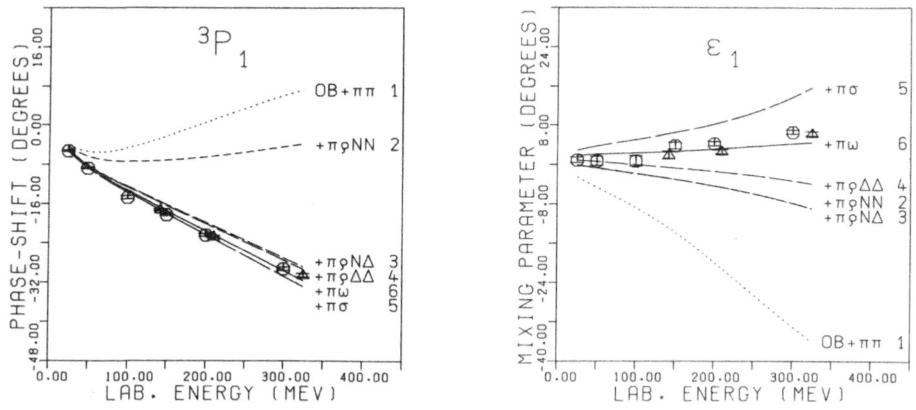

Fig. 8

Individual $\pi\rho$ as well as $\pi\sigma$ and $\pi\omega$ contributions in the $^3P_1$-phase-shift and the ε_1-mixing parameter. The individual contributions are added up successively according to the numbers given at the end of the curve labels. The abbreviated notation for the contributions is explained in fig. 7 and 10. (The curves 5 and 6 which correspond to additional 2 boson-exchanges are discussed in section 5).

The global effect of all $\pi\rho$-diagrams in other low partial waves have already been demonstrated in fig. 6 curve "$\pi\pi+\pi\rho$ 1.3" in comparison to curve "$\pi\pi$ 1.3". Clearly, the $\pi\rho$-contributions are absolutely needed for a quantitative description of low partial waves if a realistic model for the 2π-exchange is applied.

The fact that the $\pi\rho$-contributions are of relevance had been pointed out already in the work of Durso et al.[14]. In this work, however, the $\pi\rho$-contribution was suggested as an instrument to reduce the ω-coupling, which is always rather large in OBE-models if compared to the SU(3) prediction. Our findings are that this suggestion is not realistic for two reasons: First, the $\pi\rho$-contribution sometimes varies tremendously from state to state (compare e.g. the contribution in $^3P_0$ and $^3P_1$ of fig. 6) which is not the case with the ω; second, the over-attraction of the 2π-exchange in low partial waves is such that in addition to a rather large ω-coupling further repulsion is needed. Therefore, the effect of the $\pi\rho$-contribution is to function as a counterpart of the 2π-exchange contribution and not to partially provide the short range repulsion of the NN-interaction which anyhow should be about equally strong in all partial waves.

At this stage of the development of our model it is instructive to ask the question, what the fictitious scalar isoscalar σ-boson of the former one-boson-exchange (OBE) model[15] stood for. The old belief that it replaced the sum of all correlated and uncorrelated irreducible 2π-exchange contributions is true only for high partial waves (where the $\pi\rho$-contributions are negligible because of their short-range); in low partial waves the 2π-exchange contribution appears by no means to be isoscalar, whereas after its smoothening out by the $\pi\rho$-exchanges an approximation by a one-σ-exchange appears possible, as demonstrated in fig. 9.

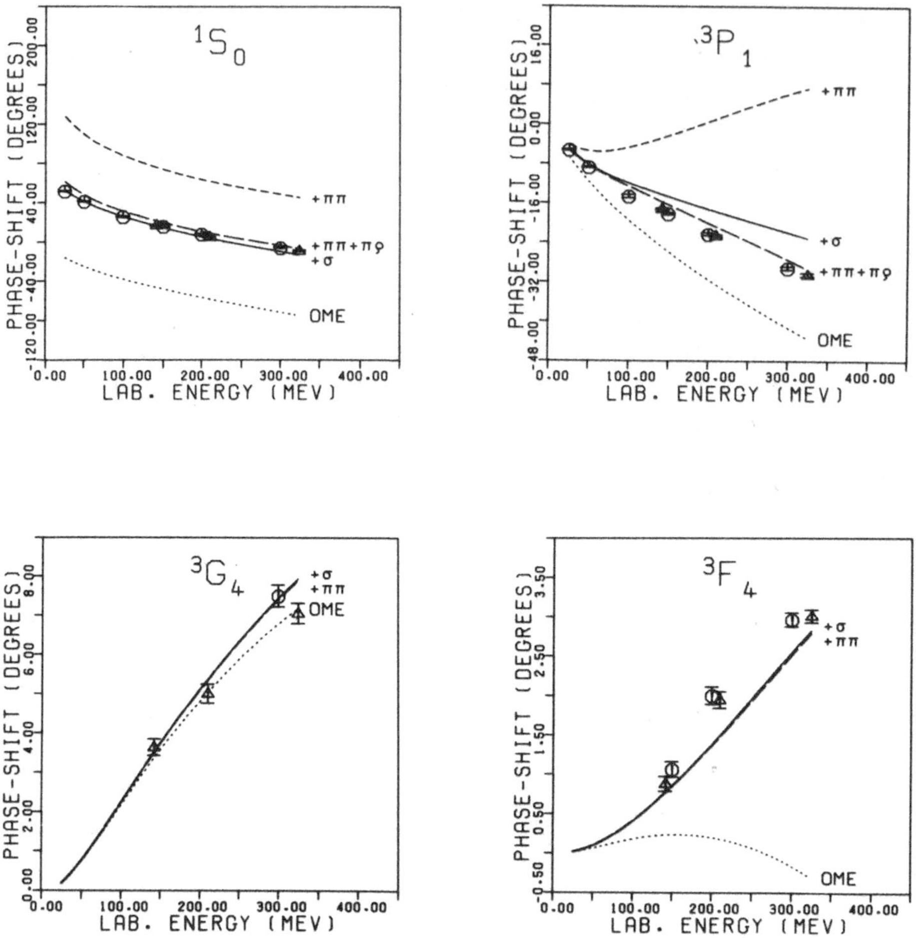

Fig. 9.

The approximation of the 2π-exchange plus πρ-exchange by a fictitious σ-boson (g_σ^2 = 11.12, m_σ = 575 MeV) in two high and two low partial waves. The dotted curve contains all iterated one-meson-exchange (OME) contributions (i.e. π,ω,δ,ρ). The other contributions, displayed with an obvious notation, are added on top of the OME. In higher partial waves the πρ-contribution is left out because it is negligible.

5. Further irreducible contributions of fourth order

Guided by our rule that we are going to consider all diagrams up to an exchanged mass of 1 GeV which seems sensible as the cutoffs applied in our model are of the order 1.0-1.5GeV we have to take into account further diagrams. One irreducible contribution of that kind will certainly be the $\pi\omega$-contribution (fig. 10). However, from the last section we know that the antagonist structure of the one-boson--exchanges may continue in higher irreducible meson-exchanges. The counterpart to the ω is the intermediate-range attraction, which in our explicit meson-exchange model is provided by the 2π-exchange, damped, however, in an essential way by the $\pi\rho$-diagrams. Therefore, we have to consider a $\pi\sigma$-contribution as well (fig. 10). The σ stands for the sum of all $\pi\pi$- and $\pi\rho$-diagrams as explained in the last section. (The σ-approximation is used by us only in this higher diagram).

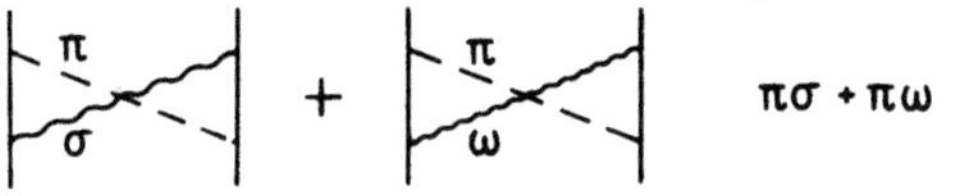

Fig. 10

$\pi\sigma$- and $\pi\omega$-contributions to the NN-interaction included in our model.

These two contributions almost completely cancel each other in most partial waves, except for the ε_1-mixing parameter where we observe a sizeable increase (see fig. 8), and it is for this reason why a rather low cutoff-mass for the pion-nucleon vertex could be chosen, namely 1.3 GeV, to be compared to 1.5 GeV in usual meson-exchange models. In this way more consistency between the cutoffs of the πNN- and $\pi N\Delta$--vertices could be achieved; the latter has a cutoff-mass of 1.2 GeV in our model.

Table 1

Meson parameters applied in the present model and from other sources.

vertex	meson-mass, m_α (MeV)	$g_\alpha^2(t=-m_\alpha^2)$	$g_\alpha^2(t=0)$;(f/g)	coupling constant from other sources or comments	cutoff-mass Λ_α (GeV)	n_α
NN π	138.03	14.4	14.08	14.28±0.18 πN-scattering, ref. 16	1.3	1
				14.52±0.40 $p\bar{p}$ forward dispersion relation, ref. 17		
				NN-phase-shift analyses: 14.25 ref. 18 (Bugg) 14.5 ref. 19 (Arndt)		
ρ	769.	0.75	0.41;(6.1)	0.55±0.06 (6.1±0.6) fit to $N\bar{N}\to\pi\pi$ partial waves, ref. 20	1.5	1
ω	782.6	20.	10.6	12.0 ref. 21 8.1±1.5 ref. 22	1.5	1
δ	983.	2.93	1.69	-	2.0	1
σ'	600.	9.0	5.33	$\pi\pi$-S-wave interaction, compare ref. 10	1.25	1
NΔ π	138.03	0.224	0.218	quark model with g_π^2 = 14.4 : 0.224	1.2	1
ρ	769.	15.51	4.58	quark model with g_ρ^2 = 0.55 : 13.4	1.5	2

$g_\alpha(t) := g_0 \left\{ \frac{\Lambda_\alpha^2 - m_\alpha^2}{\Lambda_\alpha^2 + t} \right\}^{n_\alpha}$; nucleon mass: 938.926 MeV;
mass of Δ-isobar: 1232. MeV.

Table 2

Deuteron and low energy parameters predicted by our model (theory) and by experiment (experiment).

	Theory	Experiment
Deuteron:		
binding energy, E_B (MeV)	2.22469	2.224644±0.000046 ref. 23
D-state probability, P_D (%)	4.27	5±2
quadrupole-moment, $Q_D(fm^2)$	0.281	0.2860 ± 0.0015 ref. 24
asymptotic S-state, A_S	0.9043	0.8846 ± 0.0008 ref. 25
asymptotic D/S-state, η	0.0267	0.0271 ± 0.0004 ref. 25
root-mean-square radius (fm)	2.0026	1.9635 ± 0.0045 ref. 26
ΔΔ-probability (%)	0.5	
np low energy scattering:		
singlet: a_s(fm)	-23.745	-23.748 ± 0.010
r_s(fm)	2.755	2.75 ± 0.05
triplet: a_t(fm)	5.424	5.424 ± 0.004 ref. 27
r_t(fm)	1.760	1.759 ± 0.005

In the theoretical results quoted here the nucleonic wave function of the deuteron has been normalized to unity for simplicity. In a more refined consideration of the deuteron, Δ- and mesonic components should be separated out. They would, on the one hand, reduce the normalization of the nucleonic wave function and by that the theoretical result of most deuteron quantities cited, but, on the other hand, add meson-current contributions which will at least partly compensate the former effect. This will be discussed in detail in a future paper.

6. The meson parameters and the quantitative description of the NN data

In table 1 we give the meson-parameters used throughout this work and compare them to other sources. A rather close agreement can be observed. The description of the low energy scattering and the deuteron data is excellent as can be seen from table 2. The fit of the phase-shifts of NN-scattering are shown in fig. 11, where also a comparison of our theoretical results with those from energy-independent (error-bars) and energy-dependent (dashed curve) phase-shift analysis is performed. Further, results from a typical OBE model are also given in fig. 11 (dotted curve) demonstrating the considerable improvements of our extended meson-exchange model compared to the simplified approach (watch especially the $^1P_1$, $^3P_1$ and $^3D_2$ phase-shifts). Apart from the phase-shifts the so-called observables of NN-scattering were directly computed for the present model: They also show a superior agreement to experiment[28).

7. Nuclear Matter

Though we have not yet applied the present extended model to nuclear matter, we may estimate the main qualitative results from the experience with former models including iterative isobar-diagrams[7,29). Pauli- and dispersion effects, which in a natural way occur in a model for the NN-interaction with explicit diagrams of the fourth order, will reduce the overbinding, commonly found in nuclear matter calculations, especially when a nuclear force with a realistically weak tensor force is applied. These medium effects have proven to be essential for nuclear saturation, and cannot be taken into account by simple potential models for the nuclear force. The density-dependence of the effects improve the relation between the binding energy of light and heavy nuclei. Moreover, isobar excitations of the nucleons in nuclear matter are a natural feature of our model. We expect the probability to be in the order of former calculations[29) which was found to be about 10%. This may be a possible explanation for recent structure function measurements in iron by the EMC-group[5).

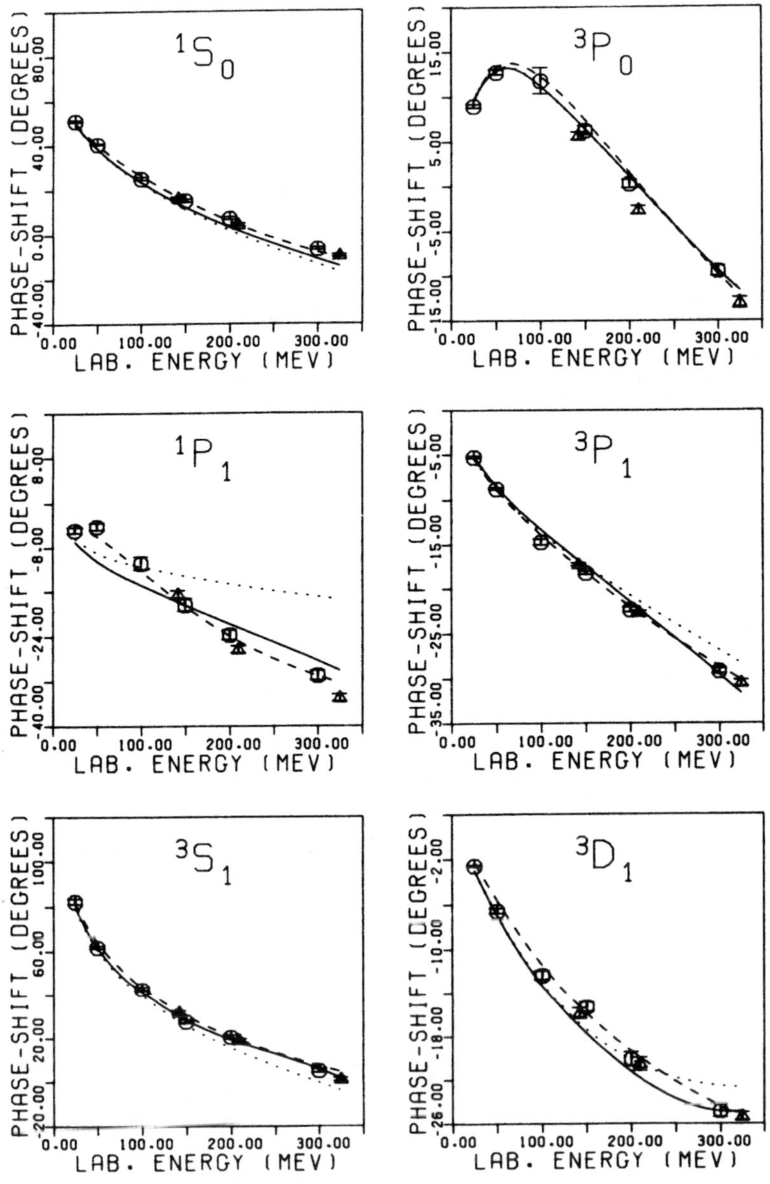

Fig. 11: Partial-wave phase-shifts of NN-scattering. The full line represents the results from our model. The dotted line refers to a typical OBEP. The energy-dependent phase-shift analysis of Arndt et. al.[19] is displayed in the dashed curve; energy-independent analyses are given from the latter group (octagon) and from Bugg and coworkers[18] (triangle).

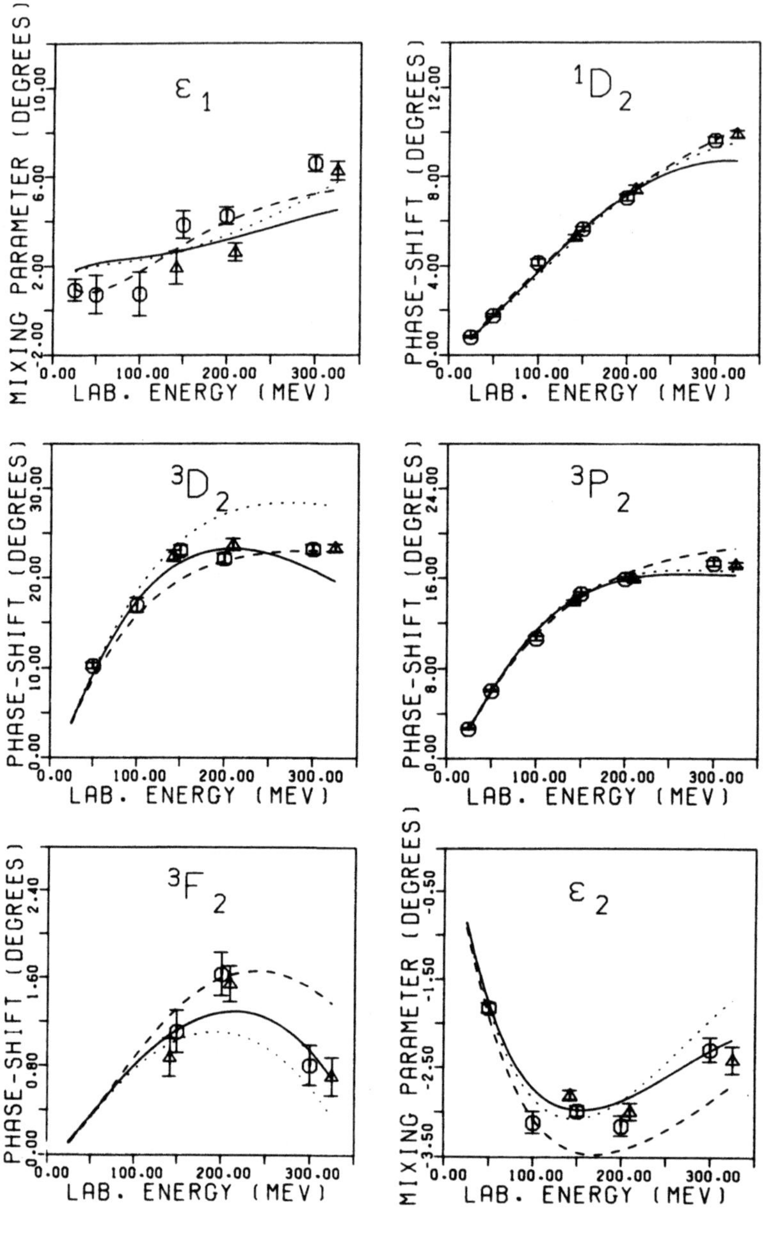

ε 1
MIXING PARAMETER (DEGREES)
LAB. ENERGY (MEV)
1D 2
PHASE-SHIFT (DEGREES)
LAB. ENERGY (MEV)
3D 2
PHASE-SHIFT (DEGREES)
LAB. ENERGY (MEV)
3P 2
PHASE-SHIFT (DEGREES)
LAB. ENERGY (MEV)
3F 2
PHASE-SHIFT (DEGREES)
LAB. ENERGY (MEV)
ε 2
MIXING PARAMETER (DEGREES)
LAB. ENERGY (MEV)

Fig. 11, cont'd

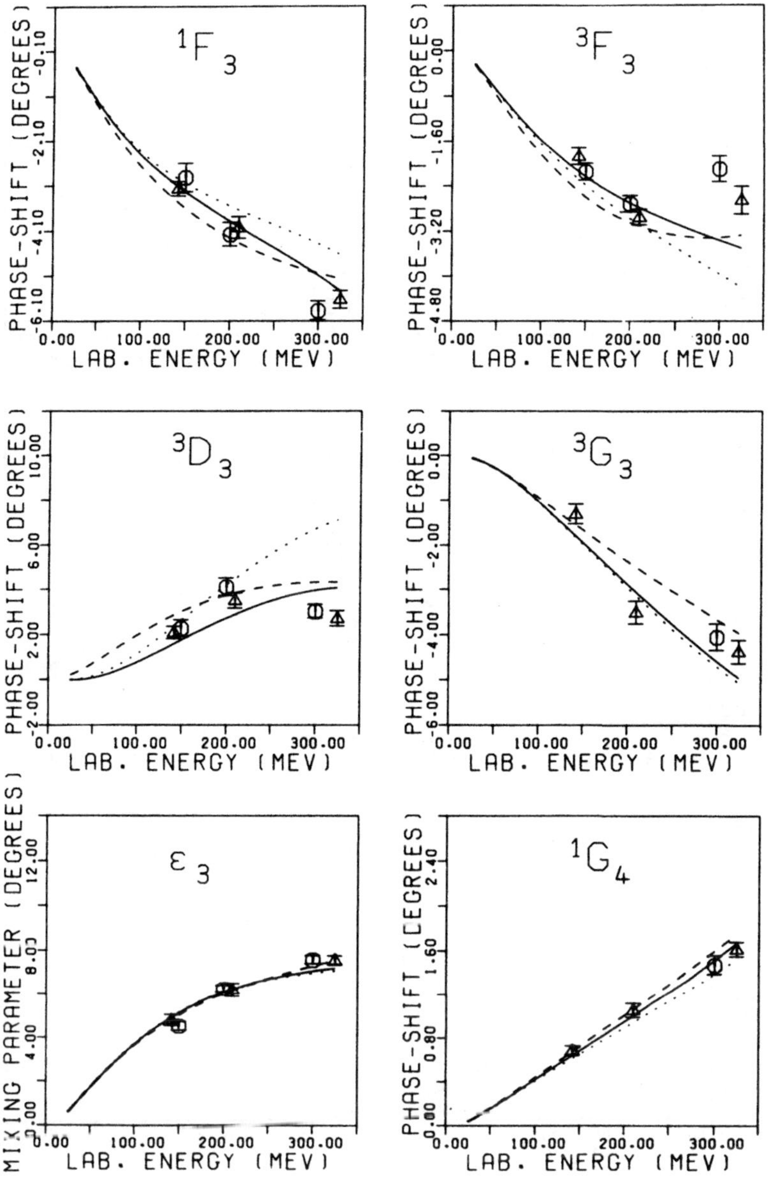

$^1F_3$
$^3F_3$
$^3D_3$
$^3G_3$
ε_3
$^1G_4$
PHASE-SHIFT (DEGREES)
MIXING PARAMETER (DEGREES)
LAB. ENERGY (MEV)
0.00
100.00
200.00
300.00

Fig. 11, cont'd

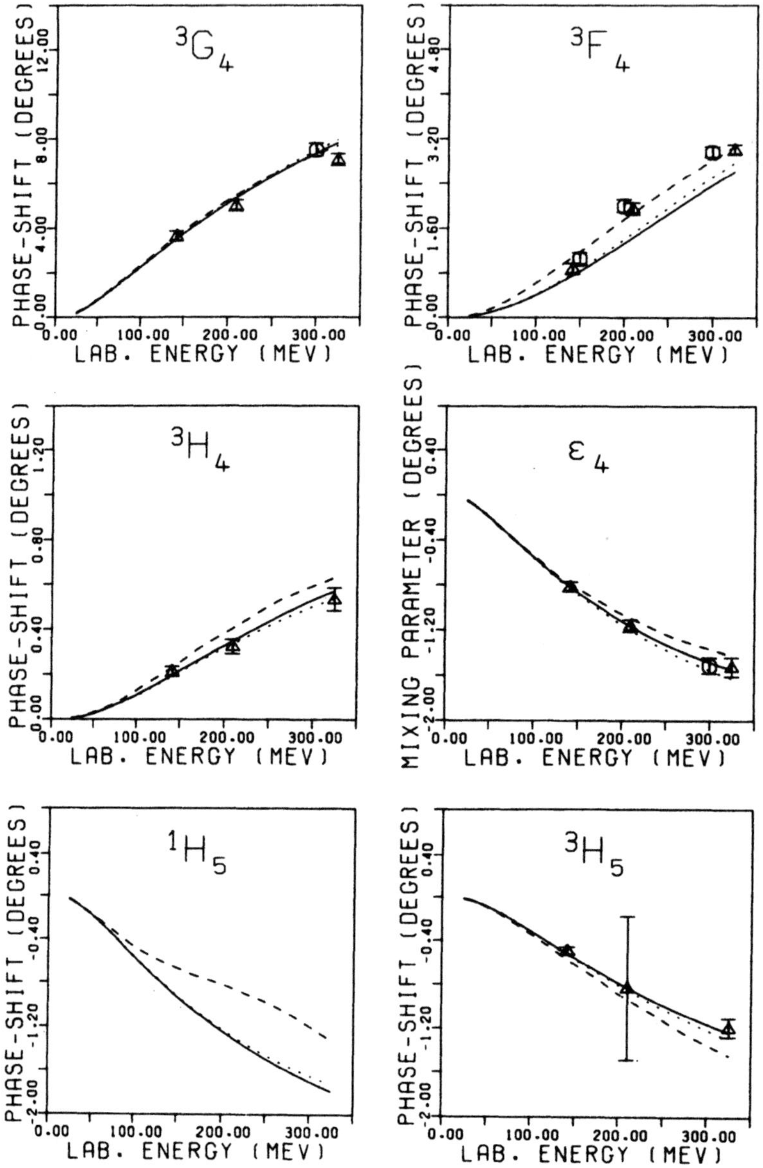

$^3G_4$
$^3F_4$
$^3H_4$
ε_4
$^1H_5$
$^3H_5$
PHASE-SHIFT (DEGREES)
MIXING PARAMETER (DEGREES)
LAB. ENERGY (MEV)
0.00
100.00
200.00
300.00
4.00
8.00
12.00
1.60
3.20
4.80
0.40
0.80
1.20
-0.40
-1.20
-2.00

Fig. 11, cont'd

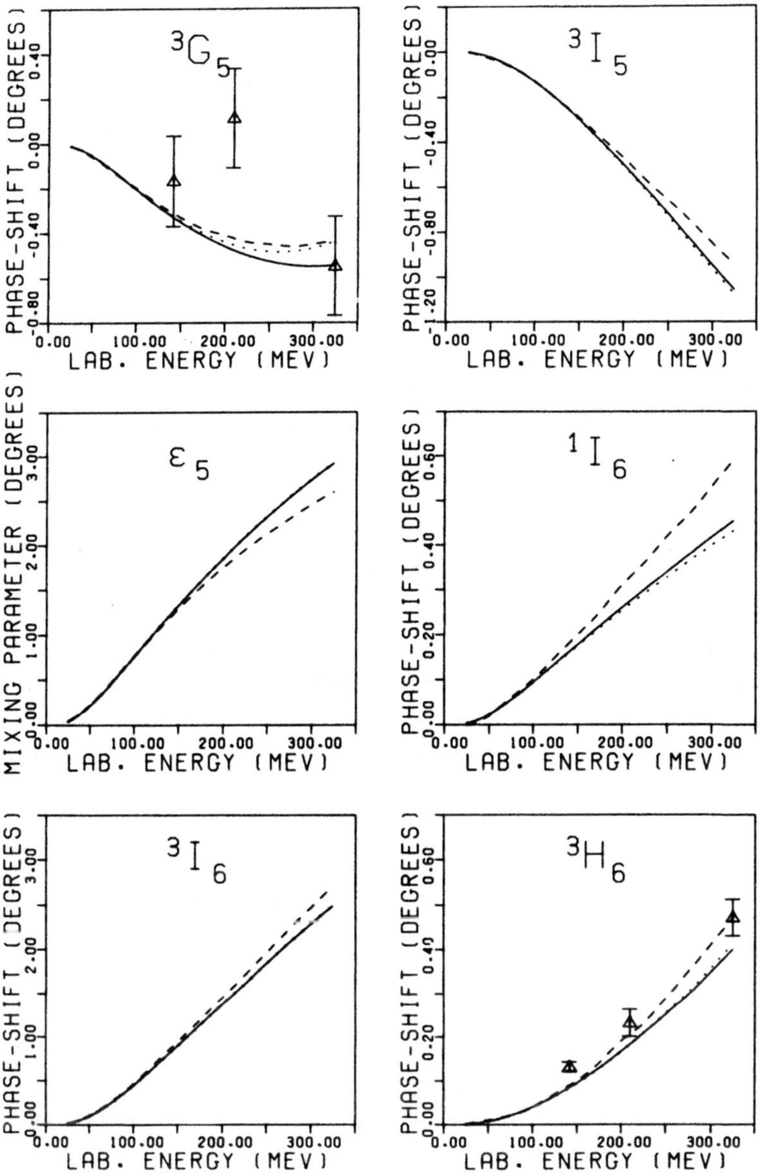

$^3G_5$
$^3I_5$
ε_5
$^1I_6$
$^3I_6$
$^3H_6$
PHASE-SHIFT (DEGREES)
MIXING PARAMETER (DEGREES)
LAB. ENERGY (MEV)

Fig. 11, cont'd

Summary and conclusions

For the NN-interaction we consider meson-exchange as the possibly adequate approximation scheme describing effectively the complicated underlying multi-quark processes. Based on that principle idea we construct a meson-exchange model for the NN-interaction which includes irreducible diagrams up to fourth order. From the quark-bag structure of hadrons it is evident that for the meson-baryon vertices a form-factor is needed. Quantitative consideration within our model lead us to a value for the cutoff mass of such a form-factor of 1.2 to 1.5 GeV. We conclude from this fact that for a consistent meson-exchange model it is necessary to take into account diagrams with a total mass of the exchanged mesons up to ≃ 1 GeV, i.e. up to almost the cutoff region.

Part of our model represents an explicit determination of the 2π-exchange contribution including virtual Δ-excitation and the direct $\pi\pi$-interaction. A quantitative check shows a remarkable agreement of our model with results from dispersion theory which are based on the analysis of πN- and $\pi\pi$-scattering data. With the same part of our model we also achieve a rather satisfactory description of higher partial waves of NN-scattering.

For lower partial waves it turns out that the 2π-exchange is by far too attractive in order to account for the empirical data. However, according to the rules for a consistent model which we stated above further diagrams including heavy boson exchange have to be considered anyhow. Following this line we find that the irreducible diagrams of π- and ρ-exchange (containing also virtual Δ-excitation) add a strong repulsion in low partial waves, and by that compensating the over-attraction of the 2π-exchange in exactly that way which is needed for a quantitative interpretation of low partial waves.

We take further irreducible diagrams of $\pi\sigma$- and $\pi\omega$-exchange into account (where the σ represents the sum of all $\pi\pi$- and $\pi\rho$-diagrams in an approximate way), observing a cancellation to a large extent in most partial waves between these two groups of fourth order diagrams. As a remarkable exception, however, the ε_1-mixing parameter is considerably increased, as the contribution to the tensor force provided by the $\pi\rho$-diagram is strongly dominant over the analogous but opposite contribution from $\pi\omega$. This net increase of the tensor force allows us to use a lower value for the cutoff mass of the πNN-vertex than in old and simplified boson-exchange models. In that way the physically necessary close agreement between the form-factor for the πNN- and the πNΔ-vertex is achieved. In all older boson models with virtual Δ-excitation the vertex-functions for the two different pion vertices used to differ tremendously.

Summarizing we may state that the overall description of the two nucleon data by our complete model is excellent. The predictions for the phase--shifts of NN-scattering and the observables up to 300 MeV laboratory energy is most satisfactory. Remarkable improvements are to be seen especially in the $^1P_1$-, $^3P_1$- and $^3D_2$-phase-shifts. The empirical low energy data, i.e. scattering length and effective range, are reproduced exactly. With the latter quantities there have always been problems within boson models in the past.

Our extended model for the NN-interaction leads to considerable alterations of the usual nuclear matter theory which uses simple potential models for the nuclear force. Due to the different treatment of intermediate states in a many-body surrounding compared to the free case, characteristic modifications occur when our model for the NN-interaction is inserted into nuclear matter. These well-known Pauli- and dispersion--effects cause a density-dependent quenching of the nuclear force. This additional repulsion which results from our model in a natural way is desperately needed for a quantitative explanation of the binding energy in nuclear matter which is usually too large in conventional potential models, especially when higher orders in the hole-line expansion are included. Due to the density-dependence of the quoted effects our nuclear matter results further improve the relation in the binding energy of light and heavy nuclei. In addition, our findings about the Δ-excitation in nuclear matter provides a possible explanation of the recent quark structure function measurements in nuclei[5)]: The Δ-admixture is found in the order to 10 - 15% in rough agreement with the analysis of Szwed[5)].

Acknowledgement

The author would like to thank Prof. K. Bleuler for the substantial support and the many illuminating discussions over the whole decade during which this work was pursued. This project has been performed together with Dr. K. Holinde with whom a fruitful and most productive collaboration is gratefully acknowledged. Thanks are also due to X. Bagnoud who contributed substantially during the last years to the work of our group and to Ch. Elster for her help with the preparation of this manuscript.

References

1) C. De Tar, Phys. Rev. D17 (1978) 302; 323;
M. Harvey, Nucl. Phys. A352 (1981) 326;
K. Holinde, Phys. Lett. 118B (1982) 266;
A. Fäßler et al., Phys. Lett. 112B (1982); see also the contributions of the last three authors to this conference
2) compare for example: M. Bander, Phys. Rep. 75 (1981) 205 and contributions by M. Bander, J. Engels and G. Schierholz in this volume
3) A.N. Mitra, Relativistic few Quark Dynamics for Hadrons, Report Internat. Centre for Theoret. Physics, Trieste IC/83/92
4) H. Yukawa, Proc. Phys. Math. Soc. Japan, 17 (1935) 48
5) J.J. Aubert et al., Phys. Lett. 123B (1983) 275
J. Szwed, Structure Functions of Nucleons inside Nuclei, Preprint Univ. of Cracow TPJU-1/83
6) D. Schütte, Nucl. Phys. A221 (1974) 450
7) K. Holinde, R. Machleidt, M.R. Anastasio, A. Fäßler and H. Müther, Phys. Rev. C18 (1978) 870;
K. Holinde, Phys. Reports 68 (1981) 121
8) G.E. Brown and A.D. Jackson, The Nucleon-Nucleon Interaction (North-Holland, Amsterdam 1976) p. 138
9) X. Bagnoud, K. Holinde and R. Machleidt, Phys. Rev. C24 (1981) 1143
10) J.W. Durso, A.D. Jackson and B.J. Verwest, Nucl. Phys. A345 (1980) 471
11) R. Vinh Mau, The Paris Nucleon-Nucleon Potential, in Mesons in Nuclei, Vol. I (North-Holland, Amsterdam, 1979) p. 151
12) X. Bagnoud, K. Holinde and R. Machleidt, "Role of ρ-exchange in isobar contributions to the NN interaction", preprint Bonn 1983 and to be published
13) K. Holinde and R. Machleidt, Nucl. Phys. A372 (1981) 349
14) J.W. Durso, M. Saarela, G.E. Brown and A.D. Jackson, Nucl. Phys. A278 (1977) 445
15) K. Erkelenz, Phys. Reports 13C (1974) 191
16) R. Koch and E. Pietarinen, Nucl. Phys. A336 (1980) 331
17) P. Kroll, Physics Data, 22-1 (1981)
18) R. Dubois et al., Nucl. Phys. A377 (1982) 554
19) R.A. Arndt et al., Phys. Rev. D28 (1983) 97

20) E. Pietarinen, Helsinki, Univ. HU-/TFT-17-77
21) W. Grein, Nucl. Phys. B131 (1977) 255
22) W. Grein and P. Kroll, Nucl. Phys. A338 (1980) 332
23) T.L. Honk, Phys. Rev. 43 (1971) 1886
24) R.V. Reid and M.L. Vaida, Phys. Rev. Lett. 34 (1975) 1064
25) T.E.O. Ericson and M. Rosa-Chlot, Nucl. Phys. A405 (1983) 497
26) R.W. Berard et al., Phys. Lett. 47B (1973) 355
27) O. Dumbrajs et al., Nucl. Phys. B216 (1983) 277
28) C. Hamburger, private communication
29) M.R. Anastasio, A. Fäßler, H. Müther, K. Holinde and R. Machleidt, Phys. Rev. C18 (1978) 2416;
R. Machleidt and K. Holinde, Nucl. Phys. A350 (1980) 396

THE SHORT-RANGE PART OF THE NN-INTERACTION QUARK-GLUON VERSUS MESON EXCHANGE+)

K. Holinde*)
Institut für Theoretische Kernphysik der Universität Bonn
Nußallee 14-16, D-5300 Bonn, West-Germany

It is shown that the concept of large nucleon bags ($R \geq 0.8$ fm) suppresses essential features of the meson exchange picture, which are needed in order to describe empirical NN data. It is investigated whether genuine quark-gluon-exchange processes can provide the characteristic features conventionally obtained from vector-meson exchange.

1. Introduction

Many examples in the literature now clearly reveal the fact that the picture of meson exchange is able to provide a convenient parametrization of the empirical low-energy ($E_{lab} \leq 300$ MeV) data. In fact, as Dr. R. Machleidt demonstrated in the foregoing talk, the latest version of the Bonn potential, which contains an explicit model for the 2π-exchange contribution consistent with what is known from dispersion theory and, in addition, corresponding $\pi\rho$-exchange diagrams, describes the NN scattering phase shifts very well, solely in terms of meson-nucleon coupling constants and cutoff-masses, i.e. physical parameters, which can (and should) be determined from the underlying quark-gluon structure of hadrons.

The advantage of keeping the meson-exchange expressions throughout the whole region, only supplementing them by form factors which are anyhow dictated by the quark structure, should be obvious: In this way, one can really demonstrate how far meson exchange works and where it breaks down. Surprisingly enough, our model works down to rather small distances.

+) Supported in part by Deutsche Forschungsgemeinschaft (Az.:Ho 730/5-1)
*) Now at MP Theory Division, LAMPF, Los Alamos National Laboratory, Los Alamos, NM 87545, USA

Furthermore, our calculations have clearly shown that some structure of the nucleon (characterized by a cutoff-mass $\Lambda \simeq 1.5$ GeV corresponding to a radius of about 0.4 fm) has to be implemented in order to account for the empirical data.

These results suggest that the meson exchange picture be essentially valid; the remaining task would be to justify the meson-nucleon vertices quark-theoretically. (We will demonstrate later that possible discrepancies inside 0.5 fm do not influence the empirical data). Also, since a rather small radius of the nucleon is implied, a meson-theoretic treatment of nuclear structure problems should then be reliable.

On the other hand, some current bag models (e.g. the Cloudy Bag Model[1)]) predict a nucleon radius of about 0.8 fm. Such a large value wipes out essential, characteristic features of the meson-exchange picture, and a reasonable description of the empirical data is not possible anymore in the conventional framework. Instead, genuine quark-gluon-exchange processes, which occur in the overlapping region of the nucleon bags, should be considered, and it is an interesting question whether such processes can provide the necessary features conventionally obtained from meson exchange. This will be the main topic of this talk.

If such rather large extensions of hadrons will be confirmed (e.g. by lattice gauge calculations), then meson-exchange provides only a phenomenological description and the physical picture is completely different. In this case, of course, an extension of meson exchange to other physical systems is at least questionable.

2. πNN vertex and NN data

In this chapter, we will shortly discuss predictions for the πNN vertex given by current bag-type models. Its general structure is

$$v_{\underline{k},i} \simeq F_R(k)\underset{\sim}{\sigma}\cdot\underset{\sim}{k}\tau_i \tag{2.1}$$

where $\underset{\sim}{k}$ is the momentum of the pion and $\underset{\sim}{\sigma}$, τ_i denote the usual (nucleon) spin and isospin operators. The form factor $F_R(k)$ is given explicitly by

$$F_R(k) = 3\,\frac{j_1(kR)}{kR} \tag{2.2}$$

R being the bag radius. To a good approximation (as we will see below), $F_R(k)$ can be written in the usual monopole-type form

$$F_R(k) \simeq \frac{\Lambda_R^2}{\Lambda_R^2+k^2} \tag{2.3}$$

where the cutoff-parameter Λ_R is related to the bag radius by $\Lambda_R = \sqrt{10}\cdot R^{-1}$.

In the Cloudy Bag Model, R has been fixed from pion-nucleon scattering to be $R \simeq 0.8$ fm, which implies $\Lambda_R \simeq 800$ MeV. In a different type of approach, Weise et al.[2)] found for the form factor

$$F_\Lambda(k) = e^{-k^2/\Lambda^2} \tag{2.4}$$

with values of Λ being in the same range as for the Cloudy Bag Model. Again, eq. (2.4) can be well approximated by eq. (2.3).

Fig. 1 shows the strong influence of the form factor with such a small cutoff mass, i.e. $\Lambda = 800$ MeV. In the important region around 1 fm, it suppresses the pion tensor interaction by almost a factor of 2. Values of $\Lambda \simeq 1.5$ GeV, which are needed in order to fit the empirical data, start to give an appreciable modification at much smaller distances.

Fig. 2 shows the influence of a strong pion cutoff on two selected partial wave phase shifts. Starting from an OBE-model which fits the scattering data and uses a cutoff-mass of $\Lambda = 1530$ MeV[3)] at all vertices, it is seen that a value of $\Lambda = 800$ MeV for the πNN-vertex destroys the fit in a characteristic way: One phase shift moves up, the other one goes down. It is important to note that such a defect cannot be cured by a change of other parameters in the model.

Two additional conclusions can be drawn from this figure: First, the forms eq. (2.2, 2.4) have essentially the same effect on the phase shifts considered; second, the monopole form, eq. (2.3), is a sufficiently good approximation.

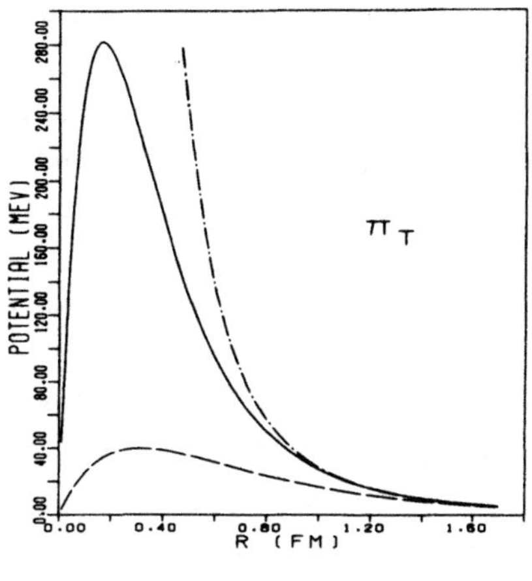

Fig. 1

One-pion-exchange tensor force (in MeV) as function of the distance R (in fm) between two nucleons. For the dash-dot curve, no form factor is used, whereas for the solid (dashed) curve, a monopole form factor, eq. (2.3), with Λ = 1530 (800) MeV is added.

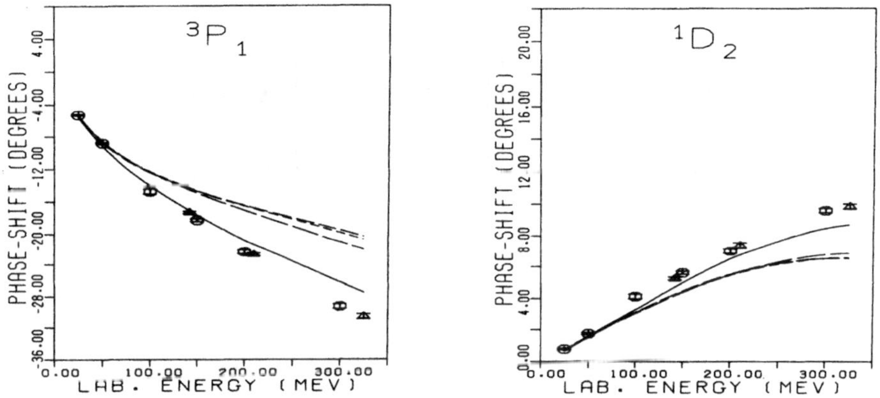

Fig. 2

NN scattering phase shifts (in degrees), in the $^3P_1$ and $^1D_2$ channel, as function of the nucleon lab energy (in MeV). The solid lines are obtained from an OBEP[3], based on a form factor, eq. (2.3), with Λ_α = 1530 MeV for all mesons exchanged. The long-dashed lines are obtained if Λ_π = 800 MeV is used in eq. (2.3). The medium-dashed curves result from the use of the form factor in eq. (2.2), with R = 0.8 fm. The short-dashed lines originate from the form factor, eq. (2.4), with Λ = 800 MeV.

Thus, in the conventional boson-exchange framework, there is a lower limit for the cutoff mass Λ still allowing a good fit to the data. It was observed already in ref.[3] that this limit is around 1200 MeV. Similar conclusions were reached by Ericson and Rosa-Clot[4] in the context of the asymptotic D/S ratio of the deuteron.

In summary, nucleon radii R > 0.8 fm, suggested e.g. by the Cloudy Bag Model[1], lead to a strong suppression of boson-exchange contributions, destroying the beautiful description provided by such models. This fact as such, however, does not rule out large bag radii. Instead, it leads to the fascinating question whether genuine quark-gluon--exchange processes, which should occur in the overlapping region of the nucleon bags and must be taken into account, might be able to provide the necessary features conventionally obtained from the meson--exchange picture.

3. Quark-gluon versus meson exchange

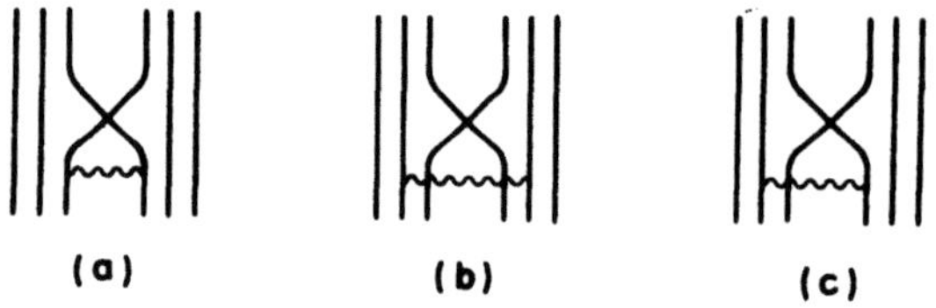

Fig. 3

Various one-gluon-exchange processes in the NN problem.

The one-gluon-exchange processes between two nucleon clusters, shown in fig. 3, represent a convenient starting point. Due to asymptotic freedom, they might be able to describe the short-range part of the NN interaction, i.e. to replace conventional vector-meson exchange, which is almost completely suppressed in case of large bag radii.

In fact, the isospin structure of the simplest one-gluon-exchange process, fig. 3 (a), is in remarkable agreement with the isospin dependence of the short-range part of the NN interaction provided by ($\omega+\rho$)-meson exchange[5]. (Note that ω provides the isoscalar, ρ the isovector part). We stress that this result is very basic, i.e. it depends only on the structure of diagram 3 (a). It is essential that the elementary (direct) quark-quark amplitude is flavour-independent and that there is no direct first-order process, i.e. quarks have to be interchanged. The result does not depend on specific model assumptions about the underlying quark-quark interaction and radial cluster wave functions. These features enter if the study is extended to the spin-dependence and to the other relevant processes (fig. 3 (b,c)). The spin-isospin structure and radial behaviour of the effective NN interaction resulting from processes in fig. 3 is suitably investigated in the Born-Oppenheimer approximation. In this treatment, the NN interaction V_{NN} is defined as the expectation value of the energy of the six-quark system minus the energy at infinite separation of the two clusters

$$V_{NN} \equiv \frac{\langle\psi|H_A+H_B+\sum_{i=1}\sum_{j=4}^{6} V(ij)|\psi\rangle}{\langle\psi|\psi\rangle} - 2\langle\phi_A|H_A|\phi_A\rangle \; . \qquad (3.1)$$

Here $|\psi\rangle$ is a totally antisymmetric six-quark state, $|\psi\rangle \equiv A|\phi_A\phi_B\rangle$ where ϕ_N is the antisymmetric and normalized nucleon cluster wave function consisting of a space, spin-isospin and colour part. $H_{A,B}$ is the internal Hamiltonian of cluster A,B, i.e.

$$H_A = \sum_{i=1}^{3} t_i + \sum_{\substack{i,j=1\\ j>i}}^{3} V(ij) - T_A^{CM}$$

with t_i being the kinetic energy operator of quark i and T_A^{CM} the corresponding c.m. energy of cluster A. V_{NN} can be split up into $V_{NN} = V_{NN}^{(1)}+V_{NN}^{(T)}+V_{NN}^{(P)}$. The first term, $V_{NN}^{(1)}$, arises from the interactions between the two clusters characterized by the processes of fig. 3. $V_{NN}^{(T,P)}$ are (non-negligible) correction terms which take the difference of the kinetic (T) and potential (P) cluster energy inside the six-quark system compared to the three-quark cluster into account.

An evaluation of eq. (3.1) has been performed[6] using an underlying quark-quark interaction

$$V(ij) = V_{OGE}(ij) + V_{conf}(ij)$$

$$V_{OGE}(ij) = \underset{\sim}{\lambda}_i \cdot \underset{\sim}{\lambda}_j \frac{\alpha_s}{4} \left[\frac{1}{r_{ij}} - \frac{2}{3} \frac{\pi}{m^2} \delta(\underset{\sim}{r}_{ij}) \underset{\sim}{\sigma}_i \cdot \underset{\sim}{\sigma}_j\right]$$

$$V_{conf}(ij) = -\underset{\sim}{\lambda}_i \cdot \underset{\sim}{\lambda}_j a r_{ij}^2$$

and

$$\varphi_A = \left(\frac{\beta^2}{\pi}\right)^{9/4} \exp\left\{- \frac{\beta^2}{2} \left[\sum_{i=1}^{3} \left(\underset{\sim}{r}_i - \frac{\underset{\sim}{R}}{2}\right)^2 \right]\right\} \quad (3.2)$$

for the radial cluster wave function, the two clusters being a distance $\underset{\sim}{R}$ apart. Two parameter sets have been used:

(i) $m = 355$ MeV, $\beta^{-1} = 0.475$ fm, $a = 34.5$ MeV$\cdot$fm^{-2}, $\alpha_s = 0.97$ given by Fäßler, Fernandez, Lübeck and Shimizu[7] (FFLS) and

(ii) $m = 300$ MeV, $\beta^{-1} = 0.6$ fm, $a = 62.5$ MeV$\cdot$fm^{-2}, $\alpha_s = 1.39$ provided by Oka and Yazaki[8] (OY).

The main results are displayed in fig. 4. The colour-magnetic spin-spin term provides the dominant, repulsive contribution to $V_{NN}^{(1)}$. Both parameter sets yield the same repulsion at $R = 0$; they differ, however, considerably (by roughly a factor of 2.5) at $R = 1$ fm due to the different size parameter β. Compared to the central and spin-spin part of conventional ω, ρ meson exchange (taken from a realistic one-boson-exchange model which fits the two-nucleon data[9]), this contribution yields relatively too weak repulsion in odd-parity states, which can be traced back to the different (negative) sign for the resulting spin-spin contribution, see ref.[6].

The situation is improved if the kinetic energy correction $V_{NN}^{(T)}$ is added. It yields a small attractive contribution in even-parity states and, on the other hand, a considerable repulsive contribution in odd-parity states. In fact, the FFLS set now provides a good agreement with vector-meson exchange around 1 fm in all channels, whereas the OY set is consistently too large.

This nice agreement for the FFLS set is spoiled to some extent by the remaining correction terms, namely $V_{NN}^{(P)}$ plus the contribution of the central term to $V_{NN}^{(1)}$. Whereas this contribution is negligible in even-parity states, and therefore does not destroy the agreement with

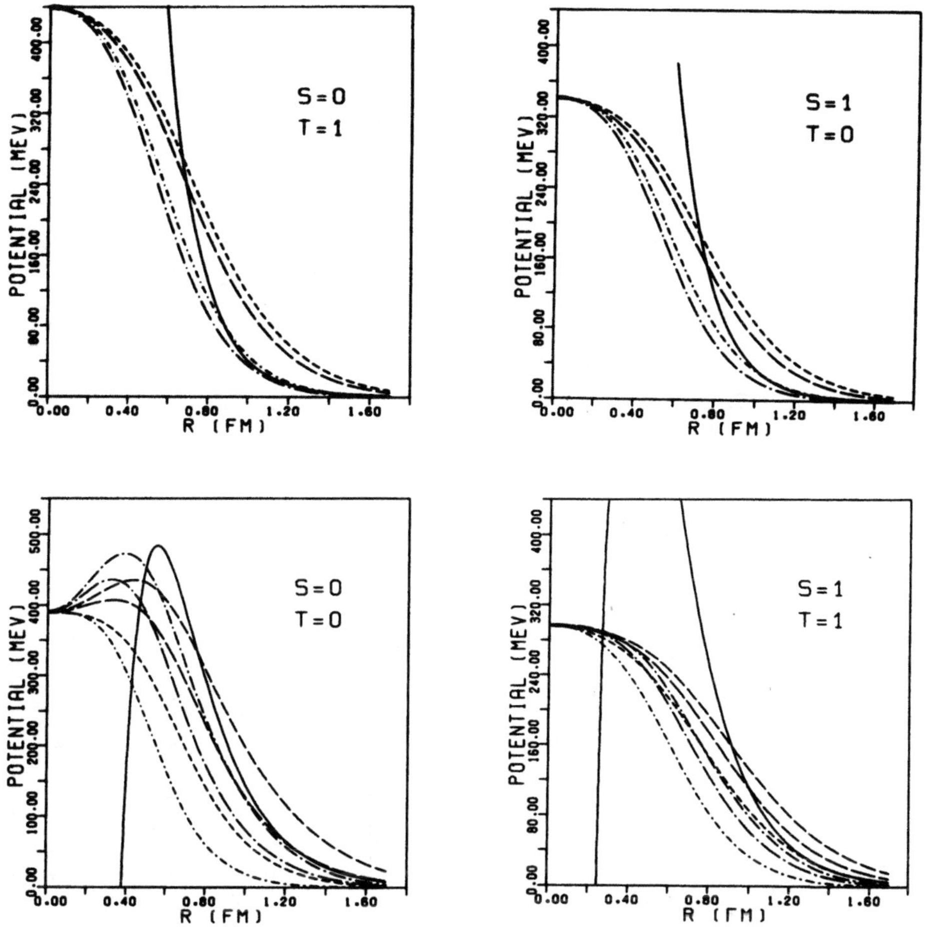

Fig. 4

V_{NN} as function of R, in the various spin-isospin channels. The solid line denotes the result provided by the central plus spin-spin part of conventional ω, ρ-meson exchange[9]. The dash-dot curves display quark--gluon-exchange results based on the FFLS parameter set[7], whereas the dashed curves originate from the OY choice[8]. The short-dashed curves denote the contribution due to the colour-magnetic term in $V_{NN}^{(1)}$, for the medium-dashed curves the kinetic-energy term $V_{NN}^{(T)}$ is added, and the long-dashed curves display the total result.

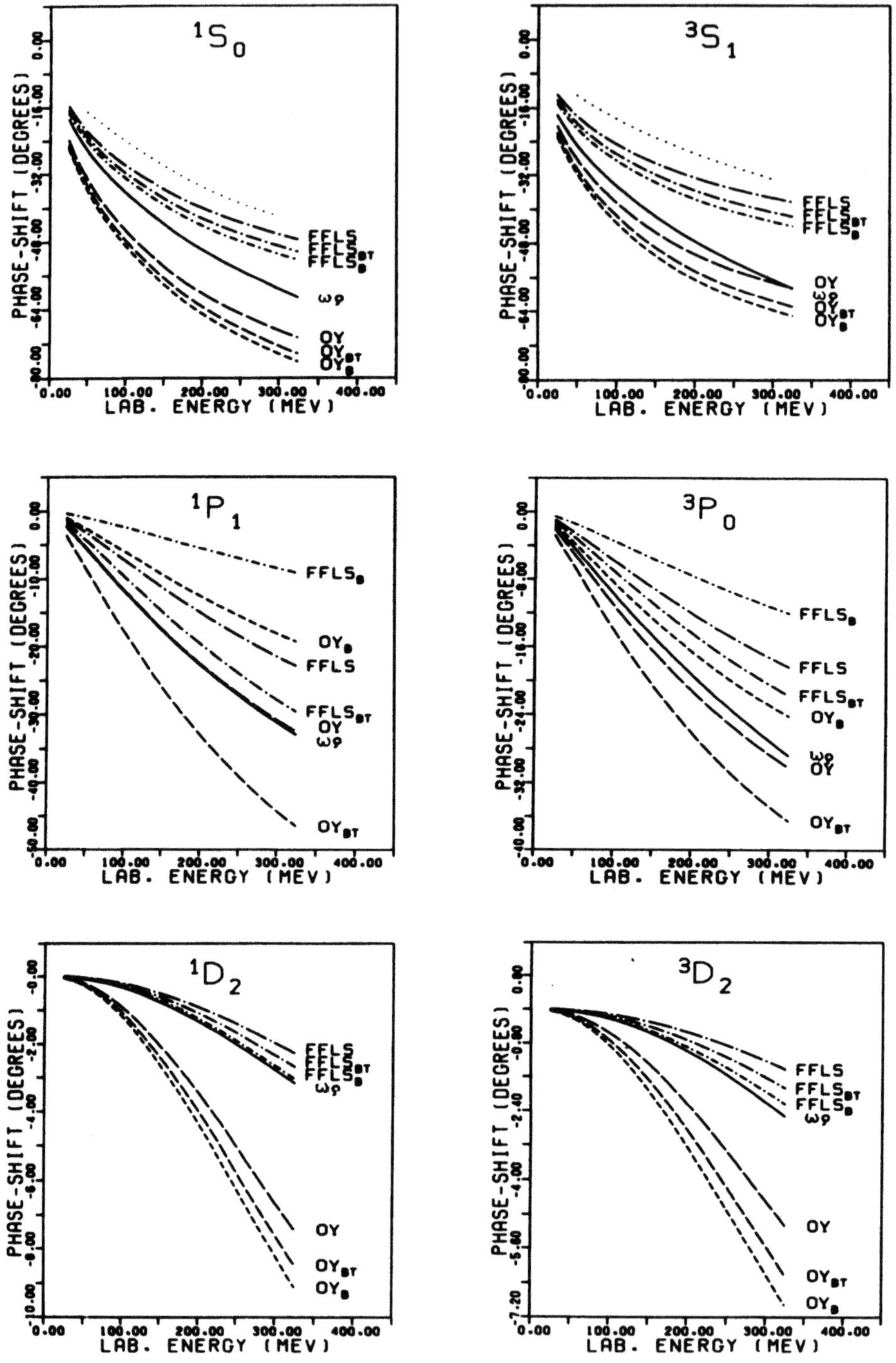

Fig. 5
(figure captions see next page)

(Fig. 5):

Nucleon-nucleon phase shifts as function of the nucleon lab energy. The solid line displays the result provided by the central plus spin-spin part of conventional ω, ρ-meson exchange[9]. The dash-dot curves display quark-gluon exchange results based on the FFLS parameter set[7], whereas the dashed curves originate from the OY choice[8]. The short-dashed curves denote the contribution due to the colour-magnetic term in $V_{NN}^{(T)}$ (B), for the medium-dashed curves the kinetic-energy term $V_{NN}^{(T)}$ is added (BT), and the long-dashed curves display the total result. The dotted curves, taken from ref.[7], result from a full resonating group calculation including all channels, based on the FFLS set.

vector-meson exchange in case of the FFLS set, it provides a strong attractive contribution in odd-parity states. Obviously, our results seem to indicate that a consistent description in all channels requires an almost complete suppression of the contribution arising from the colour-Coulomb and confining terms.

Corresponding predictions for NN scattering phase shifts, based on the effective NN interaction defined in eq. (3.1) are displayed in fig. 5. A comparison between interaction and phase shifts will clarify the role of the various contributions; especially, it will be demonstrated that the very inner part ($R < 0.5$ fm) of the NN interaction is only of minor relevance. Furthermore, our results can be compared with those from more involved resonating-group calculations. Thus the reliability of the Born-Oppenheimer scheme can be tested.

Obviously, the conclusions drawn already from the results shown for the NN interaction in fig. 4 hold also in case of the NN scattering phase shifts. Furthermore, a comparison between figs. 4,5 clearly reveals the dominant role of the region around $R = 1$ fm governing low-energy phase shifts and a correspondingly negligible role of the very inner part, especially for P- and D-waves. For example,

(i) the attractive core of (ω+ρ)-meson exchange in odd-parity states (originating from the vertex form factor, see ref.[6]) is not felt since the corresponding phase shifts in $^1P_1$, $^3P_1$ remain negative;

(ii) in case of the S-wave phase shifts, there is a strong sensitivity to the choice of parameter set, in spite of the fact that both sets provide the same repulsion at $R = 0$;

(iii) in both D-wave phase shifts, (ω+ρ)-exchange yields the weakest repulsion.

With a size parameter $\beta^{-1} \simeq 0.5$ fm, additional ($\Delta\Delta$ and hidden colour) channels should start to modify the present results at $R \simeq 2 \cdot \beta^{-1} = 1$fm; their effect should increase with decreasing distance. Consequently, we do not expect drastic changes for low-energy NN scattering phase shifts, even for S-waves. In fact, full resonating-group calculations based on the FFLS set[7] lower the repulsion in S-waves by only 20 %, see fig. 5. An even smaller effect is to be expected in higher partial waves although, in odd-parity states, the contribution from the central part in the quark-quark interaction might be sensitive to our approximate treatment, since it arises from a delicate interplay between various terms.

In summary, we feel that our results provide some indication for the possible existence of an intimate connection between the one-gluon--exchange mechanism and conventional vector-meson exchange providing the short-range part of the NN interaction. Concerning the intermediate-range attraction, this part could possibly be identified with (second-order) hidden-colour contributions (colour Van der Waals forces). It is known[10] that, at large distances, these terms provide an unrealistic contribution. However, realizing that one gluon cannot be exchanged between two separated hadrons, a cutoff factor should be introduced which is related to the overlap of the two interacting nucleons[11]. Note that, since no quark-interchange is involved in such contributions, they are exactly isoscalar, which agrees with the empirical situation. Finally, it has been suggested[12] that pion exchange between two interchanged quarks inside the two-cluster bag might account for the missing short-ranged OPE-contribution. Thus, the question whether simple quark-gluon-exchange diagrams can provide essential features of the NN interaction is a fascinating and challenging problem for the future. It might well be that both pictures, conventional meson exchange and explicit quark-gluon exchange, can provide an essentially equivalent description of the empirical situation.

Fruitful discussions with A. Fäßler and M. Harvey are gratefully acknowledged. Thanks are due to C. Elster for providing some numerical results and for preparing the figures.

References

1) A.W. Thomas, S. Théberge, G.A. Miller, Phys. Rev. D24 (1981) 216
2) R. Tegen, R. Brockmann and W. Weise, Z. Phys. A307 (1982) 339
3) K. Holinde, Phys. Repts. 68 (1981) 121
4) T.E.O. Ericson and M. Rosa-Clot, Phys. Lett. 110B (1982) 193
5) K. Holinde, Phys. Lett. 118B (1982) 266
6) K. Holinde, The Short-Range Part of the NN Interaction; Quark--Gluon versus Meson-Exchange, preprint, to appear in Nucl. Phys. A
7) A. Fäßler, F. Fernandez, G. Lübeck and K. Shimizu, Phys. Lett. 112B (1982) 201
8) M. Oka and K. Yazaki, Progr. Theor. Phys. 66 (1981) 556, 572
9) C. Elster, Diploma thesis, unpublished, and private communication
10) O.W. Greenberg and H.J. Lipkin, Nucl. Phys. A370 (1981) 349
11) C.W. Wong, Proc. Int. School of Nucl. Physics, Erice, 21-30 April 1981, Pergamon Press, Oxford 1982
12) G.A. Miller, private communication

QUARKS AND THE NN- AND $\bar{N}$N-INTERACTION

J.R. Bergervoet, P.C. van Campen, W.M. Derks,
T.A. Rijken, W.A. van der Sanden, J.J. de Swart, and P.H. Timmers

Institute for Theoretical Physics
University of Nijmegen
Nijmegen, The Netherlands

1. INTRODUCTION

One of the main problems in the nucleon-nucleon interaction today is: "What is the influence of the quark substructure of the nucleons?" Everybody is convinced that there should be quark effects, but how do we get a handle on them? One way to try to study this problem is to look at multiquark states. There are many predictions [1 - 11] of multiquark states in many different channels. But do these multiquark states really exist? What is the evidence? In our studies of the nucleon-nucleon (NN) and antinucleon-nucleon ($\bar{N}$N) interaction we have been looking at some aspects of this problem.

In the following section we will first give a quick review of some of the predictions for multiquark states. Next we indicate the construction [12, 13] of a modified effective range function, where the long range part of the interaction is treated correctly. This modified effective range function is used for the description of the $^1S_0$ phaseshift in a multienergy phaseshift analysis of the low energy pp-data below 30 MeV.

The P-matrix introduced by Jaffe and Low [14] is closely connected to this modified effective range function [15]. We made a multienergy phaseshift analysis where we parametrized this P-matrix. These P-matrix parameters can be compared with predictions for the six quark states [3,4,5]. Some other examples of very simple applications of the P-matrix are presented.

We will also look at the $\bar{N}$N-interaction. Recently a coupled channel $\bar{N}$N-potential was constructed [16] based on the Nijmegen Model D baryon-baryon potential [17]. This model D potential has been very successful in the description of the NN-data, as well as the YN-data. This $\bar{N}$N-potential gives a good fit to the presently available $\bar{N}$N scattering data. Using backward elastic $\bar{p}$p scattering data we present some evidence for the S(1940) meson, which could be a $Q^2\bar{Q}^2$-state.

Finally we look at another aspect of quarks. In the construction of NN-potentials from meson exchange one should take into account the quark substructure of the nucleons. In the last section we report on a preliminary estimate of quark structure effects in low-energy pp-scattering.

2. MULTIQUARK STATES

Specific predictions about multiquark states are almost all based on the MIT-bagmodel [18]. Let us first look at some of these predictions of the MIT-bagmodel, when all quarks Q (and antiquarks $\bar{Q}$) are in the lowest s-state of a spherical bag.

The mesonic multiquark states of the type $Q^2\bar{Q}^2$ were first considered by R.L. Jaffe [2] with quite interesting results. He found as lowest states a nonet of scalar mesons ($J^{PC} = 0^{++}$). The lowest meson of this nonet has I = 0, M = 640 MeV, and a flavor wave function $ud\bar{u}\bar{d}$. Because this state couples strongly to the $\pi\pi$-channel it will be wide. This state can be identified with the $\varepsilon(700)$. This wide ε-meson is very essential in the Nijmegen baryon-baryon potential [17,19]. In section 5 about some other applications of the P-matrix we will present once more some evidence for this meson. The other non-strange mesons of this scalar nonet are a degenerate I = 0 and I = 1 pair of states at M = 1.12 GeV. The corresponding flavor wavefunctions are for I = 0: $(u\bar{u} + d\bar{d})s\bar{s}$, and for I = 1: $u\bar{d}s\bar{s}$, $(u\bar{u} - d\bar{d})s\bar{s}$, and $d\bar{u}s\bar{s}$. These states couple strongly to the $K\bar{K}$-channel and can be identified with the S*(980) and $\delta(980)$ mesons. The mass shift from the bagmass M = 1.12 GeV to the observed meson mass M = 0.98 GeV ($K\bar{K}$ threshold) can be understood. The recognition of the states ε, S*, and δ as crypto-exotic states did solve a long standing puzzle in the old quark model, where these states were thought to be $^3P_0$ $Q\bar{Q}$-states. Now we believe [11] these $^3P_0$-mesons to appear around M = 1.28 GeV. We will not discuss here the baryonic multiquark states [1] of the type $Q^4\bar{Q}$, but only mention that also for these states there is some evidence [20].

Very important was the calculation [3] by R.L. Jaffe of the Q^6-states in a spherical bag [4,5]. The lowest state is an SU(3) flavor-singlet with $J^P = 0^+$, M = 2.167 GeV and a flavor wave function $u^2d^2s^2$. This state should appear as a bound state in the $^1S_0$ $\Lambda\Lambda$-channel about 64 MeV below the $\Lambda\Lambda$-threshold. Such a state can decay only via the weak interactions. A recent discussion about this state can be found in [21].

In the I = 1 NN channel Q^6-states are predicted with $J^P = 0^+$, M = 2.24 GeV and $J^P = 2^+$, M = 2.36 GeV. There is experimental evidence [22] for a $^1D_2$ resonance in pp-scattering at M ≃ 2.17 GeV. The mass shift from the predicted bagmass to the neighborhood of the NΔ-threshold (2.17 GeV) can be understood. However, it is unclear, if this is the predicted Q^6-state. Other interpretations [23] are also offered. Evidence for the $^1S_0$ bag state comes from P-matrix analyses. In section 4 we will give again some of this evidence.

The predictions of the multiquark states, when all quarks are in the lowest s-state of the spherical bag, are probably the strongest predictions. More speculative are the predictions of the orbitally excited states [6 - 11], which are based on the model of a fast rotating string [24]. The quarks (and antiquarks) are situated in two oppositely colored clusters each at one end of this string. The color charges of these clusters have to be complementary, such that the whole string is colorless. The string itself is formed by the color electromagnetic fields, which hold the colored clusters together.

When this model is applied to the baryon resonances, considered as a Q-Q^2 string, the results are very good [10]. This gives then some confidence for the applications to more speculative situations as the Q^2-$\bar{Q}^2$ strings [7,8,11] and Q^2-Q^4-strings [6]. With the model of the Q^2-$\bar{Q}^2$ string a very rich spectrum of baryonium states is predicted. There has lately been a very negative mood [25] about the experimental evidence for such states in $\bar{p}p$ scattering. In section 6 we will present some evidence for the S(1940) state. Even if this state exists, the exact nature of the state is of course unclear.

3. MODIFIED EFFECTIVE RANGE EXPANSION

An extensive literature [12,13,26] exists on effective range expansions in two body scattering. We will outline the method [12,27], but for clarity we will restrict ourselves to simple potentials. We assume that the potential between the two particles can be written as

$$V = V_S + V_L \quad ,$$

where the long range part V_L of the interaction is well-known. To avoid inessential complications we assume that V_L is finite at the origin ($r = 0$), and when $r \to \infty$

$$r\, V_L \sim e^{-\mu_L r} \quad .$$

We need to study the solutions of the radial Schrödinger equation for V_L

$$\left(\frac{d^2}{dr^2} - \frac{\ell(\ell+1)}{r^2} + k^2 - 2m\, V_L\right) \chi(r) = 0 \quad ,$$

where k is the relative c.m. momentum and m the reduced mass. This equation has two independent solutions.

The regular solution $\chi_R(r)$ is defined by the boundary condition at the origin

$$\chi_R(r) = \frac{r^{\ell+1}}{(2\ell+1)!!} \sum_{n=0}^{\infty} a_n\, r^n \qquad \text{with } a_0 = 1 \quad .$$

The asymptotic behavior of χ_R for $r \to \infty$ is

$$\chi_R(r) \sim \frac{N_L}{k^{\ell+1}} \sin\left(kr - \ell\frac{\pi}{2} + \delta_L\right) \quad .$$

This way we can determine for the long range potential V_L the phase shift $\delta_L(k)$ and the normalization constant $N_L(k)$. We omit in our notation the dependence of δ_L and N_L on the orbital angular momentum ℓ. The other solution, called irregular solution $\chi_I(r)$ will be defined by the boundary condition at the origin

$$\chi_I(r) = -\frac{(2\ell-1)!!}{r^\ell} \sum_{n=0}^{\infty} b_n r^n \quad ,$$

with $b_0 = 1$ and $b_{2\ell+1} = 0$. Strictly speaking this solution $\chi_I(r)$ is only irregular at the origin for $\ell \neq 0$. The asymptotic behavior of this irregular solution for $r \to \infty$ can be shown to be

$$\chi_I(r) \sim -\frac{k^\ell}{N_L} \cos\left(kr - \ell\frac{\pi}{2} + \delta_L\right) + H_L\, \chi_R(r) \quad .$$

This defines the function $H_L(k)$. We can now define the effective range function

$$X(k) = \frac{1}{N_L^2} k^{2\ell+1} \cot(\delta - \delta_L) + H_L \quad ,$$

where $\delta(k)$ is the phaseshift belonging to V. This effective range function X(k) has nice analyticity properties in the complex k^2-plane or in the E-plane. The function X(E) is regular in a large region containing the origin. It is a real analytic function, so

$$X(E) = X^*(E^*) \quad .$$

The closest righthand singularity appears for $E = E_R$ where the first inelastic threshold opens up. The closest lefthand singularity E_L is determined by the range of the potential V_S. When we assume that $r\,V_S \sim e^{-\mu_S r}$ when $r \to \infty$, then the first lefthand singularity appears for $k^2 = -(\mu_S/2)^2$ or $E_L = -(\mu_S^2/8m)$.

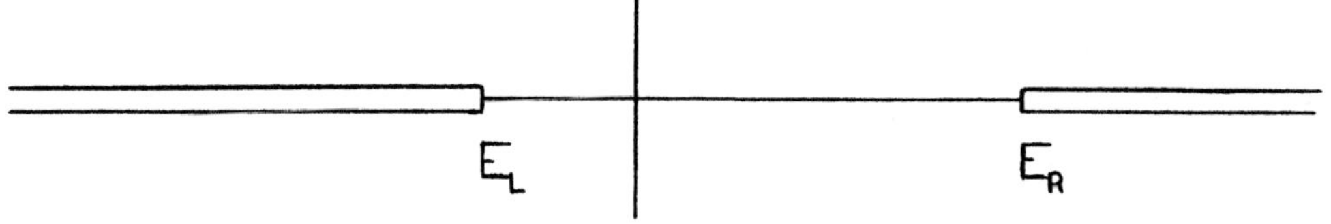

Fig. 1. Region of analyticity of X(E) in the complex E-plane.

The simplest and most well-known example of an effective range function is obtained when we take $V_L = 0$. Then we get $\delta_L = H_L = 0$, $N_L = 1$ and

$$X(k) = k^{2\ell+1} \cot \delta \quad .$$

Also well-known are the cases $V_L = V_C$ and $V_L = V_C + V_{VP}$, where V_C is the Coulomb potential and V_{VP} the vacuum polarization potential.

We have applied this method to the low energy pp-scattering in the $^1S_0$ state. We took

$$V_L = \tilde{V}_C + V_{VP} + V_\pi \quad .$$

Here $\tilde{V}_C$ is the improved Coulomb potential [28], and V_π the static one-pion-exchange potential. We used the relativistic connection between k^2 and E. When we make for $X(k^2)$ the expansion

$$X(k^2) = -\frac{1}{a_\pi} + \frac{1}{2} r_\pi k^2 + \frac{pk^4}{1 - qk^2} \quad ,$$

then we get a very good fit to the $^1S_0$ phaseshift from $T_L = 0$ to 300 MeV. This method allows a determination of the π^0-proton coupling constant.

We made [29] a multienergy phaseshift analysis of 556 low energy pp-scattering data below $T_{lab} = 30$ MeV. We used 11 parameters to describe the various partial waves. The parameter q in the modified effective range expansion was fixed by [30] $\delta(^1S_0) = 0$ at 253 MeV. We searched also for the π^0-proton coupling constant. We found $g_0^2/4\pi = 14.1 \pm 1.0$ and $\chi^2_{min} = 504.1$. From the this way determined values of a_π, r_π, p, q, and g_0^2 we calculated the more customary [31] used scattering length a_E and effective range r_E. We found $a_E = -7.804 \pm 0.002$ fm and $r_E = 2.783 \pm 0.007$ fm in agreement with earlier determined values [31].

4. THE P-MATRIX

In the P-matrix approach [14] one divides the interaction region into two parts. In the region $r > b$ one assumes that the interaction is well-known and can be described by a potential V. The interaction in the region $r < b$ will be described by a boundary condition on the radial wave function $\chi(r)$ at $r = b$. This boundary condition is called the P-matrix, where

$$P(k;b) = \left[\frac{r}{\chi} \frac{d\chi}{dr} \right]_{r=b} \quad .$$

This is essentially the boundary condition model of Lomon and Feshbach [32].

This P-matrix approach can be related to the treatment [15] of the effective range function X(k) when we choose

$$V_L = 0 \quad \text{for } r < b \quad \text{and}$$
$$V_L = V \quad \text{for } r > b \quad .$$

In this special case one can write down directly the regular and irregular solutions of V_L in the region $r < b$. They are

$$\chi_R(r) = \frac{1}{k^{\ell+1}} J(kr) \quad \text{and} \quad \chi_I(r) = k^{\ell} N(kr) \quad ,$$

where $J(x) = x\, j_\ell(x)$, $N(x) = x\, n_\ell(x)$ with $j_\ell(x)$ and $n_\ell(x)$ the spherical Bessel and Neumann functions, and $J'(x) = dJ/dx$, $N'(x) = dN/dx$.

The modified effective range function X(k) and the P-matrix are related in this case by

$$X = k^{2\ell+1} \frac{PN(kb) - kbN'(kb)}{PJ(kb) - kbJ'(kb)} \quad .$$

This shows that the functions X and P have the same domain of analyticity. We have made use of this property in the following way. Instead of parametrizing the experimental data by the 4 parameters a_π, r_π, p, and q, one could just as well try to parametrize the P-matrix [33]. We used the one-pole-approximation

$$P(s;b) = c + \frac{r}{s - s_0} \quad .$$

This implies again 4 parameters: b, s_0, r, and c.

This one-pole-approximation was used [34] for the parametrization of the energy dependence of the $^1S_0$ phaseshift in an energy dependent phaseshift analysis of the 556 pp-scattering data below T_{lab} = 30 MeV. For the potential V_L in the region r > b we used the sum of the improved Coulomb potential $\tilde{V}_C$, the vacuum polarization potential V_{VP} and the static one-pion-exchange potential V_π. The higher partial waves were parametrized with 8 parameters. Because the low energy data do not determine b very well, we used some of Arndt's s-wave phaseshifts at higher energies to determine b. The other parameters were determined by

Table 1: P-matrix parameters.

	b(fm)	$\sqrt{s_0}$(GeV)	r (GeV^2)	c
P. Mulders [35]	1.18	2.34	4.8	
	1.28	2.25	5.4	
	1.38	2.17	4.9	
Y. Simonov [33]	1.44	2.11	2.7	2.95
our fit	1.61	2.08	3.36	4.51
bagmodel [14,34]	1.48	2.24	0.84	

the low energy data and gave a good fit to these data (χ^2_{min} = 504.1). This fit gives $g_0^2/4\pi$ = 14.1 ± 1.0.

In table 1 we compare our results with those of other analyses [33, 35] and with the predictions of the MIT-bagmodel. The agreement with the bagmodel predictions for the radius b and for the pole position is reasonable. Our result prefers a somewhat larger bagradius. The residue at the pole is, however, in disagreement. For a discussion, see [36].

5. SOME OTHER APPLICATIONS OF THE P-MATRIX

The relation between the P-matrix and the S-matrix is very simple when $V_L \equiv 0$. Then

$$S = - \frac{P\ H_2(kb) - kb\ H_2'(kb)}{P\ H_1(kb) - kb\ H_1'(kb)} \quad ,$$

where $H_1(x) = x\ h_\ell^{(1)}(x)$ and $H_1' = dH_1/dx$. The S-matrix has a pole when

$$P = \left[\frac{x\ H_1'(x)}{H_1(x)} \right]_{x = kb} \quad .$$

For s-waves this becomes P = ikb. When we have parametrized a set of data with the help of the P-matrix, we can then determine very easily the position E_R of the S-matrix pole. It is interesting to apply this method to some well-known cases.

(i) The phaseshifts in the $J^P = \frac{3}{2}^+$ partial wave in π^+p scattering around the Δ^{++} are well-known [37]. The best fit with V_L = 0 was obtained with the parameters b = .88 fm and $\sqrt{s_0}$ = 1.38 GeV. The pole in this P-matrix lies much higher than the standard Δ^{++}-mass M = 1.232 GeV. For the position of the S-matrix pole we find

$$E_R(\Delta^{++}) = (1.21 - i\ 0.05)\ \text{GeV} \quad .$$

This should be compared with the value [38]

$$E_R(\Delta^{++}) = (1210.6 \pm 0.5) - i(49.7 \pm 0.3)\ \text{MeV} \quad .$$

We find the agreement very good for such a simple approximation.

(ii) The phaseshifts in the I = J = 1 partial wave in $\pi^+\pi^-$-elastic scattering around the ρ are also well-known [39]. Our best fit (χ^2/datapoint ≈ 3) to these phaseshifts was obtained for b = 0.91 fm and $\sqrt{s_0}$ = 850 MeV. Again we notice that the P-matrix pole lies much higher than the standard ρ-mass M = 776 MeV. The S-matrix pole we found at

$$E_R(\rho) = (767 - i\ 65)\ \text{MeV} \quad .$$

(iii) Also the $\pi\pi$-phases in the I = J = 0 partial wave are known [39]. Discrepancies between phaseshift analyses exist. We used the Estabrooks et al. s-channel data. A best fit to these data was obtained for $b = 0.70$ fm and $\sqrt{s_0} = 850$ MeV. This is the ε-meson. The S-matrix pole we find at $E_R(\varepsilon) = (760 - i\ 200)$ MeV. The data of Biswas et al. [40] give the same result, only the width comes out smaller. The Estabrooks et al. t-channel data give very different results.

6. AN $\bar{N}N$-POTENTIAL AND THE S(1940) MESON

Recently we constructed a coupled channel $\bar{N}N$-potential [16] starting from the Nijmegen Model D baryon-baryon potential [17]. This Nijmegen D potential gives good fits to the NN-scattering data, the YN scattering data, and the binding energies of the Λ and Σ in infinite nuclear matter [41]. This meson-theoretical potential is cut off in the inner region and there a phenomenological potential has been added. The annihilation is described by purely phenomenological potentials. The total number of free parameters used in the construction of this potential is 14. We fit 970 scattering data up to $T_{lab} = 482$ MeV with χ^2/datapoint = 1.49. Especially the fit to the elastic $\bar{p}p$-scattering data is good at about all angles and energies. Very interesting is the fit to the $\bar{p}p$ backward elastic cross section [42] between $p_{lab} = 406$ MeV/c and $p_{lab} = 922$ MeV/c. This fit is shown in Fig. 2. The total χ^2 of the fit to these 30 datapoints is $\chi^2 = 27.0$. Although this appears to be quite an acceptable χ^2, one observes that the 6 neighboring points from $p_{lab} = 486$ MeV/c to $p_{lab} = 534$ MeV/c all lie above the theoretical curve and they contribute

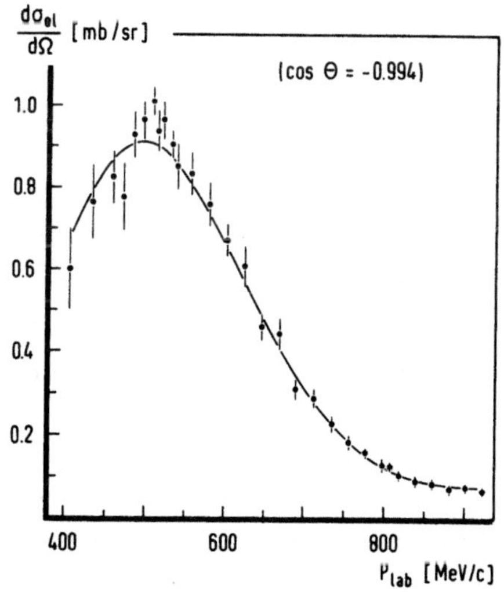

Fig. 2: The $\bar{p}p$ backward elastic cross section. Experimental points from [42].

together 13.0 to the total χ^2 of 27.0. We tried to test the statistical significance of this deviation for these 6 neighboring points as follows. First we would like to be reasonably sure that the experimental points (except for those 6) are spread around the theoretical curve according to a Gaussian probability function. The Pearson-χ^2 test [43] indicates that this is indeed the case with about 99% reliability. Assuming then such a Gaussian probability destribution for all 30 points, the probability to find a set of 6 points on a row with these (or larger) deviations from the curve is less than 0.5%. This suggests that there is indeed a statistically significant structure in these backward elastic data around p_{lab} = 509 MeV/c ($\sqrt{s}$ = 1940 MeV). When we blame this effect on the existence of a resonance, we can estimate the parameters of this resonance. We find

$$E_R = 1935 \pm 3 \text{ MeV} \quad \text{and} \quad \Gamma = 7.0 \pm 3 \text{ MeV} \quad .$$

These values for E_R and Γ are found by using a parametrization [44] of the elastic part of the S-matrix for a multichannel resonance in the presence of a background.

In a certain partial wave the elastic part of S is written as

$$S_{el} = S_{bg} + z\,(S_R - 1) \quad .$$

Here $S_{bg} = \eta\, e^{2i\delta}$ describes the elastic background,
$S_R = [E - E_R - i\ \Gamma/2]/[E - E_R + i\ \Gamma/2]$ is the Breit-Wigner form, and
$z = \frac{1}{2}\, e^{2i\delta}\,[\sqrt{(1+\eta)\Gamma_{el}/\Gamma} + i\ \sqrt{(1-\eta)\Gamma'_{in}/\Gamma}]^2$.
We write the total width Γ as

$$\Gamma = \Gamma_{el} + \Gamma'_{in} + \Gamma''_{in} \quad ,$$

where Γ_{el} is the partial width to the elastic channel, Γ'_{in} is the partial width to the (effective) channel to which also the elastic channel is coupled, and Γ''_{in} is the partial width to those (effective) channels to which the resonance is coupled, but from which the elastic channel is (effectively) decoupled.

In our model we claim that we can calculate the background (η and δ). To describe a resonance in a certain partial wave we have still four parameters (E_R, Γ, Γ_{el}, Γ'_{in}) to be adjusted. It then turns out that almost every partial wave can accommodate a resonance, such that the fit to the backward elastic cross section is much improved. We get typically $\chi^2 \lesssim 16$ for 30 datapoints. The values of E_R and Γ are almost independent of the partial wave in which we assume the resonance, whereas the values of Γ_{el} and Γ'_{in} vary widely.

We also tried to fit the structure in the backward cross-section using a coupled channels model. This consisted of the before mentioned

$\bar{N}N$-coupled channels model, supplemented with one additional confined channel. The idea behind this approach is, that $\bar{N}N$-scattering might show the existence of Q^2-$\bar{Q}^2$ states [7,8,11]. From the model of the fast rotating string for such states we take over the idea of a linear confining potential of the form

$$V = A + Br$$

with B = 972 MeV/fm, as well as an estimate M = 730 MeV/c^2 for the mass of the diquark system. Moreover we require an orbital angular momentum L unequal to zero for these Q^2-$\bar{Q}^2$ states.

The reaction $\bar{N}N \rightarrow \bar{Q}^2$-$Q^2$ can be described by the quark pair creation or $^3P_0$-model. This $^3P_0$-model suggests a transition potential between the $\bar{N}N$-channel and the $\bar{Q}^2$-Q^2 channel of the form

$$V_S(r) \sim r\, e^{-(mr)^2} \quad .$$

This form for the transition potential we have taken over in our simple model. Also in this case the data are not sufficient to constrain the resonance to a particular partial wave. For example we coupled the $\bar{N}N$ $^{11}P_1$-wave to a confined channel with relative orbital angular momentum L = 2. The constant term A in the potential is chosen such that the L = 2 bound state appears at 1940 MeV.

In this example the χ^2 for the 30 points is lowered from 27.0 to 15.9, σ_T decreases somewhat (maximally about 2.1 mb), and σ_{CE} increases (at maximum 0.32 mb). These changes in the cross section are about at the level of the current experimental error. In the future we can expect much more accurate results from LEAR, which might then well resolve the question whether the S-meson exists or not.

7. QUARK-ANTISYMMETRIZATION CORRECTIONS TO THE OBE-POTENTIALS

In low energy pp-scattering there are some discrepancies between recent phase shift analyses and the predictions of the better potential models [45]. For example the Nijmegen soft core OBE-model [19] fits the central combination Δ_C of the ^3P-phase shifts excellently, but for the tensor combination Δ_T a discrepancy exists between the model and experiment [46]. It appeared that this problem could not be solved by a simple refit of the model due to constraints imposed by the $^1S_0$- and the Δ_C-phase shifts.

In looking for an explanation of the discrepancies in $\Delta_T(^3P)$ in terms of quark structure effects we found that we can account for the differences by calculating the effects of the following processes

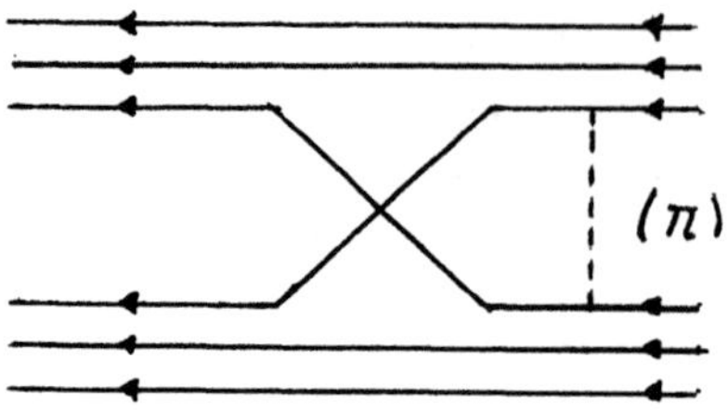

(a)

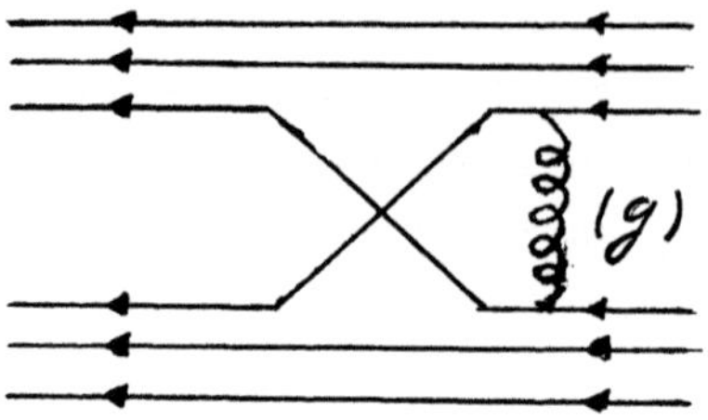

(b)

Process (a) is a correction to the OPE-potential which improves $\Delta_T(^3P)$. Process (b) is the contribution of gluon exchange, which in our calculation serves to cancel the central potential contribution from (a) to preserve $\delta(^1S)$ and $\Delta_C(^3P)$.

In this section we report briefly on our estimate of quark structure effects in low energy pp scattering. A more detailed version will be published elsewhere [47].

In the evaluation of the effect of (a) and (b) we use the resonating-group-method [48] formalism in a form as it has been applied to nucleon-nucleon scattering by Ribeiro [49] and Oka and Yazaki [50]. These authors calculated the contribution from (b). In diagram (a) and (b) the solid lines denote constituent quarks and their distribution in each nucleon is supposed to be provided by the nonrelativistic harmonic oscillator ground state wave function. The gluon and also the bosons π, ρ, ε, ... are supposed to couple directly to the quarks in the nucleons. We describe the nucleon at position $\underline{r}$ by

$$\phi(\underline{r}) = \int \Big(\prod_{i=1}^{3} d^3 r_i \Big) \, \delta^3\Big(\underline{r} - \frac{1}{3} \sum_{i=1}^{3} \underline{r}_i\Big) \, \varphi(\underline{r}_1, \underline{r}_2, \underline{r}_3) \, \varphi_{FS} \, \varphi_C \quad ,$$

where φ_{FS} and φ_C are the flavor-spin (SU(4)) and color (SU(3)) wave-functions respectively. The quark distribution is described by

$$\varphi(\underline{r}_1, \underline{r}_2, \underline{r}_3) = (\pi \sqrt{3} R^2)^{-3/2} \exp \Big[- \frac{1}{6R^2} \sum_{i<j} (\underline{r}_i - \underline{r}_j)^2 \Big] \quad ,$$

where R is the radius of the nucleon.
In the resonating group method the two-nucleon wave-function is

$$\psi(\underline{r}_1, \underline{r}_2) = \quad \{ \chi(\underline{r}_1 - \underline{r}_2) \, \phi_1(\underline{r}_1) \, \phi_2(\underline{r}_2) \}$$

where   is the antisymmetrizer (see [49]) and $\chi(\underline{r})$, with $\underline{r} = \underline{r}_1 - \underline{r}_2$, is the relative wave function. Variation with respect to the relative wave function leads to the integro-differential equation [51]

$$(-\Delta/2\mu + V(\underline{r}))\,\chi(\underline{r}) + \int K(\underline{r},\underline{r}')\,\chi(\underline{r}')\,d^3r' = E\,\chi(r)$$

where $\mu = M/2$, M = nucleon mass, and E = kinetic energy in cm-frame. In this equation $V(\underline{r})$ stands for the local potential (OBE-potential) and the non-local kernel $K(\underline{r},\underline{r}')$ contains all terms from the overlap

$$K(\underline{r}',\underline{r}) = \langle\psi'(\underline{r}_1',\underline{r}_2')|A(H-E)|\psi(\underline{r}_1,\underline{r}_2)\rangle$$

where A =   - 1 . The energy overlap $\langle\psi'|A(H_0 - E)|\psi\rangle$ is small in particular at low energies and we neglect it henceforth. So

$$K(\underline{r}',\underline{r}) \simeq \langle\psi'|AV|\psi\rangle$$

which defines the effects due to the quark structure of the nucleons in the present treatment. In fact one can view $K(\underline{r}',\underline{r})$ as a correction to the OBE-potential plus the effect of one-gluon exchange. In principle $K(\underline{r}',\underline{r})$ contains besides (a) also contributions from ρ, ω, ε, etc. It appeared that at low energies these latter contributions are much smaller than those from π-exchange and we neglect them. Also one gets contributions where the gluon and the mesons couple to quarks that are not operated on by A, but these are much smaller than (a) and (b) and so we neglect these too. In the calculation of $K(\underline{r}',\underline{r})$ for (a) and (b) we use the following quark-quark interactions.

$$V_{ji}(\pi) = (f^2/4\pi)(\underline{\tau}_j\cdot\underline{\tau}_i)\left[\tfrac{1}{3}\,(\underline{\sigma}_j\cdot\underline{\sigma}_i)\,\phi(\underline{r}_{ij}) + S_{ij}\,\chi(r_{ij})\right]$$

$$V_{ji}(g) = (\lambda_j\cdot\lambda_i)\left[a\,\underline{r}_{ij}^2 + b\,(\underline{\sigma}_j\cdot\underline{\sigma}_i)\right] \quad .$$

In $V_{ji}(\pi)$ $\phi(r)$ and $\chi(r)$ are the usual spin-spin and tensor pion-exchange potential functions $\phi(r) = \exp(-mr)/mr$ and $\chi(r) = \frac{1}{3}\left(1 + \frac{3}{mr} + \frac{3}{(mr)^2}\right)\phi(r)$. In $V_{ji}(g)$ the central and spin-spin term are taken similar as those of Ribeiro [49]. For the evaluation of $K(\underline{r}',\underline{r})$ we make the following approximation in the pion propagator

$$(\underline{K}^2 + m_\pi^2)^{-1} \simeq \sum_{\alpha=1}^{2} a_\alpha \exp(-\underline{K}^2/m_\alpha^2)$$

in order to be able to carry out analytically the internal integration, which occurs implicitly in $K(\underline{r}',\underline{r})$. Here the parameters are $a_1 = 3/4$, $m_1 = m_\pi$ and $a_2 = 1/4$, $m_2 = 3\,m_\pi$. Then we find for the nonlocal kernel

(i) π-exchange (a):

$$K_{ji}^{(\pi)}(\underline{r}',\underline{r}) = \sum_{\alpha=1}^{2} a_\alpha \frac{3^{9/2}}{2^8\,\pi^2}\left(\frac{f^2}{4\pi}\,\frac{m_\alpha^3}{m_\pi^2\,R^3}\right)(\tau_j\cdot\tau_i)P_{ji}\Big[(1 + 3\,m_\alpha^2\,r_{12}^2)$$

$$\times\,(\underline{\sigma}_j\cdot\underline{\sigma}_i) + 9\,m_\alpha^2\,((\underline{\sigma}_j\cdot\underline{r}_{12})(\underline{\sigma}_i\cdot\underline{r}_{12}) - \tfrac{1}{3}\,\underline{r}_{12}^2\,(\underline{\sigma}_j\cdot\underline{\sigma}_i))\Big]$$

$$\times\,\exp\left[-(15 + 9\,m_\alpha^2\,R^2)\,\frac{r'^2 + r^2}{16\,R^2}\right]\,\exp\left[\frac{9}{8}\,(1 + m_\alpha^2\,R^2)\,\frac{\underline{r}'\cdot\underline{r}}{R^2}\right]$$

where $\underline{r}_{12} = \underline{r}' - \underline{r}$ and P_{ji} is the quark interchange operator.
(ii) g-exchange (b):

$$K_{ji}^{(g)}(\underline{r}',\underline{r}) = -\left(\frac{3}{4\pi^{1/3}}\right)^{9/2} \frac{1}{R^3} (\lambda_j \cdot \lambda_i)\, P_{ji} \left[-\frac{9}{4}\, a\, r_{12}^2 + b\, (\underline{\sigma}_j \cdot \underline{\sigma}_i)\right]$$

$$\times \exp\left[-\frac{15}{16 R^2}(r'^2 + r^2)\right] \exp\left[\frac{9}{8 R^2}(\underline{r}' \cdot \underline{r})\right] \quad .$$

The total kernel is given by

$$K_{ji}(\underline{r}',\underline{r}) = K_{ji}^{(\pi)}(\underline{r}',\underline{r}) + K_{ji}^{(g)}(\underline{r}',\underline{r}) \quad .$$

The final steps in the evaluation of the corrections to the phase shifts are (i) the performance of the radial integrals using Coulomb wave functions, (ii) the evaluation of the $SU(4)_{FS}$ and $SU(3)_C$ matrix elements of the operators $(\lambda_j \cdot \lambda_i)P_{ji}$ etc. occurring in $K_{ji}^{(g)}$ and $K_{ji}^{(\pi)}$, (iii) partial wave expansion. We skip here the details of (i) - (iii) and will only give the final results.

As to (ii) we note that in the "intermediate configurations", which occur between the gluon or pion exchange and the quark pair exchange in (a) and (b), we must include all isospin, spin, and color configurations. We did not solve the integro-differential equation in our estimate discussed here, but evaluate the effect of the kernel $K(\underline{r}',\underline{r})$ in the Distorted-Wave-Born-Approximation (DWBA). This is reasonable at low energies. The correction to the phase shift is

$$\delta\, \Delta_\ell = -\frac{M}{p} \int_0^\infty dr_1 \left[r_1\, F(pr_1) \int_0^\infty dr_2\, K(r_1,r_2)\, F(pr_2)\right]$$

where p denotes the cm-momentum and $F_\ell(pr)$ are the Coulomb wave functions.

We find the following results for the tensor phase shifts $\Delta_T(^3P)$:

T_L(MeV)	6.14	10.00	16.00
$\Delta_T(\text{OBE})^{(a)}$	- 0.537	- 0.989	- 1.640
$\delta\, \Delta_T^{(b)}$	+ 0.050	+ 0.104	+ 0.202
$\Delta_T(\text{tot})$	- 0.487	- 0.885	- 1.438
$\Delta_T(\exp)^{(c)}$	- 0.506	- 0.837	- 1.549
	± 0.160	± 0.054	± 0.051

(a) Nijmegen potential [19]
(b) $a = 100$ MeV fm^{-2}, $b = 67.5$ MeV
(c) phase shift analysis [46].

We observe a movement into the right direction.

Finally we note that we have treated b as an adjustable parameter in order to bring about the approximate cancellation of (a) and (b) in the central phase shifts. We found a value which is considerable smaller than that in [49]. It could be argued that this is a signal of the relevance of other quark cluster wave functions than those contained in $\psi(\underline{r}_1,\underline{r}_2)$ (see above) [52] for the description of scattering processes.

Acknowledgements

Part of this work was included in the research program of the Stichting voor Fundamenteel Onderzoek der Materie (F.O.M.) with financial support from the Nederlandse Organisatie voor Zuiver-Wetenschappelijk Onderzoek (Z.W.O.).

References

1. R.L. Jaffe and K. Johnson, Phys.Lett. 60B (1976) 201.
2. R.L. Jaffe, Phys.Rev. D15 (1977) 267 and 281.
3. R.L. Jaffe, Phys.Rev.Lett. 38 (1977) 195, E 38 (1977) 617.
4. A.T. Aerts, P.J. Mulders, and J.J. de Swart, Phys.Rev. D17 (1978) 260.
5. V. Matveev and P. Sorba, Lett.Nuov.Cim. 20 (1977) 435.
6. P.J. Mulders, A.T. Aerts, and J.J. de Swart, Phys.Rev.Lett. 40 (1978) 1543 and Phys.Rev. D21 (1980) 2653.
7. R.L. Jaffe, Phys.Rev. D17 (1978) 1444.
8. Chan H.-M. and H. Høgaasen, Nucl.Phys. B136 (1978) 401.
9. H. Høgaasen and P. Sorba, Nucl.Phys. B145 (1978) 119.
10. P.J. Mulders, A.T. Aerts and J.J. de Swart, Phys.Rev. D19 (1979) 2635.
11. A.T. Aerts, P.J. Mulders and J.J. de Swart, Phys.Rev. D21 (1980) 1370.
12. G.J. Austen, Ph.D. thesis, University of Nijmegen, March 1982.
13. H. van Haeringen, L.P. Kok, Phys.Rev. A26 (1982) 1218.
14. R.L. Jaffe, F.E. Low, Phys.Rev. D19 (1979) 2105.
15. J.J. de Swart, Lecture Notes, University of Nijmegen, 1981, THEF-NYM-81.7 (unpublished).
16. P.H. Timmers, W.A. van der Sanden, and J.J. de Swart, Nijmegen preprint THEF-NYM-83.06.
17. M.M. Nagels, T.A. Rijken, and J.J. de Swart, Phys.Rev. D12 (1975) 744 and Phys.Rev. D15 (1977) 2547.
18. A. Chodos et al., Phys.Rev. D9 (1974) 3471 and Phys.Rev. D10 (1974) 2599; T. DeGrand et al., Phys.Rev. D12 (1975) 2060.
19. M.M. Nagels, T.A. Rijken, and J.J. de Swart, Phys.Rev. D17 (1978) 768.
20. J.J. de Swart, P.J. Mulders, and L.J. Somers, Proceedings of the IVth International Conference on Baryon Resonances (ed. N. Isgur), Toronto (Canada), 14 - 16 July 1980, pg. 405.
21. A.T.M. Aerts and C.B. Dover, Phys.Rev. D28 (1983) 450 and Phys.Rev. Lett. 49 (1982) 1752.
22. A. Yokosawa, Phys.Rep. 64C (1980) 47; N. Hoshizaki, Progr.Theor.Phys. 60 (1978) 1796 and Progr.Theor.Phys. 61 (1978) 129; R.A. Arndt et al., Virginia preprint VPISA-2 (82) (1982).
23. W.M. Kloet and R.R. Silbar, Nucl.Phys. A338 (1980) 281, 317 and Phys. Rev.Lett. 45 (1980) 970; M. Araki, Y. Koike and T. Ueda, Nucl.Phys. A389 (1982) 605; M. Araki and T. Ueda, Nucl.Phys. A379 (1982) 449; Y. Avishai and T. Mizutani, Phys.Rev. C27 (1983) 312; T. Mizutani et al., Phys.Lett. 107B (1981) 177; B. Blankleider and I.R. Afnan, Phys. Rev. C24 (1981) 1752.

24. K. Johnson and C.B. Thorn, Phys.Rev. D13 (1976) 1934.
25. Proceedings of the 5th European Symposium on Nucleon-Antinucleon Interactions, Bressanone (Italy), 23 - 28 June 1980.
26. L. Heller, Phys.Rev. 120 - 2 (1960) 627; J. Blatt and D. Jackson, Phys.Rev. 76 (1949) 18; H.A. Bethe, Phys.Rev. 76 (1949) 38.
27. W.A. van der Sanden, G.J. Austen, and J.J. de Swart (in preparation).
28. G.J. Austen and J.J. de Swart, Phys.Rev.Lett. 50 (1983) 2039.
29. W.A. van der Sanden, et al. (in preparation).
30. R.A. Arndt, R.H. Hackmann, and R.D. Roper, Phys.Rev. C15 (1977) 1002.
31. W.A. van der Sanden, A.H. Emmen, and J.J. de Swart, Nijmegen preprint THEF-NYM-83.11.
32. E.L. Lomon, H. Feshbach, Ann.Phys. (NY) 29 (1964) 19.
33. Y.A. Simonov, Phys.Lett. B107 (1981) 1.
34. J. Bergervoet, P. van Campen, W. van der Sanden, and J.J. de Swart (in preparation).
35. P.J. Mulders, Phys.Rev. D28 (1983) 443.
36. P.J. Mulders, Proceedings of the "10th International Conference on Few Body Problems in Physics", Karlsruhe (Germany), 21 - 27 August 1983 (to be published).
37. G. Höhler et al., Handbook of Pion-Nucleon Scattering, Physics Data 12 - 1 (1979) 97.
38. PDG, Phys.Lett. 111B, April 1982.
39. P. Estabrooks et al., Nucl.Phys. B79 (1974) 301.
40. N.N. Biswas et al., Phys.Rev.Lett. 47 (1981) 1378.
41. J. Rozynek and J. Dabrowski, Phys.Rev. C20 (1979) 1612 and C23 (1981) 1706;C.B. Dover and A. Gal, BNL-preprint 30124 (1981); H. Bandō, Progr.Theor.Phys. 66 (1981) 1349.
42. M. Alston-Garnjost et al., Phys.Rev.Lett. 43 (1979) 1901.
43. L.G. Parratt, Probability and Experimental Errors in Science, Wiley & Sons.
44. W.A. van der Sanden and P.H. Timmers, Parametrization of a Resonance in the Presence of a Multichannel Background (in preparation).
45. J.J. de Swart, W.A. van der Sanden, and W. Derks, Proceedings of the "10th International Conference on Few Body Problems in Physics" Karlsruhe (Germany), 21 - 27 August 1983 (to be published).
46. W.A. van der Sanden et al., Comments on the P-wave parameters Δ_{LS} and Δ_T in p-p scattering below 30 MeV (in preparation).
47. T.A. Rijken and W. Derks (in preparation).
48. J. Wheeler, Phys.Rev. 52 (1937) 1083, 1107.
49. J.E.F.T. Ribeiro, Z.Physik C 5 (1980) 27.
50. M. Oka and K. Yazaki, Progr.Theor.Phys. 66 (1981) 556.
51. See for example the excellent review: K. Wildermuth and W. McClure, Springer Tracts in Modern Physics, 41 (Springer, 1966).
52. M. Harvey, Nucl.Phys. A 352 (1980) 301, 326; A. Faessler, F. Fernandez, and G. Lübeck, Phys.Lett. 112B (1982) 201.

BOSON-MEDIATED INTERACTIONS BETWEEN STATIC SOURCES

M. Bolsterli
T Division, MS B279, Los Alamos National Laboratory
Los Alamos, New Mexico 87545, USA

As we have seen in previous presentations to this symposium, today's framework for nuclear theory includes a short-range interaction dominated by quarks and gluons, which obey the laws of quantum chromodynamics, and a long-range region in which the interaction is carried by pions. The long-range interaction may or may not be carried also by heavier mesons; this point is controversial, and it will be some time before the role of the heavier mesons is settled. Dr. Machleidt's talk[1] showed that the trend is toward the replacement of at least some of the heavy-meson exchange interaction by multi-pion exchange interactions of various types. At present it seems that the pionic interaction will continue to be an essential part of the interaction between nucleons. It is important to understand the pionic nucleon-nucleon interaction quantitatively so that we can determine the nature of whatever other interaction is required by the nucleon-nucleon data; presumably the nonpionic interactions must be due to quarks and gluons (and possibly heavy-meson exchange). The exact role of the delta resonance is also a question; uncertainty here leads to corresponding uncertainty in the theoretical treatment of pionic interactions. The talk by Dr. Miller[2] showed that it is most attractive to treat the delta as a distinct state of the three quarks that constitute the nucleon, although any model of the delta naturally includes a substantial nucleon-plus-pion field component.

The theoretical treatment of mesonic nuclear interaction has been of two types, one that is indirect, and one that is direct. The indirect treatment first converts the meson field theory into a nucleon-nucleon potential (and a nucleon-nucleon-nucleon three-nucleon potential, a four-nucleon potential, etc.) and then uses the potentials to compute nuclear properties like nucleon-nucleon scattering. This approach is exemplified by the work of the Bonn,[1] Nijmegen,[3] and Paris[4] groups. It has the advantage that the parameters of the theory are severely restricted by being required to reproduce the extremely accurate two-nucleon data in a reasonable way.

In multinucleon systems a more direct approach has been used, in which the meson quantum field is treated as an essential component of the state vector of the system. The underlying Hamiltonian is taken to be of the form

$$
\begin{aligned}
H &= T_F + T_B + H_Y + H_Y^+ \\
T_F &= \int t(\vec{p})\, \psi^+(\vec{p})\psi(\vec{p})\, d\vec{p} \\
T_B &= \sum_\lambda \int \omega(k)\, a_\lambda^+(\vec{k})a_\lambda(\vec{k})\, d\vec{k} \qquad (1) \\
H_Y &= -\, g \sum_\lambda \int a_\lambda^+(\vec{k})\, \psi^+(\vec{p})W_\lambda(\vec{k},\tfrac{\vec{p}+\vec{q}}{2})\psi(\vec{q})\, \delta(\vec{p}+\vec{q}-\vec{k})\, d\vec{p}\, d\vec{q}\, d\vec{k} \quad .
\end{aligned}
$$

Since the nucleon is no longer to be regarded as an elementary particle it is reasonable to use a noncovariant form for the Hamiltonian. Thus, the form factor $W_\lambda(\vec{k},\vec{K})$ can be chosen to be such that there are no infinities in the system. From the talks that we have heard on lattice gauge theories, it seems likely that in the reasonably near future the pion-nucleon form factors $W_\lambda(\vec{k},\vec{K})$ will be provided to us by the computational results of those theories. In treating Hamiltonians like that given above, the Stanford[5] and Brooklyn[6] groups have used what is basically a mean-field technique to treat scalar and vector meson fields. Since there is no pion mean field in even nuclei, a generalization of the mean-field theory is required in order to treat the pion case; this is provided by the recently developed[7] "meson-nucleon shell model." The direct techniques have the advantage of being able to deal with strong coupling of the meson fields, but they have not yet been applied to the two-nucleon system, so that the parameters of the direct approach have not been constrained by the two-nucleon data.

The aim of the present work is to apply the techniques of the direct approach to meson-nucleon states to the treatment of the potential between static sources of meson field, with a view to testing the validity of some of the approximations used in the calculatons of the indirect potentials between nonstatic sources. In the case of static sources, new methods have been developed that permit a reasonably exact treatment of the meson field for any coupling strength, and comparison with one-meson-exchange interactions can be made explicit.

The static sources have the distributions $\rho_o(\vec{r}-\vec{R}_1)$ and $\rho_o(\vec{r}-\vec{R}_2)$, where $\rho_o(r)$ is the intrinsic density distribution of a single source, taken to be spherically symmetric. Then for isovector scalar meson field interacting with the sources, the Hamiltonian of (1) becomes

$$H = T_B + H_I + H_I^\dagger \quad , \tag{2}$$

where T_B is given in Eq. (1) and

$$H_I = -g \sum_{p=1}^{2} \sum_{\lambda} \tau_\lambda^p \int a_\lambda^\dagger(\vec{k}) \frac{\tilde{\rho}_o(k)}{[16\pi^3\omega(k)]^{1/2}} e^{-i\vec{k}\cdot\vec{R}_p} d\vec{k} \quad , \tag{3}$$

where $\tilde{\rho}_o(k)$ is the Fourier transform of $\rho_o(x)$, and τ^p is the isospin operator for source p.

The ground-state energy $E_g(R)$ of the Hamiltonian depends on the magnitude R of the source separation $\vec{R}$,

$$\vec{R} = \vec{R}_1 - \vec{R}_2 \quad . \tag{4}$$

Once $E_g(R)$ is determined, the static meson potential is given by

$$V(R) = E_g(R) - 2E_1 \quad , \tag{5}$$

where E_1 is the ground-state energy of a single source interacting with the meson field. As will be seen, the magnitude of V(R) is often of the order of up to a few percent of the magnitude of $E_g(R)$ over much of the range of R values of interest, so it is important to have accurate methods for calculating E_g and E_1. The case of isovector-scalar meson field may be regarded as a testing ground for developing accurate techniques that can be applied later to the more difficult case of pion mediating field.

The static meson potential computations involve five basic ideas: they are (1) variational calculations based on (2) decomposition of the meson field operator into mutually commuting internal and external parts; (3) a restricted set of one or two internal modes is chosen in a particular way so as to give (4) correct asymptotic behavior of the static meson potential; in all cases (5) accurate strong-coupling techniques are used to generate the trial variational states.

The decompositon of the meson field operator $a(\vec{k})$ into internal and external parts

$$a(\vec{k}) = a_{int}(\vec{k}) + a_{ext}(\vec{k}) \tag{6}$$

with

$$\begin{aligned} a_{int}(\vec{k}) &= \sum_{1}^{N} A_i \, \phi_i(\vec{k}) \\ (\phi_i, a_{ext}) &= 0 \; , \; i = 1,2,\ldots N \\ (\phi_i, \phi_j) &= \delta_{ij} \\ [A_i, A_j^+] &= \delta_{ij} \end{aligned} \tag{7}$$

has been discussed at some length in references 7-10. The L_2 functions $\phi_i(\vec{k})$ are the internal mode functions or wave functions of the meson field; some specific forms are given below. The variational calculations are carried out within a subspace of the full Hilbert space; the subspace is generated by the internal-mode creation actors A_i^+ acting on the bare source state. The effective Hamiltonian in this subspace is H_{int}, where H_{int} is just H with each $a(\vec{k})$ or $a^+(\vec{k})$ replaced by $a_{int}(\vec{k})$ or $a_{int}^+(\vec{k})$. For the case of a single source, it was shown in reference 11 that a single meson mode really does give an adequately accurate approximation to the single-source ground state. Procedures for computing corrections due to the external meson field have been given in references 8 and 10.

When a single mode is used for the single source and two modes are used for the case of two sources, it has recently been shown[12] that the correct asymptotic behavior

$$V(R) \to - \frac{g_R^2}{4\pi} \tau_1 \cdot \tau_2 \int \rho_1(\vec{r}_1) \frac{e^{-m|\vec{r}_1 - \vec{r}_2|}}{|\vec{r}_1 - \vec{r}_2|} \rho_2(\vec{r}_2) \, d\vec{r}_1 \, d\vec{r}_2 \tag{8}$$

is obtained as R becomes large. This demonstration shows that the static meson potentials as computed are valid for large R and supports the use of only two modes for other values of R. The same technique that gives the correct asymptotic potential also gives an interesting variational "distorted-field" approximation (DFA) to the potential at any value of R.

In a previous computation of static meson potentials[13] the asymptotic behavior of the potential was not treated correctly; an unjustified ad hoc procedure was used to determine the zero of the potential.

The strong-coupling techniques are based on the use of coherent meson pairs[11] in the variational state vectors; this method allows an accurate energy to be computed with a fraction of the number of state components that would otherwise be required. Strong-coupling techniques are required for treating the single source and therefore also for two sources even in the asymptotic region. In the region of small R, special strong-coupling states appropriate to two isospin-$\frac{1}{2}$ sources have been used.

For the Hamiltonian of Eq. (3), the appropriate mode for the case of a single source at the origin has been shown in references 8 and 14 to be

$$\phi(k) = \frac{1}{G\omega(k)} \frac{g\tilde{\rho}_o(k)}{[16\pi^3\omega(k)]^{1/2}} , \tag{9}$$

where G is both the normalization constant for the mode ϕ and a "normalized coupling constant," since with (9) it follows that

$$H_I = - G \sum_{\lambda p} \tau_\lambda^p \int \omega(k)\, a_\lambda^+(\vec{k})\, \phi(k)\, e^{-i\vec{k}\cdot\vec{R}_p}\, d\vec{k} \quad ; \tag{10}$$

comparison with T_B shows that G measures the strength of H_I relative to the strength of T_B. The static-model calculations of the 1950's used the form

$$\frac{1}{G(\omega(k) + \lambda)} \frac{g\tilde{\rho}_o(k)}{[16\pi^3\omega(k)]^{1/2}} , \tag{11}$$

but it has since been shown that the appropriate value of λ in this case is zero.[8,10,15,16]

When there are two sources present, the minimum number of modes required to obtain the independent-source modes asymptotically is two, one symmetric and one antisymmetric. These are assumed to have the forms

$$\phi_\pm(\vec{k}) = \frac{1}{n_\pm\sqrt{2}}\,\phi(k)\,[\,e^{-i\vec{k}\cdot\vec{R}_1} \pm e^{-i\vec{k}\cdot\vec{R}_2}\,] \quad , \tag{12}$$

where $\phi(k)$ is the single-source mode function of (9), and the mode normalization functions $n_\pm(R)$ are given by

$$\begin{aligned} n_\pm(R) &= 1 \pm \int |\phi(k)|^2 \cos\vec{k}\cdot\vec{R}\, d\vec{k} \\ &= 1 \pm \frac{1}{R}\int \sin kR\, |\phi(k)|^2\, kdk \quad . \end{aligned} \tag{13}$$

Corresponding to the modes $\phi_\pm$ there are annihilation operators $A_\pm$. It is also useful to introduce the $\tau_\pm$ operators

$$\tau_\pm = \frac{1}{\sqrt{2}}(\tau^{(1)} \pm \tau^{(2)}) \quad . \tag{14}$$

In the asymptotic large-R region, the distorted normal modes that are orthogonal and go over into single-source modes as $R\to\infty$ are ϕ_1 and ϕ_2, where

$$\phi_\pm(\vec{k}) = \frac{1}{\sqrt{2}}(\phi_1(\vec{k}) \pm \phi_2(\vec{k})) \quad . \tag{15}$$

The mode ϕ_1 differs from $\phi(k)e^{-i\vec{k}\cdot\vec{R}_1}$ because of the factors $n_\pm$ in Eq. (12). Corresponding to ϕ_1 and ϕ_2 are the annihilation operators A_1 and A_2:

$$A_\pm = \frac{1}{\sqrt{2}}(A_1 \pm A_2) \quad . \tag{16}$$

For the source form factor $\rho_o(k)$ it is possible to make various choices. The particular choice

$$\rho_o(k) = \frac{1}{(1+\frac{k^2}{\Lambda^2})} \tag{17}$$

has been used in the work presented here. This choice is suggested by the nonrelativistic limit of the covariant scalar source density. It is also possible to use a cloudy-bag source function or a function with a sharp cutoff. The requirement is only that no divergent integrals be obtained. As was mentioned earlier, it is expected that the relevant form factors will eventually be computed from lattice gauge theory.

The above choices of meson modes and internal field lead to corresponding internal Hamiltonians. For the single source, the internal Hamiltonian is

$$H_{int}^{(1)} = W \left[A_1^{+}\cdot A_1 - G\tau^{(1)}\cdot(A_1^{+} + A_1) \right] , \tag{18}$$

where W is the expectation value of the meson energy $\omega(k)$ in the meson mode $\phi(k)$,

$$W = \int \omega(k) \; |\phi(k)|^2 \, d\vec{k} \quad . \tag{19}$$

Despite the apparent simplicity of $H_{int}^{(1)}$, its eigenvalues have so far only been determined numerically. When there are two sources, the Hamiltonian is most simply written in terms of $A_{\pm}$:

$$\begin{aligned} H_{int} &= \sum_{s=\pm} W_s \left[A_s^{+}\cdot A_s - G n_s \tau_s\cdot(A_s^{+}+A_s) \right] \\ W_s &= \int \omega(k) \; |\phi_s(\vec{k})|^2 \, d\vec{k} \quad , \end{aligned} \tag{20}$$

but for the asymptotic region the description in terms of A_1 and A_2 is more useful:

$$\begin{aligned} H_{int} &= \sum_{p=1}^{2} H^{p}(R) \; + H_{12}(R) \\ H^{p}(R) &= \frac{W_+ + W_-}{2} \left[A_p^{+}\cdot A_p - G(R)\tau_p\cdot(A_p^{+}+A_p) \right] \\ G(R) &= G\,\frac{n_+W_+ + n_-W_-}{W_+ + W_-} \\ H_{12} &= \frac{W_+ - W_-}{2}\left[A_1^{+}\cdot A_2 + A_2^{+}\cdot A_1\right] - G\,\frac{n_+W_+ - n_-W_-}{2}\left[\tau_1\cdot(A_2^{+}+A_2) + \tau_2\cdot(A_1^{+}+A_1)\right] \quad . \end{aligned} \tag{21}$$

Since the modes $\phi_{\pm}$ depend on R, it follows that $W_{\pm}$ and H_{int} also depend on R.

For a given separation of the two sources, H_{int} can be diagonalized over a selected subspace of the Hilbert space. Figures 1 and 2 show the results of some calculations for the case that $g^2/4\pi$ is 1.0 and the cutoff Λ in (17) is taken to be 7m, where m is the meson mass. The computed value of G is .735; the renormalized value of $g^2/4\pi$ is .245. The value of E_1 is −5.88m, so that $2E_1$ takes the rather large value of −11.76m. In both of these figures, the solid curve shows the result of using the form (21) for H_{int} with a single state that is the product of the ground states of $H^{(1)}(R)$ and $H^{(2)}(R)$; this is the "distorted-field" approximation to the potential (DFAP). The dashed curve in both figures shows the short-range potential (SRP), which is the result of diagonalizing H_{int} in the form given by Eq. (20) over subspaces with three states for T=0 and six states for T=1. Each state in the subspace is a coherent-pair state[11] constructed on a basic state. The basic states include all basic states with up to two mesons. Coherent pairs are used for both the + and − modes. The dot-dashed curve in both figures is the renormalized one-meson-exchange potential (ROMEP). As was shown in reference 12, the DFAP is asymptotically equal to the ROMEP.

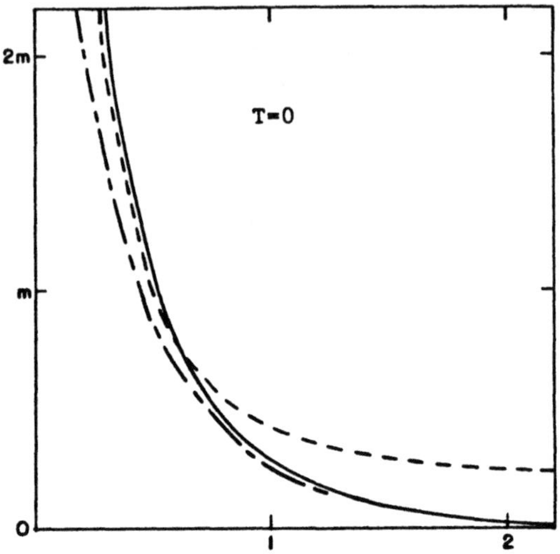

Figure 1

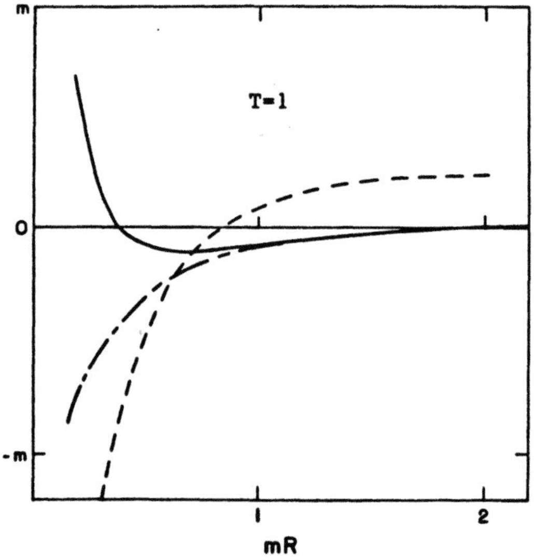

Figure 2

Note that the ROMEP is not a variational result, so that it is not a candidate for the potential at any R. At each R, the calculated potential is the lesser of DFAP and SRP. As would be expected, the SRP is lower than the DFAP for small values of R. For large values of R the SRP cannot treat the asymptotic single-source meson clouds correctly; only an approximation based on the DFA can be used for large R. As R decreases, the difference between the T=0 potential and the T=1 potential grows; at the value R_c the difference becomes equal to the meson mass. For $R < R_c$ the lowest T=0 state is not the one that gives the T=0 SRP, but rather its energy is bounded by the energy of the lowest T=1 state plus the energy m of a meson at infinity.[13]

Figure 3 shows a plot of the combined lowest potentials. The cusp in the T=1 potential shows that the DFAP needs to be improved by using more states in the subspace in which H_{int} of Eq. (21) is diagonalized. The T=0 potential in the vicinity of R_c can be improved by adding to the subspace states with one external-meson creation operator, so that the lowest T=0 state becomes

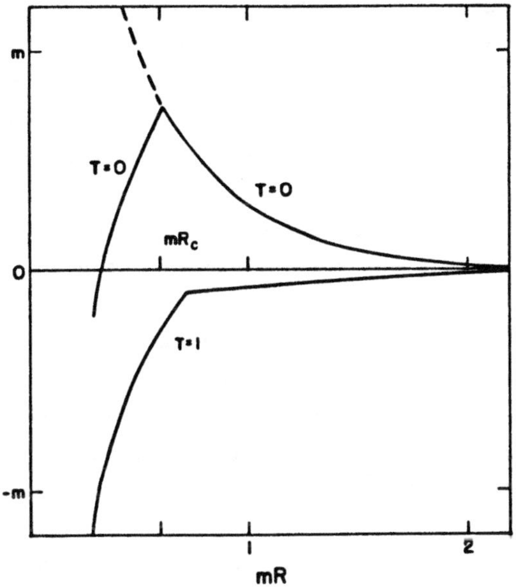

Figure 3

$$\alpha|T=0\rangle + \int f(\vec{k})\ \{a^+_{ext}(\vec{k}), |T=1\rangle\}^{T=0}\ d\vec{k} \quad ; \tag{22}$$

with this ansatz, the equations for α and $f(\vec{k})$ are like the equations for the solution of the Lee model in the V sector and are easily solved (this is not meant to imply that this has actually been done as yet). The result must be to round the cusp in the T=0 potential and to move the actual threshold for conversion of the T=0 state into the T=1 state plus a meson at infinity to a value of R smaller than R_c.

For small enough values of R both the T=0 and T=1 potentials are attractive. It is expected that this is a general feature of mesonic interactions when the source operators are nonabelian (like τ or σ). In the Abelian case it is known[13] that the full static potential is identical to the OMEP (there is no coupling-constant renormalization in this case), so that there can be a repulsive potential at short distances. The fact that nonabelian mesonic interactions always give attraction at short distances means that it is difficult to achieve saturation of nuclear forces in the usual old-fashioned way by using meson-exchange potentials with appropriate mixture of various nonabelian terms.

Summary and prospects

The techniques are now available for doing accurate computations of static potentials arising from the exchange of virtual mesons. Such computations must take account of the fact that different approximation methods must be used in the regions where R is large and where R is small. In the asymptotic region, the distorted-field approximation provides an appropriate starting-point, but it must be improved before trustworthy results are obtained for all but the largest values of R. In the region of small R, accurate strong-coupling methods are based on the use of states with coherent meson pairs. For small R, it is also important to take account of the possibility of meson emission or near-emission.

Current work is aimed at applying the techniques described here to the case of static sources interacting via pion field. In particular, it will be interesting to see how sensitive the potential is to the value of the cutoff Λ. Other areas of application are the study of the effects of nonlinearity[17] and models of quark-quark and quark-antiquark potentials.

This work was performed under the auspices of the U. S. Department of Energy.

References

1. H. Machleidt, invited paper at this symposium.
2. G. A. MIller, invited paper at this symposium.
3. J. de Swart, invited paper at this symposium.
4. R. Vinh Mau, in Mesons in Nuclei, edited by M. Rho and D. Wilkinson (North-Holland, New York, 1979), Vol. I, p. 151.
5. B. D. Serot and J. D. Walecka, Phys. Lett. 87B, 172 (1979) and references therein.
6. M. R. Anastasio, L. S. Celenza, and C. M. Shakin, Phys. Rev. C 23, 2258 (1981) and references therein.
7. M. Bolsterli, Phys. Rev. C 25, 2728 (1982).
8. M. Bolsterli, Phys. Rev. D 25, 1095 (1982).
9. M. Bolsterli, Phys. Rev. D 27, 2725 (1983).
10. M. Bolsterli, Phys. Rev. D 27, 2940 (1983).
11. M. Bolsterli, Phys. Rev. D 24, 400 (1981).
12. M. Bolsterli, Phys. Rev. D, in press.
13. M. Bolsterli, Phys. Rev. D 22, 437 (1980).
14. M. Bolsterli, Phys. Rev. C 26,2397 (1982).
15. R. J. Drachman, Phys. Rev. 125, 1758 (1962).
16. P. Hoodbhoy, Phys. Rev. D 26, 3235 (1982).
17. M. Bolsterli, Phys. Rev. D 27, 349 (1983).

EPS Europhysics Journal

Zeitschrift für Physik C Particles and Fields

ISSN 0170-9739 Title No. 288

Aims and Scope:

Zeitschrift für Physik C, **Particles and Fields** is devoted to the experimental and theoretical investigation of elementary particles. In view of the steadily growing interplay of theory and experiment in this field, particular emphasis is given to a clear and complete presentation of research.

The topics covered include:
- Experimental and theoretical particle physics
- Structure of elementary particles
- High energy processes
- Strong, electromagnetic and weak interactions
- Symmetry principles
- Unification schemes
- S-matrix theory
- Quantum field theory
- Lattice field theory

Special features: Rapid publication, no page charge. Language of publications is English.

Zeitschrift für Physik appears in three parts: A: Atoms and Nuclei; B: Condensed Matter; C: Particles and Fields

Each part may be ordered separately.

Coordinating editor for Zeitschrift für Physik, Parts A, B and C: O. Haxel, Heidelberg

Subscription information and/or **sample copies** are available from your bookseller or directly from Springer-Verlag, Journal Promotion Dept., P.O.Box 105280, D-6900 Heidelberg, FRG

Springer-Verlag
Berlin
Heidelberg
New York
Tokyo

Lecture Notes in Physics

Vol. 173: Stochastic Processes in Quantum Theory and Statistical Physics. Proceedings, 1981. Edited by S. Albeverio, Ph. Combe, and M. Sirugue-Collin. VIII, 337 pages. 1982.

Vol. 174: A. Kadić, D.G.B. Edelen, A Gauge Theory of Dislocations and Disclinations. VII, 290 pages. 1983.

Vol. 175: Defect Complexes in Semiconductor Structures. Proceedings, 1982. Edited by J. Giber, F. Beleznay, J. C. Szép, and J. László. VI, 308 pages. 1983.

Vol. 176: Gauge Theory and Gravitation. Proceedings, 1982. Edited by K. Kikkawa, N. Nakanishi, and H. Nariai. X, 316 pages. 1983.

Vol. 177: Application of High Magnetic Fields in Semiconductor Physics. Proceedings, 1982. Edited by G. Landwehr. XII, 552 pages. 1983.

Vol. 178: Detectors in Heavy-Ion Reactions. Proceedings, 1982. Edited by W. von Oertzen. VIII, 258 pages. 1983.

Vol. 179: Dynamical Systems and Chaos. Proceedings, 1982. Edited by L. Garrido. XIV, 298 pages. 1983.

Vol. 180: Group Theoretical Methods in Physics. Proceedings, 1982. Edited by M. Serdaroğlu and E. İnönü. XI, 569 pages. 1983.

Vol. 181: Gauge Theories of the Eighties. Proceedings, 1982. Edited by R. Raitio and J. Lindfors. V, 644 pages. 1983.

Vol. 182: Laser Physics. Proceedings, 1983. Edited by J. D. Harvey and D. F. Walls. V, 263 pages. 1983.

Vol. 183: J. D. Gunton, M. Droz, Introduction to the Theory of Metastable and Unstable States. VI, 140 pages. 1983.

Vol. 184: Stochastic Processes – Formalism and Applications. Proceedings, 1982. Edited by G. S. Agarwal and S. Dattagupta. VI, 324 pages. 1983.

Vol. 185: H. N. Shirer, R. Wells, Mathematical Structure of the Singularities at the Transitions between Steady States in Hydrodynamic Systems. XI, 276 pages. 1983.

Vol. 186: Critical Phenomena. Proceedings, 1982. Edited by F.J.W. Hahne. VII, 353 pages. 1983.

Vol. 187: Density Functional Theory. Edited by J. Keller and J.L. Gázquez. V, 301 pages. 1983.

Vol. 188: A.P. Balachandran, G. Marmo, B.-S. Skagerstam, A. Stern, Gauge Symmetries and Fibre Bundles. IV, 140 pages. 1983.

Vol. 189: Nonlinear Phenomena. Proceedings, 1982. Edited by K. B. Wolf. XII, 453 pages. 1983.

Vol. 190: K. Kraus, States, Effects, and Operations. Edited by A. Böhm, J.W. Dollard and W.H. Wootters. IX, 151 pages. 1983.

Vol. 191: Photon Photon Collisions. Proceedings, 1983. Edited by Ch. Berger. V, 417 pages. 1983.

Vol. 192: Heidelberg Colloquium on Spin Glasses. Proceedings, 1983. Edited by J. L. van Hemmen and I. Morgenstern. VII, 356 pages. 1983.

Vol. 193: Cool Stars, Stellar Systems, and the Sun. Proceedings, 1983. Edited by S. L. Balliunas and L. Hartmann. VII, 364 pages. 1984.

Vol. 194: P. Pascual, R. Tarrach, QCD: Renormalization for the Practitioner. V, 277 pages. 1984.

Vol. 195: Trends and Applications of Pure Mathematics to Mechanics. Proceedings, 1983. Edited by P. G. Ciarlet and M. Roseau. V, 422 pages. 1984.

Vol. 196: WOPPLOT 83. Parallel Processing: Logic, Organization and Technology. Proceedings, 1983. Edited by J. Becker and I. Eisele. V, 189 pages. 1984.

Vol. 197: Quarks and Nuclear Structure. Proceedings, 1983. Edited by K. Bleuler. VIII, 414 pages. 1984.

Selected Issues from

Lecture Notes in Mathematics

Vol. 836: Differential Geometrical Methods in Mathematical Physics. Proceedings, 1979. Edited by P. L. García, A. Pérez-Rendón, and J. M. Souriau. XII, 538 pages. 1980.

Vol. 837: J. Meixner, F. W. Schäfke and G. Wolf, Mathieu Functions and Spheroidal Functions and their Mathematical Foundations Further Studies. VII, 126 pages. 1980.

Vol. 838: Global Differential Geometry and Global Analysis. Proceedings 1979. Edited by D. Ferus et al. XI, 299 pages. 1981.

Vol. 840: D. Henry, Geometric Theory of Semilinear Parabolic Equations. IV, 348 pages. 1981.

Vol. 841: A. Haraux, Nonlinear Evolution Equations- Global Behaviour of Solutions. XII, 313 pages. 1981.

Vol. 842: Séminaire Bourbaki vol. 1979/80. Exposés 543–560. IV, 317 pages. 1981.

Vol. 843: Functional Analysis, Holomorphy, and Approximation Theory. Proceedings. Edited by S. Machado. VI, 636 pages. 1981.

Vol. 845: A. Tannenbaum, Invariance and System Theory: Algebraic and Geometric Aspects. X, 161 pages. 1981.

Vol. 849: P. Major, Multiple Wiener-Itô Integrals. VII, 127 pages. 1981.

Vol. 851: Stochastic Integrals. Proceedings, 1980. Edited by D. Williams. IX, 540 pages. 1981.

Vol. 852: L. Schwartz, Geometry and Probability in Banach Spaces. X, 101 pages. 1981.

Vol. 856: R. Lascar, Propagation des Singularités des Solutions d'Equations Pseudo-Différentielles à Caractéristiques de Multiplicités Variables. VIII, 237 pages. 1981.

Vol. 858: E. A. Coddington, H. S. V. de Snoo: Regular Boundary Value Problems Associated with Pairs of Ordinary Differential Expressions. V, 225 pages. 1981.

Vol. 861: Analytical Methods in Probability Theory. Proceedings 1980. Edited by D. Dugué, E. Lukacs, V. K. Rohatgi. X, 183 pages. 1981.

Vol. 866: J.-M. Bismut, Mécanique Aléatoire. XVI, 563 pages. 1981.

Vol. 878: Numerical Solution of Nonlinear Equations. Proceedings, 1980. Edited by E. L. Allgower, K. Glashoff, and H.-O. Peitgen. XIV, 440 pages. 1981.

Vol. 881: R. Lutz, M. Goze, Nonstandard Analysis. XIV, 261 pages. 1981.

Vol. 888: Padé Approximation and its Applications. Proceedings, 1980. Edited by M. G. de Bruin and H. van Rossum. VI, 383 pages. 1981.

Vol. 898: Dynamical Systems and Turbulence, Warwick, 1980. Proceedings. Edited by D. Rand and L.-S. Young. VI, 390 pages. 1981.

Vol. 901: Séminaire Bourbaki vol. 1980/81 Exposés 561–578. III, 299 pages. 1981.

Vol. 904: K. Donner, Extension of Positive Operators and Korovkin Theorems. XII, 182 pages. 1982.

Vol. 905: Differential Geometric Methods in Mathematical Physics. Proceedings, 1980. Edited by S.J. Andersson, H.-D. Doebner, and H.R. Petry. VI, 309 pages. 1982.

Vol. 909: Numerical Analysis. Proceedings, 1981. Edited by J.P. Hennart. VII, 247 pages. 1982.

Vol. 912: Numerical Analysis. Proceedings, 1981. Edited by G. A. Watson. XIII, 245 pages. 1982.

Vol. 920: Séminaire de Probabilités XVI, 1980/81. Proceedings. Edité par J. Azéma et M. Yor. V, 622 pages. 1982.

Vol. 921: Séminaire de Probabilités XVI, 1980–81 Supplément: Géométrie Différentielle Stochastique. Proceedings. Edité par J. Azéma et M. Yor. III, 285 pages. 1982.

Vol. 922: B. Dacorogna, Weak Continuity and Weak Lower Semicontinuity of Non-Linear Functionals. V, 120 pages. 1982.

Vol. 923: Functional Analysis in Markov Processes. Proceedings, 1981. Edited by M. Fukushima. V, 307 pages. 1982.

Vol. 926: Geometric Techniques in Gauge Theories. Proceedings, 1981. Edited by R. Martini and E.M.de Jager. IX 219 pages. 1982.

Vol. 927: Y. Z. Flicker, The Trace Formula and Base Change for GL (3). XII, 204 pages. 1982.

Vol. 928: Probability Measures on Groups. Proceedings 1981. Edited by H. Heyer. X, 477 pages. 1982.

Vol. 929: Ecole d'Eté de Probabilités de Saint-Flour X – 1980. Proceedings, 1980. Edited by P.L. Hennequin. X, 313 pages. 1982.

Vol. 930: P. Berthelot, L. Breen, et W. Messing, Théorie de Dieudonné Cristalline II. XI, 261 pages. 1982.

Vol. 931: D.M. Arnold, Finite Rank Torsion Free Abelian Groups and Rings. VII, 191 pages. 1982.

Vol. 932: Analytic Theory of Continued Fractions. Proceedings, 1981. Edited by W.B. Jones, W.J. Thron, and H. Waadeland. VI, 240 pages. 1982.

Vol. 934: M. Sakai, Quadrature Domains. IV, 133 pages. 1982.

Vol. 935: R. Sot, Simple Morphisms in Algebraic Geometry. IV, 146 pages. 1982.

Vol. 936: S.M. Khaleelulla, Counterexamples in Topological Vector Spaces. XXI, 179 pages. 1982.

Vol. 937: E. Combet, Intégrales Exponentielles. VIII, 114 pages. 1982.

Vol. 938: Number Theory. Proceedings, 1981. Edited by K. Alladi. IX, 177 pages. 1982.

Vol. 942: Theory and Applications of Singular Perturbations. Proceedings, 1981. Edited by W. Eckhaus and E.M. de Jager. V, 363 pages. 1982.

Vol. 953: Iterative Solution of Nonlinear Systems of Equations. Proceedings, 1982. Edited by R. Ansorge, Th. Meis, and W. Törnig. VII, 202 pages. 1982.

Vol. 956: Group Actions and Vector Fields. Proceedings, 1981. Edited by J.B. Carrell. V, 144 pages. 1982.

Vol. 957: Differential Equations. Proceedings, 1981. Edited by D.G. de Figueiredo. VIII, 301 pages. 1982.

Vol. 963: R. Nottrot, Optimal Processes on Manifolds. VI, 124 pages. 1982.

Vol. 964: Ordinary and Partial Differential Equations. Proceedings, 1982. Edited by W.N. Everitt and B.D. Sleeman. XVIII, 726 pages. 1982.

Vol. 968: Numerical Integration of Differential Equations and Large Linear Systems. Proceedings, 1980. Edited by J. Hinze. VI, 412 pages. 1982.

Vol. 970: Twistor Geometry and Non-Linear Systems. Proceedings, 1980. Edited by H.-D. Doebner and T.D. Palev. V, 216 pages. 1982.

Vol. 972: Nonlinear Filtering and Stochastic Control. Proceedings, 1981. Edited by S.K. Mitter and A. Moro. VIII, 297 pages. 1983.

Vol. 978: J. Ławrynowicz, J. Krzyż, Quasiconformal Mappings in the Plane. VI, 177 pages. 1983.

Vol. 979: Mathematical Theories of Optimization. Proceedings, 1981. Edited by J.P. Cecconi and T. Zolezzi. V, 268 pages. 1983.

GPSR Compliance

The European Union's (EU) General Product Safety Regulation (GPSR) is a set of rules that requires consumer products to be safe and our obligations to ensure this.

If you have any concerns about our products, you can contact us on ProductSafety@springernature.com

In case Publisher is established outside the EU, the EU authorized representative is:

Springer Nature Customer Service Center GmbH
Europaplatz 3
69115 Heidelberg, Germany

Batch number: 10838062

Printed by Printforce, the Netherlands